IMIDIC POLYMERS AND GREEN POLYMER CHEMISTRY

New Technology and Developments in Process and Product

IMIDIC POLYMERS AND GREEN POLYMER CHEMISTRY

New Technology and Developments in Process and Product

Edited by

Andreea Irina Barzic, PhD
Neha Kanwar Rawat, PhD
A. K. Haghi, PhD

First edition published 2021

Apple Academic Press Inc.
1265 Goldenrod Circle, NE,
Palm Bay, FL 32905 USA

4164 Lakeshore Road, Burlington,
ON, L7L 1A4 Canada

CRC Press
6000 Broken Sound Parkway NW,
Suite 300, Boca Raton, FL 33487-2742 USA

2 Park Square, Milton Park,
Abingdon, Oxon, OX14 4RN UK

First issued in paperback 2021

Library and Archives Canada Cataloguing in Publication

Title: Imidic polymers and green polymer chemistry : new technology and developments in process and product / edited by Andreea Irina Barzic, PhD, Neha Kanwar Rawat, PhD, A.K. Haghi, PhD.
Names: Barzic, Andreea Irina, editor. | Rawat, Neha Kanwar, editor. | Haghi, A. K., editor.
Description: Includes bibliographical references and index.
Identifiers: Canadiana (print) 20200360396 | Canadiana (ebook) 20200360450 | ISBN 9781771889032 (hardcover) | ISBN 9781003057918 (ebook)
Subjects: LCSH: Polyimides. | LCSH: Polyimides—Industrial applications. | LCSH: Polymers. | LCSH: Green chemistry.
Classification: LCC TP1180.P66 I45 2021 | DDC 668.9—dc23

Library of Congress Cataloging-in-Publication Data

....

CIP data on file with US Library of Congress

....

ISBN: 978-1-77188-903-2 (hbk)
ISBN: 978-1-77463-767-8 (pbk)
ISBN: 978-1-00305-791-8 (ebk)

About the Editors

Andreea Irina Barzic, PhD
Researcher, Department of Physical Chemistry of Polymers,
"Petru Poni" Institute of Macromolecular Chemistry, Romania

Andreea Irina Barzic, PhD, is Researcher in the Department of Physical Chemistry of Polymers at "Petru Poni" Institute of Macromolecular Chemistry in Romania. Her area of scientific activity is focused on textured polyimide surfaces, liquid crystals, nanocomposites, rheology, optics, and transparent dielectrics for solar cells. She has published over 40 papers in peer-reviewed ISI journals and has contributed as author and/or editor to 12 books and book chapters. She was a leader and member of several research grants as well as a member of the organizing committees of scientific conferences. She has also been a reviewer for a number of prestigious journals in the field of physics and polymer science. Dr. Barzic received her PhD from the Faculty of Physics from "Alexandru Ioan Cuza" University, Iasi, Romania. She received her PhD in Chemistry on thermodynamics and morphology of imidic polymers. Afterwards, she joined a postdoctoral fellowship program, where she expanded her research skills toward bio-applications of imidic polymers.

Neha Kanwar Rawat, PhD
Researcher, Materials Science Division, CSIR-National Aerospace
Laboratories, Bangalore, India

Neha Kanwar Rawat, PhD, is a recipient of the prestigious DST Young Scientist Postdoctoral Fellowship and is presently a researcher in the Materials Science Division, CSIR–National Aerospace Laboratories, Bangalore, India. She received her PhD in Chemistry from Jamia Millia Islamia (A Central University), India. She has published numerous peer-reviewed research articles in journals of high repute. Her contributions have led to many chapters in international books published with Apple Academic Press, and many others in progress. She has worked on multiple prestigious research and academic fellowships in her career. She is a member of many groups, including the Royal Society of Chemistry and the American Chemical Society (USA) and is a life member of the Asian Polymer Association.

A. K. Haghi, PhD
Professor Emeritus of Engineering Sciences, Former Editor-in-Chief,
*International Journal of Chemoinformatics and Chemical Engineering
and Polymers Research Journal*; Member, Canadian Research and
Development Center of Sciences and Culture

A. K. Haghi, PhD, is the author and editor of over 200 books and more than
1000 published papers in various journals and conference proceedings. Dr.
Haghi has received several grants, consulted for a number of major corpora-
tions, and is a frequent speaker to national and international audiences. Since
1983, he served as professor at several universities. He is former Editor-
in-Chief of the *International Journal of Chemoinformatics and Chemical
Engineering* and *Polymers Research Journal* and is on the editorial boards of
many international journals. He is also a member of the Canadian Research
and Development Center of Sciences and Cultures (CRDCSC), Montreal,
Quebec, Canada.

Contents

Contributors

Raluca Marinica Albu
Petru Poni" Institute of Macromolecular Chemistry, Laboratory of Physical Chemistry of Polymers, 41A Grigore Ghica Voda Alley, 700487 Iasi, Romania

Alexander Alentiev
A. V. Topchiev Institute of Petrochemical Synthesis, Moscow 119991, Russia

Devrim Balkose
Department of Chemical Engineering Gulbahce, Izmir Institute of Technology, Urla, Izmir, Turkey

Susanta Banerjee
Materials Science Centre, Indian Institute of Technology, Kharagpur 721302, India

Razvan Florin Barzic
"Petru Poni" Institute of Macromolecular Chemistry, Laboratory of Physical Chemistry of Polymers, 41A Grigore Ghica Voda Alley, 700487 Iasi, Romania

Mariana Cristea
"Petru Poni" Institute of Macromolecular Chemistry, 41A Grigore Ghica Voda Alley, 700487, Iasi, Romania

Mariana-Dana Damaceanu
"Petru Poni" Institute of Macromolecular Chemistry, 41A Grigore Ghica Voda Alley, 700487, Iasi, Romania

Hüseyin Deligöz
Faculty of Engineering, Department of Chemical Engineering, İstanbul University-Cerrahpaşa, Avcılar 34320, Istanbul, Turkey

Göknur Dönmez
Faculty of Engineering, Department of Chemical Engineering, İstanbul University-Cerrahpaşa, Avcılar 34320, Istanbul, Turkey

Ayça Ergün
Faculty of Engineering, Department of Chemical Engineering, İstanbul University-Cerrahpaşa, Avcılar 34320, Istanbul, Turkey

Manuel Garbea
Microsin SRL, Bucharest, Romania

Barnali Dasgupta Ghosh
Department of Chemistry, Birla Institute of Technology, Mesra, Ranchi 835215, India

Mehmet Gonen
Chemical Engineering Department, Suleyman Demirel University, Isparta, Turkey

Corneliu Hamciuc
"Petru Poni" Institute of Macromolecular Chemistry, Laboratory of Polycondensation and Thermostable Polymers, 41A Grigore Ghica Voda Alley, Iasi 700487, Romania

Elena Hamciuc
"Petru Poni" Institute of Macromolecular Chemistry, Laboratory of Polycondensation and
Thermostable Polymers, 41A Grigore Ghica Voda Alley, Iasi 700487, Romania

Camelia Hulubei
"Petru Poni" Institute of Macromolecular Chemistry, Laboratory of Polycondensation and
Thermostable Polymers, 41A Grigore Ghica Voda Alley, Iasi 700487, Romania

Daniela Ionita
"Petru Poni" Institute of Macromolecular Chemistry, 41A Grigore Ghica Voda Alley, 700487,
Iasi, Romania

Shama Islam
Department of Physics, Jamia Millia Islamia, New Delhi 110025, India

Purnima Jain
Department of Chemistry, Netaji Subhas University of Technology (Erstwhile Netaji Subhas
Institute of Technology, University of Delhi) Dwarka, New Delhi 110078, India

Athira John
Centre for Biopolymer Science and Technology (CBPST), CIPET, Kochi, Kerala, India

Hana Khan
Department of Physics, Jamia Millia Islamia, New Delhi 110025, India

Jolanta Konieczkowska
Centre of Polymer and Carbon Materials Polish Academy of Sciences, 34 M. Curie-Sklodowska Str.,
41-819 Zabrze, Poland

Merve Okutan
Faculty of Engineering, Department of Chemical Engineering, Hitit University, Çorum 19030, Turkey

Deepak Poddar
Department of Chemistry, Netaji Subhas University of Technology (Erstwhile Netaji Subhas
Institute of Technology, University of Delhi) Dwarka, New Delhi 110078, India

Ananthu Prasad
International and Inter University Centre for Nanosicence and Nanotechnology (IIUCNN),
Mahatma Gandhi University, Kottayam, Kerala, India

Inga Ronova
A. N. Nesmeyanov Institute of Organoelement Compounds, Moscow 119991, Russia

Sevdiye Atakul Savrık
A.Ş Gaziemir, Akzo Nobel Boya, Izmir, Turkey

Ankita Singh
Department of Chemistry, Netaji Subhas University of Technology (Erstwhile Netaji Subhas
Institute of Technology, University of Delhi) Dwarka, New Delhi 110078, India

Iuliana Stoica
"Petru Poni" Institute of Macromolecular Chemistry, Laboratory of Polymeric Materials Physics,
41A Grigore Ghica Voda Alley, 700487, Iasi, Romania

Sanjeeve Thakur
Department of Chemistry, Netaji Subhas University of Technology (Erstwhile Netaji Subhas
Institute of Technology, University of Delhi) Dwarka, New Delhi 110078, India

Sabu Thomas
School of Chemical Sciences, Mahatma Gandhi University, Kottayam, Kerala, India

Yuri Yampolskii
A. V. Topchiev Institute of Petrochemical Synthesis, Moscow 119991, Russia

Senem Yetgin
Food Engineering Department, Kastamonu University, Kastomonu, Turkey

M. Zulfequar
Department of Physics, Jamia Millia Islamia, New Delhi 110025, India

Abbreviations

AC	activated carbon
AFM	atomic force microscopy
AgNW	silver nanowire
AMOLEDs	active-matrix organic light-emitting diodes
BA	biogenic amines
BD	Brownian dynamics
BMI	bis(maleimide)s
BN	boron nitride
BNC	bionanocomposite
BPDA	biphenyl-3,3′,4,4′-tetracarboxylic dianhydride
BT	bismaleimide-triazine
BTDA	benzophenone tetracarboxylic dianhidryde
CA	cellulose acetate
CB	carbon black
CE	cyanate ester
CF	carbon fiber
CLCP	crosslinked LCP
CNT	carbon nanotube
CNTs	carbon nanotubes
Co-PIs	co-polyimides
CPI	cross-linked PI
CTC	charge transfer complex
CV	cyclic voltammetry
DDFT	dynamic density functional theory
DMAc	dimethylacetamide
DMF	dimethylformamide
DMSO	dimethyl sulfoxide
DPD	dissipative particle dynamics
DPL	dynamic plowing lithography
DSC	differential scanning calorimetry
DX	1,4-dioxane
ECDs	electrochromic devices
EDLCs	electric double-layer capacitors
FETs	field-effect transistors

FFT	Fourier transform images
FG	fluorographene
FTIR	Fourier transform infrared
GO	graphene oxide
HBPSi	hyper-branched polysiloxane
HNMR	hydrogenated acrylic butadiene rubber
HPMA	hyper-branched polysiloxane containing maleimide
HSP	Hansen solubility parameter
HT	hydrothermal
IGZO	In–Ga–Zn–O
IL	ionic liquids
ITO	indium tin oxide
LB	lattice Boltzmann
LCDs	liquid crystal displays
LCE	liquid crystal elastomer
LCN	LCP network
LCP	liquid crystal polymer
LIB	lithium-ion batteries
MC	Monte Carlo
MD	molecular dynamics
MeIm-BF4	1-methyl imidazolium tetrafluoroborate
MWCNT	multi-wall carbon nanotube
MOF	metal–organic frameworks
MOSFET	metal-oxide-semiconductor FET
mPD	m-phenylene diamine
MZA	modified zeolite 4A
NF	nanofiller
OFETs	Organic field-effect transistors
OLED	organic light emitting diodes
OMPS	octa(maleimidophenyl)silsesquioxanes
OSN	organic solvent nanofiltration
OTFTs	organic thin-film transistors
PAA	poly(amic acid)
PANI	polyaniline
PBI	polybenzimidazole
PC	propylene carbonate
PCM	phase change materials
PE	polyethylene
PEDOT	poly(3,4-ethylenedioxythiophene)

PEEK	polyetheretherketone
PEGME	poly(ethylene glycol) monomethyl ether
PEI	polyethyleneimine
PEM	proton-exchange membranes
PEO	poly(ethylene oxide)
PIs	polyimides
PLA	polylactic acid
PLC	polymeric liquid crystal
PMDA	pyromellitic dianhydride
POA	photoinduced optical anisotropy
POSS	polyhedral oligomeric silsesquioxane
PPPI	poly(*p*-phenyl pyromellitimide)
PS	polystyrene
PSPI	photosensitive polyimide
PT	polythiophene
PUI	poly(urethane-imide)
RF	radio frequency
RGO	reduced graphene oxide
QM	quantum mechanics
QSPR	quantitative structure–activity relationship
SAMs	self-assembled monolayers
Sbi	surface bearing index
Sci	core fluid retention index
SCLC	space charge limited current
SEM	scanning electron microscopy
SGR	surface relief grating
SNP	silica nanoparticle
sPI	sulfonated
SPL	static plowing lithography
Sq	root mean square roughness
Stdi	texture direction index of the surface
SU	structural unit
Svi	valley fluid retention index
TDGL	time-dependent Ginzburg-Landau
TFTs	thin-film transistors
TG	thermogravimetry
TGA	thermal gravimetric analysis
TPA	triphenylamine
UMZA	unmodified zeolite 4A

Vmc	core material volume
Vmp	peak material volume
Vvc	core void volume
Vvv	valley void volume

Preface

This book reviews the latest research, development, and future potential of polyimides (PIs) and green polymer chemistry.

Polymers with imidic structure, known as PIs, are widely investigated owing to their practical implications in numerous industrial sectors. This volume explains why PIs offer versatility unparalleled in most other classes of macromolecules. Polymers can be prepared from a variety of starting materials, by a variety of synthetic routes. They can be tailor-made to suit specific applications. This book will serve as a valuable reference for those with an interest in synthesis of PIs and the chemistry and physical chemistry of polyimide compounds. It is intended as a summary of the current status of polyimide research for the specialist as well as a reference book for graduate studies in polymer chemistry.

Green polymer chemistry is an extension of green chemistry to polymer science and engineering. Developments in this area have been stimulated by health and environmental concerns, interest in sustainability, desire to decrease the dependence on petroleum, and opportunities to design and produce "green" products and processes. Major advances include new uses of green processing methodologies and green polymeric products.

This book combines the major interdisciplinary research in these fields and is targeted for scientists, engineers, and students, who are involved or interested in green polymer chemistry and imidic polymers.

Aromatic PIs, as heterocyclic polymers, are recognized as high-performance systems that combine properties including high thermal and chemical stability, high radiation resistance, high mechanical and insulating properties, and in particular, inherent high refractive index. These polymers have been applied in many engineering fields, such as microelectronics, adhesives, gas separation membranes, and advanced optical technologies after several decades of technological evaluations. The main problem associated with the superior properties of the aromatic PIs is the strong intramolecular and intermolecular charge-transfer complex (CTC) between the electron-donating diamine and the electron-accepting dianhydride moieties, which makes them insoluble and difficult to process in their fully imidized form. Aromatic-alicyclic PIs offer a solution to remove these current inconveniences caused by a high CTC, from low solubility and low processability to strong visible radiation absorption and high dielectric constant (over 3). In this context, Chapter 1 provides a basic presentation of this class of polymers, including their synthesis and processing methodologies. In addition, some new polyimide structures and their applications are discussed.

Aromatic PIs are an important class of polymers used in modern technology, including membrane gas separation. Synthesis and study of PIs with bulky groups in the elementary unit, high free volume, and attractive transport characteristics are modern trends in membrane materials science. Chapter 2 describes the methods of synthesis of PIs used to obtain new materials with bulky or pendant, noncoplanar, kink, spiro, cardo, branched substituents, flexible or unsymmetrical linkages in the backbone. Particular attention is paid to the gas separation characteristics of PIs of this class. In permeability–selectivity diagrams, PIs with bulky groups are located above the upper bound of the distribution. The reasons for this behavior are an increase in the free volume and the growth of chain conformational rigidity due to hindering rotation of the fragments of the main chain, which is proved by numerical methods. Conformational factors lead to an increase in the permeability and overall selectivity of PIs of this class and to a decrease in their permittivity. Thus, PIs with bulky groups are promising materials for various applications.

Azobenzene-containing polymers called azopolymers are a wide group of materials interesting from the point of view their potential utilization in the fields of photonics, optoelectronics, and biology as cell culture substrates and biomaterial. They are utilized as thin films on the substrate or free-standing foils. This class of polymers contains light-sensitive linkages -N=N- in their chemical structure. Azobenzene derivatives occur in the form of two isomers: an energetically stable *trans*-isomer and a metastable *cis*-isomer. Under the irradiation of azochromophores with a specific wavelength, the reversible *trans–cis–trans* isomerization is observed. The stability of *cis*-isomer in the solid-state is crucial in some azopolymers applications, which can be tailored by changing the azochromophore structure, kinds of azomolecules incorporation to the polymer backbone (covalent or noncovalent), and polymer molar masses. The knowledge about the influence of mentioned factors supports designing azopolymers with desired properties adjusted to a particular device. Chapter 3 is devoted to showing the impact of azopolyimide structure (main-chain, side-chain or T-type polymers, "guest-host" systems) on *trans–cis* and *cis–trans* isomerization process in the solid-state based on literature review and investigations published in our previous works.

Chapter 4 presents that PIs prepared from a dianhydride and a diamine generally stand out in many applications due to their superior properties such as high temperature stability, low dielectric constant, tensile and high compressive strength, good resistance to solvents and moisture, good adhesion properties for inorganic, metallic, and dielectric materials, and easy casting of thin or thick films. Due to these unique properties, PIs are prevalently used in microelectronics, films, adhesives, and membranes. In addition to the use of PIs as an insulating material, their electrical properties can also be controlled from semiconducting to conducting state using some different approaches including the addition of metallic particles, introduction of some functional groups into the PI backbone, or doping it with the ionic liquids. These research efforts pave the way for widely using of PIs not only in the insulating materials but also in the industries where semiconducting/conducting material properties are required. As a consequence, it is expected that these multipurpose materials will be used more widely in various technological applications of microelectronic, separation, energy, and aviation in the near future.

Imidic polymers are materials of great importance in various applicative fields. The macromolecular architecture is an essential parameter that affects the physico-chemical properties of PIs. Chapter 6 describes the need of molecular modeling in imidic polymer investigation, highlighting the importance of understanding the correlation between chemical structure and the performance of the final product. The main computational techniques applied to this type of polymers are presented. The connection between theoretical information extracted from modeling and experimental data is also discussed, including the applicative potential.

PIs are considered a special type of thermostable polymers, with excellent mechanical resistance combined with good dielectric, morphological, and optical features. For this reason, this category of high-performance plastics is widely exploited in electronic industry, particularly in transistor or biosensor manufacturing. Basic aspects regarding transistors or biosensors are shortly presented for a better understanding of the polyimide role in these devices. Chapter 7 also describes the current advances in transistors based on PIs by highlighting their impact on the device performance when used as substrates or gate dielectrics. Moreover, the latest developments concerning the processing of PIs for several types of biosensors are reviewed.

Natural fibers have been used by humans for centuries. The intervention of new synthetic materials replaced natural fibers in many applications. But, due to ecological concerns and better awareness of environmental pollution, natural fibers are attaining attention now. Natural fibers and derivatives of natural fibers are eco-friendly, cheap, and renewable. Large industrial sectors like automobile industry are promoting the use of natural fibers as reinforcements. However, when compared with synthetic fibers, natural fibers have several disadvantages like water absorption, low compatibility with matrix, etc. Extensive research is being carried out to overcome these limitations. In Chapter 8, we can study natural fibers, its limitations, the methods adopted to overcome these limitations, and various applications.

The characterization of a commercial particulate sample labeled as "polyethylene wax" used in pigment masterbatch preparation was performed in Chapter 9 by comparing its properties with similar materials; a paraffin wax, sorbitan monostearate (Span 60), and myristic acid. The samples were characterized by Fourier transform infrared spectroscopy, x-ray diffraction, scanning electron microscopy, differential scanning calorimetry, and thermal gravimetric analysis. Infrared spectroscopy showed that there were vibrations of CH_2 and CH_3 groups in paraffin wax, CH_2, CH_3, C-O, and C=O groups in commercial wax and Span 60 and myristic acid. It was found that commercial wax has a melting point of 54.9°C and heat of fusion of 115 J g^{-1}. The particles were in spherical form with 170 μm average diameter and mixed with cylindrical fibers of 5.45 μm diameter and 75 μm in length. There were particles having 560 μm diameter with holes of 145 μm on their surface. In these large particles, there was a crust of 32 μm in thickness coating aglomerated small particles. It was found that the commercial wax was not "polyethylene wax" as it was labeled. The melting point, functional groups, and the crystal structure of commercial wax had closer similarity to Span 60 than that of paraffin wax and myristic acid. However, its onset temperature of mass loss was much lower than that of Span 60 and the heat of fusion of Span 60 was much lower than that of commercial wax. Thus, it was concluded that further studies with more advanced methods are necessary to make a complete identification of commercial polyethylene wax sample.

In Chapter 10, it is shown that supercapacitors are the most promising candidates in future generations for energy storage devices. The very and capable new energy devices are desirable because of the power and surrounding crisis is at hazard to growth the power source in squat exaggeration beat have skillful an incredible enthusiasm for electrochemical capacitors, additionally given as supercapacitors. In the present period, supercapacitors stipulate the high essence and comprehensive lastingness needful for a few current efficiency devices in thermoelectric vehicles, backup ascend for a few electrical devices and constant desirable quality group. Novel challenging applications in usage counting microautonomous robots and in mobile/portable energy storage, distributed sensors, and other devices to convene the supplies of high power density and long durability supercapacitor devices are responsible for a search of new and exciting materials. To extend superior electrode materials for supercapacitors, an imperative approach and methods are selected to fabricate materials. Graphene-based materials have been used for electric vehicles to provide improved resources for storing electricity and the remarkable enhancement of moveable electronics. To develop the electrodes materials, various carbon–metal oxide composite have been prepared by mixing of metal oxides in the matrix of carbon nanostructures counting zero-dimensional carbon nanoparticles, one-dimensional nanostructures (carbon nanotubes and carbon nanofibers), two-dimensional nanosheets (graphene and reduced graphene oxides), as well as three-dimensional porous carbon nanoarchitecture. In the present chapter, the efforts have been enthusiastic to achieved lightweight, thin supercapacitors with enhanced supercacitance performance. In any case, upright classify supercapacitors are a liability to establish second-hand generalship polymer to show starving reliable behavior among the way lump, exceptionally conducting polymers, for example, polyaniline, polypyrroles, and polyethenedioxythiophene with admirable electrical conductivity and moderate pseudo electrical capacity have a lively extensive interest as electrode materials for supercapacitors applications.

PI has found extensive use in the fabrication of polymeric membranes because of their high selectivity and permeability. Besides these are advantageous for they have high-temperature resistance, flexibility, high chemical and mechanical stability, secure handling and inexpensive nature compared to the inorganic material and ceramics used in filtration membranes and even readily accepts the physical and chemical modification to improve its properties. PI has been used in various separation techniques, especially gas-phase separation processes. Chapter 11 mainly focuses on the greener approach to synthesize the polyimide polymer and membrane fabrication at the same time. Various methods for synthesis have been discussed and elaborated with examples. The use of the eco-friendly solvent media has been discussed in detail with the particular emphasis on the selection of the solvent as a medium to be used in synthesis methods and various parameters such as Hansen parameter has also been discussed to provide an insight to the students to choose solvent effectively.

The authors have tried to provide the content of the chapter in a concise yet understandable manner and those interested in apprehending more can go through the references provided at the end of the chapter.

The formation of PIs by the combination of dianhydrides with diamines results in a large diversity of structures having various mobilities, inter/intrachain interactions, and topologies. Generally, the processing of PIs in the imidized form is complicated even because of molecular rigidity and strong interchain interactions that entails poor organosolubility. The molecular dynamics of PIs are the result of the balance between molecular packing and chain motions. The phenomenon has effects also on the application of polyimide, for example in the membrane field. Dynamic mechanical analysis (DMA) is a convenient method to investigate molecular dynamic in PIs. The secondary relaxations of PIs can be accurately defined by DMA, along with their influence on the whole behavior of the polymer. As a polyimide is prepared by thermal imidization of its precursor, poly(amic acid), the result of a temperature sweep DMA also gives information on the evolution of the system during heating. In addition, the accurate assessment of the glass transition region cannot be established without considering the occurrence of imidization by increasing temperature. Even more challenging is the situation when, during imidization, the analyst deals with crosslinking, thermal rearrangement processes, or solvent evaporation. Chapter 12 will frame all the aspects above by considering a couple of examples for illustration.

CHAPTER 1

Aromatic-Alicyclic Polyimides: From Basic Aspects Toward High Technologies

CAMELIA HULUBEI*, ELENA HAMCIUC, and CORNELIU HAMCIUC

*"Petru Poni" Institute of Macromolecular Chemistry,
Laboratory of Polycondensation and Thermostable Polymers,
41A Grigore Ghica Voda Alley, Iasi 700487, Romania*

Corresponding author. E-mail: hulubei@icmpp.ro

ABSTRACT

Aromatic polyimides (PIs), as heterocyclic polymers, are recognized as high-performance systems that combine properties including high thermal and chemical stability, high radiation resistance, high mechanical and insulating properties, and in particular, inherent high refractive index. These polymers have been applied in many engineering fields such as microelectronics, adhesives, gas separation membranes, advanced optical technologies, after several decades of technological evaluations. The main problem associated with the superior properties of the aromatic PIs is the strong intramolecular and intermolecular charge transfer complex (CTC) between the electron-donating diamine and the electron-accepting dianhydride moieties, which makes them insoluble and difficult to process in their fully imidized form. Aromatic-alicyclic PIs offer a solution to remove these current inconveniences caused by a high CTC, from low solubility and low processability to strong visible radiation absorption and high dielectric constant (over 3). In this context, the chapter deals with the basic presentation of this class of polymers, including their synthesis and processing methodologies. In addition, some new PI structures and their applications are discussed.

1.1   BASIC ASPECTS ON POLYIMIDES

Heterocyclic polymers, representative for the heat-resistant structures, are recognized as high performance engineering plastics.[1] Among them, the most investigated are the PIs. Kapton is a familiar aromatic PI, produced in the 1960s by Du Pont. After its commercialization, other such types of imidic structures with modified properties have been proposed. Wholly aromatic PIs are remarked for their excellent physicochemical properties, like chemical stability, electrical and thermal insulation, mechanical resistance, and inherent high refractive index coupled with heat and radiation resistance.[2,3]

Starting from their outstanding balance of characteristics, the PIs have acquired considerable importance in advanced technologies involved in microelectronics,[4] aerospace,[5] gas separation membranes,[6] flexible optoelectronics, and so on.[7,8] In the past years, some PI materials have shown adequate properties for biomedical applications.[9,10]

1.2   PROBLEMS ASSOCIATED WITH AROMATIC POLYIMIDES

The nature of the PIs chemical structure consisting of rigid heterocyclic imide and aromatic rings can explain their excellent characteristics.[11] Otherwise it is worth noting that the polar nature of PIs and the charge transfer complex (CTC) that these polymers can form. The CTC is a feature of the polymers which contain donor and acceptor monomer units (Fig. 1.1).

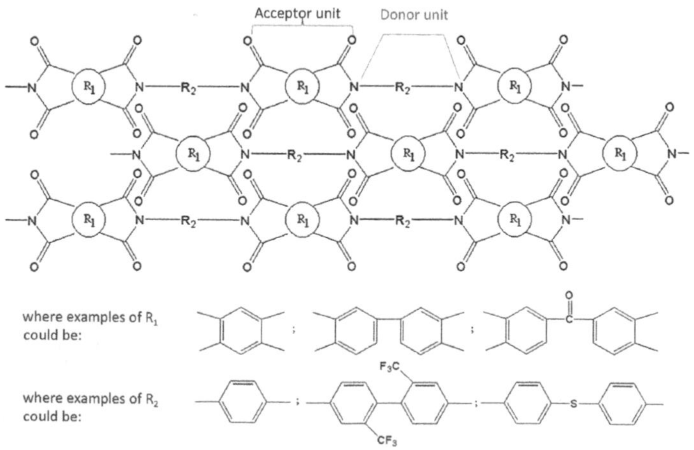

FIGURE 1.1   Schematic illustration of CTC interactions among imidic chains.

In other words, PI can be considered such type of structure, where diamine moiety, the donor, gives electrons from the nitrogen atoms to the carbonyl groups of dianhydride part, the acceptor, thus holding them tightly together. Finally, the chains of polymer become stacked in a similar manner to strips of paper, in which the donor and the acceptor groups are tied, due to the CTC interactions among the adjacent units. More than that, the macromolecular chains are held together very tightly by the CTC which does not allow them to move around very much. Thus, the lack of movement is practically trans-ferred from the molecular level to the macro level, in the whole material. Based on the effects generated by the interactions between electron-donating and the electron-accepting moieties, one may explain why PIs are so strong and display superior properties.[11] However, CTC determines inconveniences which in some cases limit the applicability of the wholly aromatic PIs in certain industrial areas. Among them it can be mentioned:

- high insolubility of structures in their fully imidized form, with a negative impact on the processability. In such cases, it is difficult to obtain PI films or fibers which are used in dielectric layers or optical waveguides;[12,13]
- great absorption of radiations from the visible domain. So, PIs cannot be used in the practical situations (e.g., solar cells encapsulant) where the colorless property is a basic requirement. The coloring phenom-enon is another result of the same strong inter and intramolecular CTC interactions;[14]
- relatively high dielectric constant generated by CTC interactions seems to be another weak point for interlayer dielectrics for multilevel high-speed electronics;[11]
- low coefficient of thermal expansion of aromatic PIs impedes their use in multilayered devices. In order to adapt this material feature (related to its ability to expand upon heating), one must control chain stiffness, chain linearity, and intermolecular interactions.[15]

Having all these in view, it can be concluded that in spite of their excel-lent properties, the problems ascribed to the CTC of aromatic PIs create some deficiencies, which do not always render the optimum properties for the majority of high tech applications. Therefore, the main challenge was to develop soluble PI systems, able to form colorless films, with high transparency, and thermal stability for the applications, where fully aromatic structures are not ideal.

1.3   AROMATIC-ALICYCLIC POLYIMIDES AS ALTERNATIVE TO CONVENTIONAL STRUCTURES

The chemical structure of a PI is the main factor which strongly impacts its processability and physical properties. Many efforts have been directed toward suppressing CTC interactions, thus improving the processability and counteracting the others PI shortcomings related to transparency, color, and dielectric constant. The literature presents some structural modifications of the PI backbone to reduce the CTC interactions, such as fluorination,[16] incorporation of flexible bridging groups,[17] bulky substituents,[18] twisted[19] or unsymmetrical[20] monomer structures, or copolymerization as synthesis method.[21]

A more convenient alternative relies on using aliphatic or alicyclic monomers[11,22,23] (organic compounds that are both aliphatic and cyclic) in PI synthesis to produce aromatic-alicyclic structures. The foundation of this last concept is based on the fact that aliphatic structures display low molecular density, low polarity, and reduced probability of undergoing inter or intra-molecular CTC. In this way, their incorporation in the PI structure induces good solubility, thermal stability, negligible birefringence, low refractive index, lack of color, a lower dielectric constant, and a higher transparency comparing with the aromatic PIs.[11]

Aromatic-alicyclic PIs are generally able to form transparent and almost colorless polymer films. However, certain conditions should be evaluated to ensure excellent optical transparency, classified by Masatoshi Hasegawa[24] as follows:

- *chemical factors*: responsible for CTC interactions and the electronic conjugation of the PI chain structures and also related to partial decomposition of terminal amino groups and aliphatic units in the main chains and the side groups;
- *physical factors*: responsible for the aggregation of the macromo-lecular chains (generated by imidization procedure and heating programs involved in film formation);
- *processing factors*: atmosphere, temperature, solvent type used for film preparation, and the purity of each reactant used in synthesis.

Therefore, careful molecular design combined with appropriate prepara-tion processes will allow to achieve highly transparent and almost colorless aromatic-alicyclic PI systems, useful as materials with target properties.

1.4 MOLECULAR DESIGN AND SYNTHESIS METHODS OF TRANSPARENT POLYIMIDES

As previously described, fully aromatic PIs with alternately linked sequences consisting of an electron donor (diamine part) and an electron acceptor (dianhydride part) generate strong inter and intramolecular CTC interactions.[25] From this perspective, in order to obtain adequate optical properties there is the possibility to diminish CTC interactions by using of aromatic diamines with electron-withdrawing $-CF_3$ or $-SO_2-$ groups[26] and aromatic tetracarboxylic dianhydrides containing electron-donating $-O-$ groups. However, in this case, the aromatic diamines having strong electron-withdrawing groups will probably perturb the formation of poly(amic acid)s (PAAs) with high molecular weights and film-forming ability. As a result, in such situations, it will be quite difficult to get ductile or flexible films.[24]

Aromatic-alicyclic PI systems can be considered a possible approach to achieve the targeted optical properties due to the chemical nature[24] and the used monomers combination. Based on their structures, such imidic polymers can be organized as follows:

- *PIs*—as combinations of alicyclic dianhydrides/aromatic diamines or aromatic dianhydrides /alicyclic diamines monomers;
- *copolyimides*—derived from various aromatic/alicyclic dianhydride and diamine monomers.

Figure 1.2 presents a basic illustration of an aromatic-alicyclic PI preparation by means of thermal or chemical imidization of a PAA prepolymer.

The incorporation of alicyclic segments in PI structure is the key solution to overcome some of the shortcomings which appear in the case of the aromatic PIs. Partial alicyclic PIs and copolyimides, which varied in chemical composition by their alicyclic content, preserve some properties referring to the thermal stability, glass transition temperature, and mechanical resistance. In the same time, such PI structures display improved optical and dielectric characteristics.[11,27] It is very important to note the high basic character of an aliphatic diamine and its association with the insoluble intermediate salts formation by the reaction with the carboxylic acid groups of the PAA[24] and, implicitly, with a modest molecular weight of these structures. Based on this aspect, when PIs are achieved by the classical two-step polycondensation method, the alicyclic diamine basicity has an important role. To reduce the salt formation, the diamine must be added very slowly to the dianhydride[2,21,28] or the two-steps synthesis approach must be changed with direct polycondensation in *m*-cresol at elevated temperatures.[29,30]

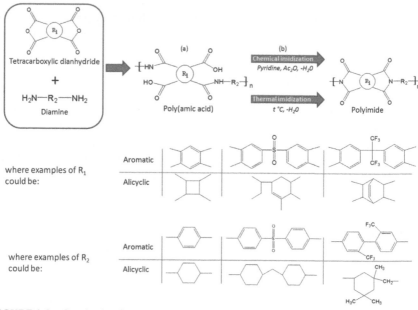

FIGURE 1.2   Synthesis of an aromatic-alicyclic PI by a two-step polycondensation reaction of diamine and dianhydride monomers in a solvent medium with: formation of a PAA prepolymer; cyclodehydration through a chemical or thermal imidization process.

The chemical imidization process can be considered as another preparation method with a positive effect regarding PI film transparency and color. The literature data[24] confirm that PI films achieved using chemical imidization have superior optical properties comparing with the films resulted using a conventional two-step process via thermal imidization. This can be explained by taking into account the instability of the terminal amino group which can be end-capped by Ac_2O during the chemical imidization stage. The PI film-forming steps by coating and drying at much lower temperatures (around 250°C) in regard with the thermal imidization process (300°C–350°C) can be viewed as another advantage.

1.5   MAIN APPROACHES OF PROCESSING

Condensation reactions represented the most practical way to obtain PIs until the early 1970s.[2,31] The synthesis involves two steps: PAA, results as intermediate polymer in the first step by a polyaddition reaction between dianhydrides and diamines (the used monomers) in a polar solvent and at low temperatures (about 50°C).[2] This pre-polymer can be employed to produce *films*,[32] *fibers*,[33] *foams*,[34] *adhesives*[35] *or impregnated fibrous reinforcements*,[36] and *particles*.[37] Afterwards, the PAA can be transformed into fully cyclized PI products, by thermal (200°C–300°C)[38] or chemical[39] imidization (using dehydrating agents).

To obtain films or fibers, a PAA with high molecular weight[40,41] is required, whereas this condition is not necessary to prepare resins matrices for composites or adhesives. Hydrolytic instability of the PAA and the volatile compounds elimination (e.g., water, methanol, and acetic acid) during the cyclodehydration process are factors that hinder processing.[42] Particularly, for achieving thick samples, careful control of the processing conditions and the use of complex heating cycles are necessary.

1.5.1 FILMS

Kapton PI is a structure obtained from pyromellitic dianhydride and 4,4'-diaminodiphenyl ether. This polymer remains the reference for the PI films production.[2] Kapton films are casted from the PAA intermediate, which subsequently are thermally cyclized to the PI form by heating at 300°C. There are several methods by which the PI films are deposited on the substrate and most often encountered ones are spin-coating,[43] solution casting followed by shearing,[44] a roll-to-roll approach,[45] and Langmuir–Blodgett.[46]

1.5.2 FIBERS

Thin fibers can be prepared by the wet spinning of thermoplastic PIs from melts or solutions.[33] An alternative is to obtain fibers from PAA prepolymer solutions and then their imidization, chemically or thermally, to the PI form. There are studies that describe this processing method in two stages,[47] being possible to use conventional wet and dry spinning equipment. The appearance of fibers defects is closely related to the solvent conductivity, polymer molecular weight, applied voltage, and other electrospinning parameters.[48] Defect-free PI fibers can be prepared by ensuring a proper concentration of the solution, where the macromolecular coils are entangled.[49] Below that critical concentration the PI fibers present beaded-shaped defects.[49] Such aspects are presented in Figure 1.3.

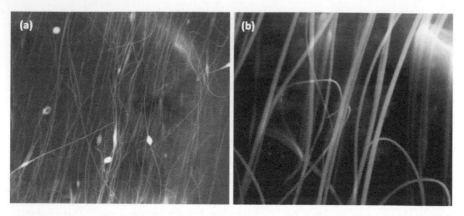

FIGURE 1.3   PI fibers with bead defects (a) and PI fibers without defects (b) obtained by electrospinning.

1.5.3   FOAMS

Foams are generally products derived from the PIs.[34] The powder resulting from drying the sprayed polyamide solution is heated in a microwave oven and thus foamed in blocks. Water or alcohol are being removed during cyclization and thus they act as blowing agents. There are also syntactic foams, which are a category of materials produced via pre-formed hollow spheres. For instance, such foams can be accomplished by mixing equal amounts of PI and silica microballoon which are further processed by heating at temperatures around 300°C and moderate pressure.

1.5.4   ADHESIVES

As adhesives, the condensation PIs are also processed from the intermediate PAA solution phase, applied to a glass cloth support. Most of the solvent is dried by subjecting the system at 100°C–150°C, after which the partially cured material is placed between the parts that need bonding and the process is finished under high temperature and pressure (e.g., 0.5 MPa at 310°C for 90 min). A posttreatment at 300°C–350°C is useful for achieving optimum properties.[50]

1.5.5 PARTICLES

There are several methods that can be applied to prepare polyimide particles (PI Ps) of different dimensions and variable porosity. In order to produce PI Ps one may need to start from:

- the PAA solution or from a solubility difference between the PI and PAA in the same solvent—for example, 1-methyl-2-pyrrolidone anhydrous, *N,N'*–dimethylacetamide, etc.—which can be good solvents for PAA, but poor solvents for PI;[51]
- the PI solution: either by cooling of solution[52] or by addition of few drops of precipitant into PI solution (water or ethanol).[53] Such methods are useful only when the PI is soluble;
- the PIs solutions with other materials to obtain composite particles (e.g., polystyrene, silica) by reprecipitation approach.[54,55] The latter is considered as a convenient technique for fabricating organic and polymer nanoparticles and/or nanocrystals in a dispersion medium.[56] The reprecipitation method is also suitable to prepare PI Ps from a soluble partial-alicyclic copolyimide based on bicyclo[2.2.2]oct-7-ene-2,3,5,6-tetracarboxylic dianhydride. Both structural forms, the polymer precursor, PAA, and the PI, respectively, were used as working solutions, together with polyvinyl alcohol, which was elected as porogen and polymer stabilizer.[55] In the scanning electron microscopy images from Figure 1.4, one may notice spherical shapes with diameters in the range of 10–700 nm for the PI Ps obtained from PAA solutions, and also shapes like sphere, ellipse or even formless PI Ps for those prepared from the PI solution. The resulting PI Ps were thermally stable around 300°C;[55]
- porous cross-linked (co)polymers particles as bead shape, by using suspension polymerization method were prepared, considering the unsaturated nature of the used polymer chemical structure. The polymer was cross-linked by radical copolymerization with a bismaleimide, in presence of two pairs of porogens. The polymer beads morphology, including aspects concerning specific and apparent densities, porosity, surface area and mean pore diameter, and pore volume, depending on various physical and chemical aspects was reported.[37] For some chosen samples thermal behavior and tendency to swell in different organic diluents were also determined. When the cross-linker amount is around 40%, the PI beads are thermally stable

above 400°C, presenting a pore volume of 1.28 mL/g, a surface area of 74.20 m²/g, and enhanced swelling abilities.

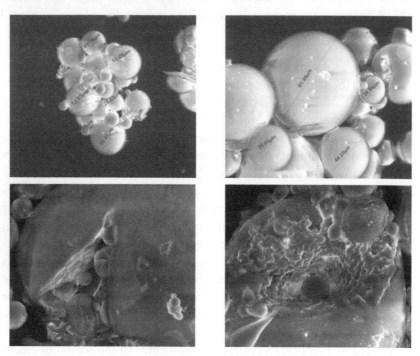

FIGURE 1.4   Scanning electron microscopy images showing examples of PI particles with various shapes.

1.6   GREEN CHEMISTRY AND POLYIMIDES

In the past decades, green chemistry concepts were used for PI synthesis in order to improve specific properties.

Luo and co-workers[57] have prepared chromophores and PIs by using with a green chemistry approach to design materials for nonlinear optical applications, particularly for high-speed electrooptic devices with very broad bandwidth (up to 150 GHz) and low driving voltages. Their research group developed chromophore-containing dendrimers that display very interesting nanostructures and electrooptic activities. The efficiency of these materials is affected by size, shape, and molecular architecture of the dendrimer. Incorporation of such dendrimers in linear polymer enhanced the poling efficiency and diminished the synthetic time for electrooptic materials. However, to achieve better thermal stability Luo and co-workers[57] have used aromatic PIs with dendronized nonlinear optic chromophores that are functionalized on a cardobisphenol linkage of a rigid imidic polymer backbone. The modified PIs having a cardo structure present high glass transition temperature, good solubility, and great thermal stability. The application of the site isolation principle to a rigid 3-D cardo-type PI led to materials with a large electro-optic coefficient, namely 71 pm/V at 1.3 µm. More than 90% of this value was maintained at 85°C for a large period of time (~650 h).

Baumgartner and collaborators[58] synthesized highly ordered PIs using the geomimetics and green polymer chemistry concepts. They applied one-step hydrothermal (HT) polymerization to a tough polymer, namely poly(*p*-phenyl pyromellitimide) (PPPI). The resulting material is highly ordered in fully imidized form and it consists of crystalline flakes and flowers at the micrometer scale. In comparison with the classical two-step procedures, the HT method does not require long reaction times and does not involve toxic solvents and catalysts. Moreover, the HT polymerization enables full conversion in only 1 h at 200°C, in hot water. Microscopy investigations revealed that PPPI aggregates form via a dissolution–polymerization–crystallization process. A PI achieved by conventional procedures is not able to recrystallize hydrothermally supporting the idea that HT polymerization and crystallization take place simultaneously. The highly ordered PI exhibits excellent thermal stability up to 600°C under nitrogen atmosphere, which arises from cumulative effects of strong, covalent polymer backbone, and interchain hydrogen bonding.

Leimhofer and co-workers[59] obtained inorganic–organic hybrid materials by green one-pot synthesis and processing. They used a semiaromatic PI derived from pyromellitic dianhydride and hexamethylene diamine. Both PI/silica have co-condensed hydrothermally, using only the respective precursors and water. Furthermore, it was proved that the polymer and the silica component were able to covalently connect under HT conditions, with the help of the compatibilizer (3-aminopropyl)-triethoxysilane. The thermoplasticity of the PI enabled processing both the matrix and its silica-based composite via sintering, which is a green method since it does not involve solvents.

Doddamani and its group[60] have also used a green chemistry approach to manufacture nonlinear optical PI materials. Their idea relied on coupling *N*-methyl-*N*-(2-hydroxyethyl)-4-amino benzaldehyde with barbituric acid, 1,3-indanedione, and 1,3-diethyl-2-thiobarbituric acid with the role of acceptors through stilbene linkage. The main advantage of this concept was that the synthesis reactions were performed in less than 10 min at room temperature and required no other solvents but water. The reaction of poly(hydroxy-imide)s with chromophores through Mitsunobu procedure led to two different side-chain PIs. The imidic polymers were mainly developed based on aromatic dianhydrides (4,4'-(hexafluoroisopropylidene) diphthalic anhydride or pyromellitic dianhydride) and 4,4'-diamino-4''-hydroxytriphenylmethane. The resulting materials have good solubility in polar aprotic solvents, being easy to process. The samples display different molecular orientations after electrical poling. The Maker fringe method allowed determining the second harmonic generation coefficients of the corona-poled PI foils, indicating values that vary between 59.33 and 77.82 pm/V. Raised thermal endurance was noticed for the PIs as a result of the extensive hydrogen bonds in the matrix. The reported materials present no decay in second harmonic generation signals under 110°C, revealing their suitability for nonlinear optical devices.

1.7 AROMATIC-ALICYCLIC POLYIMIDES AND THEIR PRACTICAL IMPORTANCE

The aromatic-alicyclic PIs can be easily processed under the form of transparent coatings and films. They mainly posses low dielectric constant, high transparency, good adhesion, and mechanical resistance combined with high heat resistance. PIs containing alicyclic segments are known to be promising materials for a variety of industries, namely in:

- *electronics* and *optoelectronics*: PIs are used as a substrate of flexible printed circuits or boards, heat resistant insulators, buffer coatings, passivation layers, liquid crystal (LC) alignment substrates, and electrochromic layers with very high contrast ratio;[61]
- *nonlinear optics:* PIs have good nonlinear optical responsibility;[62]
- *membranes*: depending on their molecular architecture PIs are useful as gas and chemical separation membranes;[63]
- *energy production devices*: transparent PI coatings with high refractive index are suitable materials for shielding solar cells;[64]
- *biomedicine*: PIs with good optical clarity allow observation of cell growth, while in some cases PIs are hemocompatible materials. In addition, functionalized PIs have the ability to link on their surface biocide agents or to become hemocompatible/cytocompatible.[65–68]

Figure 1.5 illustrates the main application fields of aromatic-alicyclic PI materials. Partially alicyclic PI-based micro and nanoparticles have received special attention for their large specific surface area and outstanding thermal performances which are adequate for applications as high-temperature thermal insulators.[37] Besides their widely known advantageous properties, PIs containing alicyclic units could be modified to possess a variety of functionalities, like photo-reactivity, molecular recognition ability, light emitting ability, thus widening the range of applications.[2,11,29]

In some cases, the bulk properties can be exploited to achieve high-performance interlevel dielectrics[69], permeation membranes[3], encapsulants[70] for optoelectronic devices,[71] and so on. In other situations, the imidic polymer film requires a pretreatment to adapt to the surface features for the pursued application. For instance, when using the PI film as orientation layer[72] of nematic molecules one may need to induce a surface anisotropy to accomplish the uniform alignment of the LC. For other purposes, when a biocide or a metal must be linked to the PI foil, the surface should be treated by chemical or physical methods to enhance the polarity, roughness, and interfacial adhesion.[73]

In the following section, the main practical uses of aromatic-alicyclic PIs are shortly reviewed.

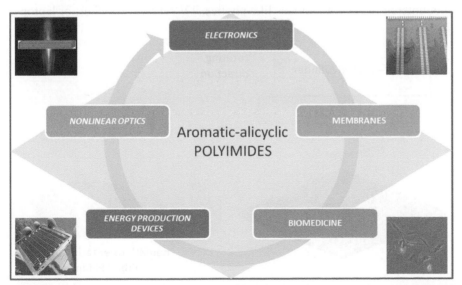

FIGURE 1.5 Schematic presentation of main application fields of aromatic-alicyclic PI materials.

1.7.1   SURFACE-RELATED APPLICATIONS

The modification of PI film surface for a pursued application is widely used, when designing self-cleaning surfaces or LC orientation layers.

The introduction of alicyclic sequences in PI structure enhances optical transparency and interfacial adhesion of PI with nematic molecules. Rubbing with clothes, like velvet, is a classical method applied in the LC display industry.[74] The contact of polymer film surface with textile fibers changes the surface by formation of microgrooves rendering an anisotropic character. Figure 1.6 displays the basic procedure of PI surface rubbing. This relies on using a velvet-covered cylinder, which rotates in close contact with the imidic polymer film surface. Depending on the chemical nature of the fibers from which the rubbing material is made, one may notice microgrooves of different dimensions and various degrees of uniformity.[74] An alternative to rubbing is photo-alignment but this procedure is restricted for photo-sensitive PI structures.[75] Regarding this topic, Song et al.[75] prepared PIs based on 1,2,3,4-cyclobutanetetracarboxylic dianhydride as the photo-alignment layers by UV exposure and thermal curing. The resulted substrates were able to induce uniform LC texture. The nematic molecules are orienting on rubbed or UV patterned PI films and tend to achieve a uniform tilt angle. This aspect is beneficial for the performance of the LC displays, which are encountered in several devices like phones, TV screens, watches, or laptops.

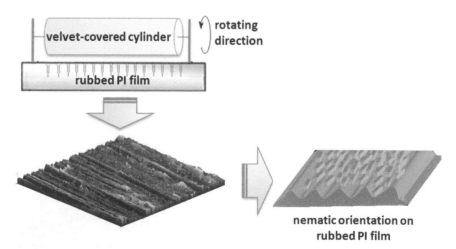

FIGURE 1.6   Schematic presentation of rubbing of the PI films.

Chemical modification of PI surface with polyhedral oligomeric silsesquioxane opened perspectives in manufacturing self-cleaning materials.[76] Oktay et al.[77] prepared segmented PI–siloxane copolymers with a variable amount of aminopropyl terminated polydimethylsiloxane for self-cleaning purposes. The PAA was subjected to electrospinning and thermal imidization to achieve the electrospun PI–siloxane mats. The resulting surfaces present nanoroughness and superhydrophobicity (contact angle ~142–165°) similar to the lotus-leaf effect recommending them for the pursued application.

1.7.2 SOLAR CELLS

PIs with good thermal stability, transparency, and high refractive index are wonderful candidates as encapsulant materials for solar cells. In such photovoltaic layered devices light reflection might take place at the interfaces if there are important differences in the refractive index of each layer, as presented in Figure 1.7. For this reason, PI encapsulant must be optically transparent and should have a refractive index close to that of the adjacent layer. Hulubei et al.[64] have synthesized semialicyclic PIs containing thioether linkages with a refractive index close to transparent conductive oxide. The presence of sulfide groups combined with alicyclic units render PI materials with high thermo-oxidative stability, small water absorption, and good transparency. The imidic polymer allows transmission of light 92–93% toward the solar cell junction after traveling through air/PI and PI/transparent conductive oxide interfaces. This aspect indicated that the reported PI coating is adequate for shielding solar cell devices.

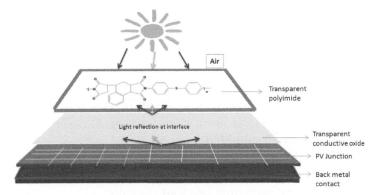

FIGURE 1.7 Schematic presentation of light reflection at interface in a solar cell shielded by a transparent PI coating.

1.7.3 BIOMEDICINE

There are some reports that describe the biomedical applicability of PIs. Therefore, such materials are useful in manufacturing blood-contacting devices or in preparation of antibacterial medical tools.

For blood-contact devices, it desirable to use hemocompatible materials. Popovici and collaborators[67] developed a series of imidic copolymers containing 1,6-diaminohexane sequences and tested their blood compatibility. Surface wettability and interfacial tension were affected by the chemical composition of the copolymer. The higher amount of the aliphatic counterpart in the PI material determines a higher rejection of the immunoglobulin G and fibrinogen. The adhesion interactions of blood cells (thrombocytes and erythrocytes) with PI surface are diminished as the aliphatic segment prevails. The rejection of platelets is indicative of thrombosis prohibition. Clinical investigations revealed that at the bio-interface there is no noticeable aggregation of platelets and no erythrocytes lyses phenomena. Optical microscopy was a useful tool to analyze if the imidic compounds affect the morphology of the blood components. These are positive aspects of hemocompatible materials,[68] which can be used in blood-contacting medical devices, such as catheters, extracorporeal circuits, or cardiac devices.

When designing antibacterial tools, one should use materials having bacteriostatic or bactericidal characteristics. Ioan et al.[65] investigated some PIs derived from alicyclic dianhydrides synthesized *via* classical two-step polycondensation. A correlation between surface wettability and antibacterial behavior was found. The antibacterial activity of the semialiphatic PI films was tested against *E. coli* and *S. aureus*. The electrostatic stacking of the PIs affected the bacterial metabolism. In this way, the specific compositions of the cell wall of Gram-negative (*E. coli*) and Gram-positive (*S. aureus*) bacteria generate distinct antimicrobial activity. The counterpart of Gram-positive bacteria cell walls is peptidoglycan, that renders the hydrophobic nature of *S. aureus*, whereas the peptidoglycan from the Gram-negative bacteria combined with lipopolysaccharides and proteins, confer the hydrophilic character of *E. coli*.[65] It was revealed that *S. aureus* is more inhibited than *E. coli* by the PI films with higher surface hydrophobicity.

The precise mechanism of the bacteria inhibition was not fully clarified, but it was clear that there is a connection between the surface wettability of the imidic polymer and bacterium cell wall structure.

1.8 CONCLUDING REMARKS AND FUTURE PERSPECTIVES

The engineering of imidic polymers is a research area in continuous development that still leads to the progress of current technologies. Aromatic-alicyclic PIs have emerged as a solution to solve classical issues of wholly aromatic ones. The advantages of this category of imidic polymers arise from their improved processability, optical, and dielectric properties. All these allow manufacturing a larger assortment of products for many industries. Depending on the used processing approach one may obtain materials with adapted properties for specific applications ranging from electronics to biomedicine.

Worldwide research activities are directed toward PIs with an optimized balance between stiffness, free volume, and polarity enabling to upgrade the functionality of the devices, like light emitting diode, organic transistors, displays, or solar cells. The future of this domain relies not only on structural optimization but also on perspective of interface engineering to achieve devices with better reliability.

Despite remarkable progress in the design of PIs, there still remain certain challenges concerning the synthesis by less polluting procedures to attain "green" products. This should be seriously taken into account by researchers considering the alarming amount of wastes produced by the electronic industry. Therefore, green chemistry concepts should be more often applied in production of PI-based products.

The relatively recent found biocompatibility of PIs must be explored more deeply when designing biomedical devices, like epidermal electronics, cardiac electrophysiological monitoring, or organic digital mechanoreceptor. Advanced procedures of physical/chemical modifications of PI surface should be researched to achieve a more stable surface functionality, which is a key factor in numerous applications.

KEYWORDS

- **PI**
- **alicyclic**
- **synthesis**
- **processability**
- **green chemistry**
- **applications**

REFERENCES

1. Lu, F. Some Heterocyclic Polymers and Polysiloxanes. *J. Macromol. Sci. Part C.* **1998,** *38*, 143–205.
2. Ghosh, M. K.; Mittal, K. L. *Polyimides: Fundamental and Applications*; Marcel Dekker: New York, 1996.
3. Favvas, E. P.; Katsaros, F. K.; Papageorgiou, S. K.; Sapalidis, A. A.; Mitropoulos, A. Ch. A Review of the Latest Development of Polyimide Based Membranes for CO_2 Separations. *React. Funct. Polym.* **2017,** *120*, 104–130.
4. Thompson, L. F.; Wilson, C. G.; Tagawa, S. *Polymers for Microelectronics: Resists and Dielectrics.* ACS Symposium Series: Washington, DC, 1994.
5. Périchaud, A. A.; Iskakov, R. M.; Kurbatov, A.; Akhmetov, T. Z.; Prokohdko, O. Y.; Razumovskaya, I. V.; Bazhenov, S. L.; Apel, P. Y.; Voytekunas, V.Yu.; Abadie, M. J. M. Auto-reparation of Polyimide Film Coatings for Aerospace Applications Challenges & Perspectives. In *High Performance Polymers - Polyimides Based - from Chemistry to Applications*, Abadie, M., Ed.; InTech: Croatia, 2012.
6. Ma, X.-H.; Yang, S.-Y. Polyimide Gas Separation Membrane. In *Advanced Polyimide Materials Synthesis, Characterization and Applications*; Yang, S.-Y., Ed.; Elsevier: Amsterdam, 2018; pp 257–322.
7. Lee, D.; Lim, Y. W.; Im, H. G.; Jeong, S.; Ji, S.; Kim, Y. H.; Choi, G. M.; Park, J. U.; Lee, J. Y.; Jin, J. et al. Bioinspired transparent laminated composite film for flexible green optoelectronics. *ACS Appl. Mater. Interfaces* **2017,** *9*, 24161–24168.
8. Kim, M.; Park, J.; Ji, S.; Shin, S. H.; Kim, S. Y.; Kim, Y. C.; Kim, J. Y.; Park, J. U. Fully-integrated, Bezel-less Transistor Arrays Using Reversibly Foldable Interconnects and Stretchable Origami Substrates. *Nanoscale* **2016,** *8*, 9504–9510.
9. Richardson, R. R. Jr.; Miller, J. A.; Reichert, W. M. Polyimides as Biomaterials: Preliminary Biocompatibility Testing. *Biomaterials* **1993,** *14*, 627–635.
10. Teo, A. J. T. Mishra, A.; Park, I.; Kim, Y.-J.; Park, W.-T.; Yoon, Y.-J. Polymeric Biomaterials for Medical Implants and Devices. *ACS Biomater. Sci. Eng.* **2016,** *2*, 454–472.
11. Mathews, A. S.; Kim, I.; Ha; C.-S. Synthesis, Characterization, and Properties of Fully Aliphatic Polyimide and their Derivatives for Microelectronics and Optoelectronics Applications. *Macromol. Res.* **2007,** *15*, 114–128.
12. Khan, M. U.; McGrath, J.; Corbett, B.; Pemble, M. Air-clad Broadband Waveguide Using Micro-molded Polyimide Combined with a Robust, Silica-based Inverted Opal Substrate. *Opt. Mater. Express.* **2017,** *7*, 3155–3161.
13. Ioan, S.; Hulubei, C.; Popovici, D.; Musteata, V. E. Origin of Dielectric Response and Conductivity of Some Alicyclic Polyimides. *Polym Eng. Sci.* **2013,** *53*, 1430–1447.
14. Ando, S.; Matsuura, T.; Sasaki, S. Coloration of Aromatic Polyimides and Electronic Properties of their Source Materials. *Polym. J.* **1997,** *29*, 69–76.
15. Ando, S.; Harada, M.; Okada, T.; Ishige, R. Effective Reduction of Volumetric Thermal Expansion of Aromatic Polyimide Films by Incorporating Interchain Crosslinking. *Polymers (Basel).* **2018,** *10*, 761.
16. Simone, C. D.; Vaccaro, E.; Scola, D. A. The Synthesis and Characterization of Highly Fluorinated Aromatic Polyimides. *J. Fluor. Chem.* **2019,** *224*, 100–112.
17. Butnaru, I.; Hamciuc, E.; Brumă, M.; Szesztay, M. Modified Aromatic Polyimides with Flexible Groups. *Rev. Roum. Chim.*, **2009,** *54*, 1023–1029.

18. Wang, X.; Li, Y.; Gong, C.; Zhang, S.; Ma, T. Synthesis and Characterization of Novel Soluble Pyridine-Containing Polyimides Based on 4-phenyl-2,6-bis 4-(4-aminophenoxy) phenyl-pyridine and Various Aromatic Dianhydrides. *J. Appl. Polym. Sci.* **2007**, *104*, 212–219.

19. Chou, C. H.; Reddy D. S.; Shu, C. F. Synthesis and Characterization of Spirobifluorene-based Polyimides. *J. Polym. Sci. Part A.* **2002**, *40*, 3615–3621.

20. Chung, I. S.; Kim, S. Y. Soluble Polyimides from Unsymmetrical Diamine with Trifluoromethyl Pendent Group. *Macromolecules.* **2000**, *33*, 3190–3193.

21. Hwang, H. J.; Li, C. H.; Wang, C. S. Dielectric and Thermal Properties of Dicyclopentadiene Containing Bismaleimide and Cyanate Ester. Part IV. *Polymer.* **2006**, 47, 1291–1299.

22. Hulubei, C.; Popovici, D. Novel Polyimides Containing Alicyclic Units. Synthesis and Characterization. *Rev. Roum. Chim.* **2011**, 56, 209–215.

23. Eichstadt, A. E.; Ward, T. C.; Bagwell, M. D.; Farr, I. V.; Dunson, D. L.; McGrath, J. E. Synthesis and Characterization of Amorphous Partially Aliphatic Polyimide Copolymers Based on Bisphenol-A dianhydride. *Macromolecules.* **2002**, *35*, 7561–7568.

24. Hasegawa, M. Development of Solution-processable, Optically Transparent Polyimides with Ultra-low Linear Coefficients of Thermal Expansion. *Polymers.* **2017**, *9*, 520.

25. Hasegawa, M.; Horie, K. Photophysics, Photochemistry, and Optical Properties of Polyimides. *Prog. Polym. Sci.* **2001**, *26*, 259–335.

26. St. Clair, A. K.; Slemp, W. S. Evaluation of Colorless Polyimide Film for Thermal Control Coating Applications. *SAMPE J.* **1985**, *21*, 28–33.

27. Mathews, A. S.; Kim, D.; Kim, Y.; Kim, I.; Ha, C.-S. Synthesis and Characterization of Soluble Polyimides Functionalized with Carbazole Moieties. *J. Polym. Sci. Part A.* **2008**, *46*, 8117–8130.

28. Bessonov, M. I.; Zubkov, V. A. *Polyamic Acid and Polyimides: Synthesis, Transformation and Structure*; CRC Press: Boca Raton, FL, 1993.

29. Ando, S.; Ueda, M.; Kakimoto, M.; Kochi, M.; Takeichi, T.; Hasegawa, M.; Yokota, R. (Eds.) *The Latest Polyimides: Fundamentals and Applications*; 2nd ed.; NTS: Tokyo, Japan, 2010.

30. Sroog, C. E. Polyimide. *Prog. Polym. Sci.* **1991**, *16*, 561–694.

31. Sroog, C. E. Polyimides. *J. Polymer Sci. Macromol. Rev.*, **1976**, *11*, 161–208.

32. Popovici, D.; Barzic, A. I.; Barzic, R. F.; Vasilescu, D. S.; Hulubei, C. Semi-alicyclic Polyimide Precursors: Structural, Optical and Biointerface Evaluations. *Polym. Bull.* **2016**, *73*, 331–344.

33. Park, S. K.; Farris, R. J. Dry-jet Wet Spinning of Aromatic Polyamic Acid Fiber Using Chemical Imidization. *Polymer.* **2001**, *42*, 10087–10093.

34. Cano, C. I.; Clark, M. L.; Kyu, T.; Pipes, B. Modeling Particle Inflation from Poly(amic acid) Powdered Precursors. III. Experimental Determination of Kinetic Parameters. *Polym. Eng. Sci.* **2008**, *48* (3), 617–626.

35. Gaw, K.; Jikei, M.; Kakimoto, M.-A.; Imai, Y.; Mochjizuki, A. Adhesion Behaviour of Polyamic Acid Cured Epoxy. *Polymer.* **1997**, *38*, 4413–4415.

36. Harito, C.; Porras, R.; Bavykin, D. V.; Walsh, F. C. Electrospinning of In Situ and Ex Situ Synthesized Polyimide Composites Reinforced by Titanate Nanotubes. *J. Appl. Polym. Sci.* **2017**, *134*, 44641.

37. Hulubei, C.; Vlad, C. D.; Stoica, I.; Popovici, D.; Lisa, G.; Nica, S. L.; Barzic, A. I. New Polyimide-based Porous Crosslinked Beads by Suspension Polymerization: Physical and Chemical Factors Affecting their Morphology. *J. Polym. Res.* **2014**, *21*, 514.

38. Chen, W.; Chen, W.; Zhang, B.; Yang, S.; Liu, C.-Y. Thermal Imidization Process of Polyimide Film: Interplay Between Solvent Evaporation and Imidization. Polymer. **2017**, *109*, 205–215.

39. Ayala, D.; Lozano, A. E.; de Abajo, J.; de la Campa, J. G. Synthesis and Characterization of Novel Polyimides with Bulky Pendant Groups. *J. Polym. Sci. A.* **1999**, *37*, 805–814.

40. Bower, G. M.; Frost, L. W. Aromatic Polyimides. *J. Polym. Sci.: Part A.* **1963**, *1*, 3135–3150.

41. Dine-Hart, R. A.; Wright, W. W. Preparation and Fabrication of Aromatic Polyimides. *J. Appl. Polym. Sci.* **1967**, *11*, 609–627.

42. Bessonov, M. I.; Koton, M. M.; Kudryavtsev, V. V.; Laius, L. A. *Polyimides. Thermally Stable Polymers*; Consultants Bureau: New York, 1987.

43. Ha, Y.; Choi, M.-C. Jo, N.; Kim, I.; Ha, C.-S.; Han, D.; Han, S.; Han, M. Polyimide Multilayer Thin Films Prepared Via Spin Coating from Poly(amic acid) and Poly(amic acid) Ammonium Salt. *Macromol. Res.* **2008**, *16*, 725–733.

44. Liu, F.; Liu, Z.; Gao, S.; You, Q.; Zoua, L.; Chena, J.; Liu, J.; Liu, X. Polyimide Film with Low Thermal Expansion and High Transparency by Self-enhancement of Polyimide/SiC Nanofibers Net. *RSC Adv.* **2018**, *8*, 19034–19040.

45. Method for Preparing Polyimide Film, U.S. 0292873 A1, 2013

46. Iwamoto, M.; Kubota, T.; Sekine, M. Electrical Properties of Polyimide Langmuir-Blodgett Films Deposited on Noble Metal Electrodes. *J. Phys D: Appl. Phys.* **1990**, *23*, 575.

47. Konkin, A. A. *Thermally Stable and Noncombustible Fibers*; Khimiya: Moscow, 1978.

48. Jacobs, V.; Anandjiwala, R. D.; Maaza, M. The Influence of Electrospinning Parameters on the Structural Morphology and Diameter of Electrospun Nanofibers. *J. Appl. Polym. Sci.* **2010**, *115*, 3130–3136.

49. Chisca, S.; Barzic, A. I.; Sava, I.; Olaru, N.; Bruma, M. Morphological and Rheological Insights on Polyimide Chain Entanglements for Electrospinning Produced Fibers. *J. Phys. Chem. B.* **2012**, *116*, 9082–9088.

50. Wright, W. W.; Hallden-Abberton, M. *Polyimides in Ullmann's Encyclopedia of Industrial Chemistry.* **2000**, DOI:10.1002/14356007.a21_253.

51. Nagata, Y.; Ohnishi, Y.; Kajiyama, T. Highly Crystalline Polyimide Particles. *Polym. J.* **1996**, *28*, 980–985.

52. Lin, T.; Stickney, K. W.; Rogers, M.; Riffle, J. S.; McGrath, J. E.; Marand, H.; Yu, T. H.; Davis, R. M. *Polymer.* **1993**, *34*, 772–777.

53. Chai, Z.; Zheng, X.; Sun, X. Preparation of Polymer Microspheres from Solutions. *J. Polym. Sci. Part B.* **2003**, *41*, 159–165.

54. Suzuki, M.; Kasai, H.; Ishizaka, T.; Miura, H.; Okada, S.; Oikawa, H.; Nihira, T.; Fukuro H.; Nakanishi, H. Fabrication of Size-controlled Polyimide Nanoparticles. *J. Nanosci. Nanotechnol.* **2007**, *7*, 2748–2752.

55. Hulubei, C.; Vlad, C. D.; Popovici, D.; Stoica, I.; Barzic, A. I.; Rusu, D. Polyimide Micro- and Nanoparticles via the Reprecipitation Method. *Rev. Roum. Chim.* **2016**, *61*, 797–804.

56. Katagi, H.; Kasai, H.; Okada, S.; Oikawa, H.; Matsuda, H.; Nakanishi, H. Preparation and Characterization of Poly-diacetylene Microcrystals. *J. Macromol. Sci Part A.* **1997,** *A34,* 2013–2024.
57. Luo, J.; Haller, M.; Li, H.; Tang, H.-Z.; Jen, A. K.-Y.; Jakka, K.; Chou, C.-H.; Shu, C.-F. A Side-chain Dendronized Nonlinear Optical Polyimide with Large and Thermally Stable Electrooptic Activity. *Macromolecules.* **2004,** *37,* 248–250.
58. Baumgartner, B.; Bojdys, M. J.; Unterlass, M. M. Geomimetics for Green Polymer Synthesis: Highly Ordered Polyimides via Hydrothermal Techniques. *Polym. Chem.* **2014,** *5,* 3771.
59. Leimhofer, L.; Baumgartner, B.; Puchberger, M.; Prochaska, T.; Konegger, T.; Unterlass, M. M. Green One-pot Synthesis and Processing of Polyimide–silica Hybrid Materials. *J. Mater. Chem. A.* **2017,** *5,* 16326–16335.
60. Doddamani, R. V.; Tasaganva, R. G.; Inamdar, S. R.; Kariduraganavar, M. Y. Synthesis of Chromophores and Polyimides with a Green Chemistry Approach for Second-order Nonlinear Optical Applications. *Polym. Adv. Technol.* **2018,** *29,* 2091–2102.
61. Zhang, Q.; Tsai, C.-Y.; Li, L.-J.; Liaw, D.-J. Colorless-to-colorful Switching Electrochromic Polyimides with Very High Contrast Ratio. *Nat. Commun.* **2019,** *10,* 1239.
62. Kim, E.-H.; Moon, I. K.; Kim, H. K.; Lee, M.-H.; Han, S.-G.; Yi, M. H.; Choi, K.-Y. Synthesis and Characterization of Novel Polyimide-based NLO Materials from Poly(hydroxy-imide)s Containing Alicyclic Units (II). *Polymer.* **1999,** *40,* 6157–6167.
63. Park, C.-Y.; Kim, E.-H.; Kim, J. H.; Lee, Y. M.; Kim, J.-H. Novel Semi-alicyclic Polyimide Membranes: Synthesis, Characterization, and Gas Separation Properties. *Polymer.* **2018,** *151,* 325–333.
64. Hulubei, C.; Albu, R. M.; Lisa, G.; Nicolescu, A.; Hamciuc, E.; Hamciuc, C.; Barzic, A. I. Antagonistic Effects in Structural Design of Sulfur-based Polyimides as Shielding Layers for Solar Cells. *Sol. Energ. Mater. Sol. Cells* **2019,** *193,* 219–230.
65. Ioan, S.; Filimon, A.; Hulubei, C.; Stoica, I.; Dunca, S. Origin of Rheological Behavior and Surface/Interfacial Properties of Some Semi-alicyclic Polyimides for Biomedical Applications. *Polym. Bull.* **2013,** *70,* 2873–2893.
66. Stoica, I.; Barzic, A. I.; Butnaru, M.; Doroftei, F.; Hulubei, C. Surface Topography Effect on Fibroblasts Population on Epiclon-based Polyimide Films. *J. Adhesion Sci. Technol.* **2015,** *29,* 2190–2207.
67. Popovici, D.; Barzic, A. I.; Stoica, I.; Butnaru, M.; Ioanid, G. E.; Vlad, S.; Hulubei, C.; Bruma, M. Plasma Modification of Surface Wettability and Morphology for Optimization of the Interactions Involved in Blood Constituents Spreading on Some Novel Copolyimide Films. *Plasma Chem. Plasma Process.* **2012,** *32,* 781–799.
68. Albu, R. M.; Hulubei, C.; Stoica, I.; Barzic, A. I. Semi-alicyclic Polyimides as Potential Membrane Oxygenators: Rheological Implications on Film Processing, Morphology and Blood Compatibility. *eXPRESS Polym. Lett.* **2019,** *13,* 349–364.
69. Chung, I. S.; Park, C. E.; Ree, M.; Kim, S. Y. Soluble Polyimides Containing Benzimidazole Rings for Interlevel Dielectrics. *Chem. Mater.* **2001,** *139,* 2801–2806.
70. Kim, T.; Lee, T.; Lee, G.; Choi, Y. W.; Kim, S. M.; Kang, D.; Choi, M. Polyimide Encapsulation of Spider-inspired Crack-based Sensors for Durability Improvement. *Appl. Sci.* **2018,** *8,* 367.
71. Tsai, C.-L.; Yen, H.-J.; Liou, G.-Sh. Highly Transparent Polyimide Hybrids for Optoelectronic Applications. *React. Funct. Polym.* **2016,** *108,* 2–30.

72. Barzic, A. I.; Hulubei, C.; Stoica, I.; Albu, R. M. Insights on Light Dispersion in Semi-alicyclic Polyimide Alignment Layers to Reduce Optical Losses in Display Devices. *Macromol. Mater. Eng.* **2018,** *303,* 1800235.

73. Nica, S.-L.; Hulubei, C.; Stoica, I.; Ioanid, E. G.; Nica, V.; Ioan, S. Electrical Resistivity Under Different Humidity Conditions for Plasma-treated and Gold-sputtered Polyimide Films. *Polym. Bull.* **2016,** 73, 1531–1544.

74. Stoica, I.; Barzic, A. I.; Hulubei, C. The Impact of Rubbing Fabric Type on Surface Roughness and Tribological Properties of Some Semi-alicyclic Polyimides Evaluated from Atomic Force Measurements. *Appl. Surf. Sci.* **2013,** *268,* 442–449.

75. Song, Y.; Yuan, L.; Wang, Z.; Yang, S. Photo-aligning of Polyimide Layers for Liquid Crystals. *Polym. Adv. Technol.* **2019,** *30,* 1243–1250.

76. Chandramohan, A.; Nagendiran, S.; Alagar, M. Synthesis and Characterization of Polyhedral Oligomeric Silsesquioxane–siloxane-modified Polyimide Hybrid Nanocomposites. *J. Compos. Mater.* **2012,** *46,* 773–781.

77. Oktay, B.; Toker, R. D.; Kayaman-Apohan, N. Superhydrophobic Behavior of Polyimide–siloxane Mats Produced by Electrospinning. *Polym. Bull.* **2015,** *72,* 2831–2842.

CHAPTER 2

Polyimides with Bulky Groups: Synthesis, Characterization, and Physical Properties

BARNALI DASGUPTA GHOSH[1], SUSANTA BANERJEE[2*],
ALEXANDER ALENTIEV[3*], INGA RONOVA[4], and YURI YAMPOLSKII[3]

[1]Department of Chemistry, Birla Institute of Technology, Mesra,
Ranchi 835215, India

[2]Materials Science Centre, Indian Institute of Technology,
Kharagpur 721302, India

[3]A. V. Topchiev Institute of Petrochemical Synthesis, Moscow 119991,
Russia

[4]A. N. Nesmeyanov Institute of Organoelement Compounds,
Moscow 119991, Russia

*Corresponding author. E-mail: susanta@matsc.iitkgp.ac.in;
alentiev1963@mail.ru

ABSTRACT

Aromatic polyimides are an important class of polymers used in modern technology, including membrane gas separation. Synthesis and study of polyimides with bulky groups in the elementary unit, high free volume, and attractive transport characteristics are modern trends in membrane materials science. The present study describes the methods of synthesis of polyimides used to obtain new materials with bulky or pendant, noncoplanar, kink, spiro, cardo, branched substituents, flexible, or unsymmetrical linkages in the backbone. Particular attention is paid to the gas separation characteristics of polyimides of this class. In permeability–selectivity diagrams, polyimides with bulky groups are located above the upper bound of the distribution. The reason for this behavior is an increase in the free volume and the growth of chain conformational rigidity due to hindering rotation of the fragments of the main chain, which is proved by numerical methods. Conformational factors lead to an increase in the permeability and overall selectivity of polyimides of this class and to a decrease in their permittivity. Thus, polyimides with bulky groups are promising materials for various applications.

2.1   INTRODUCTION

Aromatic polyimides (PIs) are known as high-performance polymers. This class of polymers attracted great attention because of their excellent set of physical and chemical properties. Though synthesis of aromatic PI was first reported by M. T. Bogert and R. R. Renshawas early as in 1908[1]; real interest to them appeared only in 1950 after the development of the two-step PI synthesis in DuPont Co.[2] Prepared polymers displayed outstanding thermal, mechanical, and electrical properties and found numerous applications in aerospace and electronic industries as adhesives and matrixes for composites.[3-6] Also, these polymers were gifted with high chemical and solvent resistant and excellent film forming ability that allowed their uses in different membrane-based applications.[7-10] Nonporous PI membranes characterized by high selectivity of gas separation.[7,8] It is for these reasons that more than 40% of the presented polymers (about 400) belong to this class[11] in the TIPS RAS Database on the transport properties of glassy polymers.[12] The relationship between chemical structure and membrane properties of PIs is discussed in a number of reviews and books.[7,8,13-15] On the basis of methods of computer modeling[16,17] and additive schemes of prediction of transport characteristics[18-21], the basic elements of the chemical structure of the elementary link[7,8,14-21] and the structure of the chain[8,17] responsible for the gas separation characteristics of PIs are revealed. Thus, it is known that the features of PI packaging, providing high selectivity of gas separation and, at the same time, good mechanical properties and high thermal stability, are determined by the increased cohesion energy due to the interaction of flat phenylimide cycles (so-called stacking). However, the same features of the packaging lead to a decrease in gas permeability. A well-known design element for obtaining highly permeable PIs is the use of structural elements with voluminous groups, or rigid kinks, which prevent the chain rotation and lead to the destruction of the stacking interaction. Thus, the most permeable in this group of polymers are PIs with structural elements characteristic of the so-called polymers "with intrinsic microporosity".[22] Less pronounced effects, but similar in conformational features of the chain, are observed for

PIs with rigid diamines with voluminous substituents.[23] On the other hand, poor solubility of PIs in organic solvents and exceptionally high glass transition or melting temperatures restrict their processing in melt and solution routes. Numerous approaches have been adopted to improve the processing characteristics of intractable PIs. One of the successful approaches to ease the processing characteristics of PIs is the incorporation of bulky side groups or bulky units in the polymer backbone.[9,23,24]

2.2 SYNTHESIS OF THE POLYIMIDES

PIs are prepared by condensation polymerization reaction of organic diamines with organic dianhydrides by one-step and two-step polymerization process.[3–5,9,25,26] A comprehensive route of synthesis of PIs from an aromatic dianhydride and an aromatic diamine is shown in Figure 2.1.

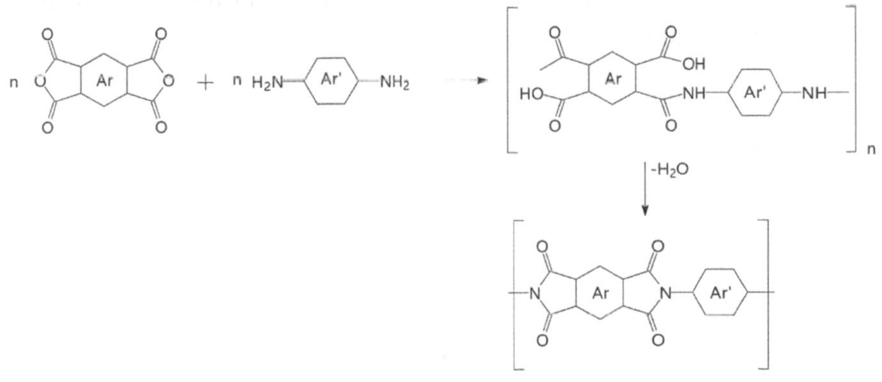

FIGURE 2.1 General reaction scheme of synthesis of polyimides.

2.2.1 ONE-STEP POLYMERIZATION

In this method, completely cyclized PIs are obtained directly from a molar equivalent mixture of dianhydrides and diamines, which are completely soluble in organic solvents at the polymerization temperature. In this method, the calculated amount of monomers is heated in a solvent having high boiling point or in a mixture of solvents in a temperature range of 140°C–250°C, that is, in high-temperature range, when the imidization reaction proceeds rapidly. The polymerization reaction is conducted in high boiling solvents such asp-chlorophenol, α-chloronaphthalene, m-cresol, nitrobenzene, o-dichlorobenzene, and dipolar aprotic amide solvents and their mixtures.[27–29] The PIs can also be synthesized using molten carboxylic acid, benzoic acid, salicylic acid, etc.[29–33] The water generated during the course of the reaction is continually removed by azeotropic distillation. The high temperature solution polymerization is often performed in the presence of catalysts, such as quinoline, tertiary amines, alkali metals, and zinc salts of carboxylic acids.[34–35] This direct polymerization is conducted typically when the final PI is soluble in the polymerizing solvent. The imidization also proceeds via the amic acid route; however, the concentration of amic acid at any time is very small. The amic acid group rapidly converts to an imide or reverts back to amine and dianhydride. This method of polymerization normally takes longer time (more than 18 h), uses carcinogenic solvents and often in polymerization the concentration must be less than 10% w/v that affects the quality final products (e.g., films and fibers). Representative examples of this type of polymer are the condensation product of 4,4′-(4,4-isopropylidenediphenoxy)bis(phthalic anhydride) (BPADA) and m-phenylene diamine (mPD) given in Figure 2.2.

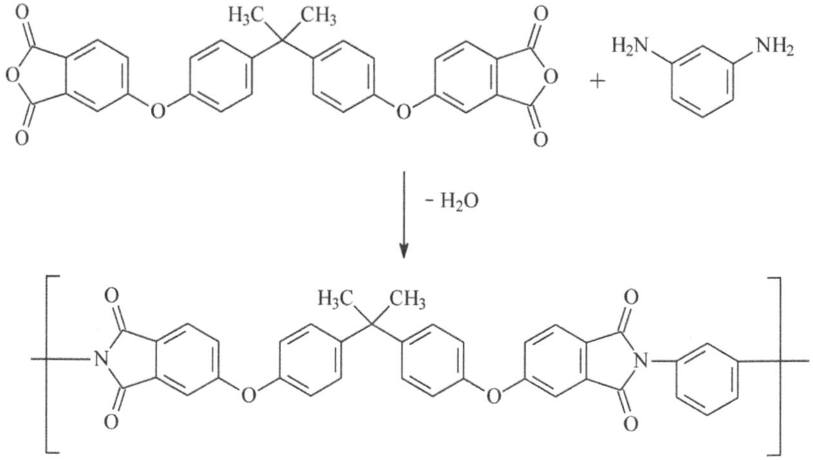

FIGURE 2.2   Synthesis of polyimide by single-step method.

2.2.2   *TWO-STEP POLYMERIZATION*

The two-step polymerization process is very frequently used procedure for the preparation of PIs. In this method, equimolar mixture of dianhydride and diamine reacts in a polar aprotic solvent, such as *N,N*-dimethylformamide (DMF), *N,N*-dimethylacetamide (DMAc), etc. to form the polyamic acid at initial stage. In Figure 2.3, general reaction pathway for the synthesis of polyamic acid is shown. In this process, the nitrogen atom of the amino group participates in the nucleophilic attack to the electrophilic carbonyl carbon atom of the anhydride group. Consequently, the anhydride ring opens to form an amic acid group.[36] Some of the important factors that regulate the formation of polyamic acid are the reactivity, stoichiometric ratio, purity of the monomers, choice of solvent, and reaction condition.

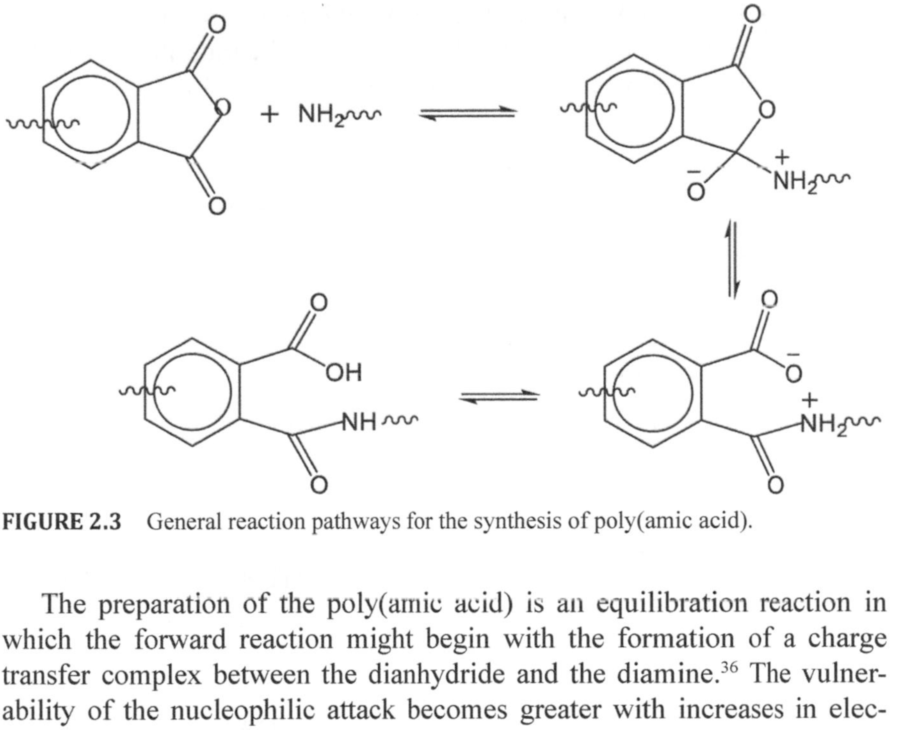

FIGURE 2.3 General reaction pathways for the synthesis of poly(amic acid).

The preparation of the poly(amic acid) is an equilibration reaction in which the forward reaction might begin with the formation of a charge transfer complex between the dianhydride and the diamine.[36] The vulnerability of the nucleophilic attack becomes greater with increases in electrophilicity of the dianhydride group. Therefore, the electron affinity of the dianhydride monomer is directly related to its reactivity. Higher the electron affinity, greater is the reactivity of the dianhydride.[37] Attachment of strong electron-withdrawing groups with the dianhydride triggers the nucleophilic attack on the anhydride carbonyl group.

The electron affinity values for some dianhydrides are given in Table 2.1. From the Table 2.1, the highest value of electron affinity (1.90 eV) is obtained for PMDA, whereas BPADA shows the lower value of 1.12 eV and EDA shows the lowest value of 1.10 eV.[38] The electron-withdrawing bridge group such as SO_2, CO, etc. enhances the E_a value substantially, while electron-donating group such as ester reduces the value. In BTDA, an electron-withdrawing carbonyl group is bridged, which decreases the electron density through the π-orbitals. So, the anhydride carbons experience a greater positive environment, which facilitates attack by a nucleophile. On the other hand, ODPA is linked by an oxygen atom that is able to donate electrons into the ring thereby decreasing the anhydride carbons' affinity for incoming electrons from an attacking nucleophile. Moreover, BPADA is linked to two such oxygen atoms and –CH_3 groups, which donate electrons into the ring, accordingly reducing the anhydride carbons' affinity for incoming electrons from an attacking nucleophile. The ether bridged dianhydride, that is, HQDA, BPADA, and EDA have low E_a value in the range 1.10–1.19 eV.

TABLE 2.1 Electron Affinity Values for Different Anhydrides.

Structure of the dianhydride	Dianhydride name	Electron affinity (eV)
	BPADA	1.12
	BPDA	1.38
	BTDA	1.55
	DSDA	1.57
	HQDA	1.19

TABLE 2.1   *(Continued)*

Structure of the dianhydride	Dianhydride name	Electron affinity (eV)
	ODPA	1.30
	PMDA	1.90
	EDA	1.10

On the other hand, the reactivity of the diamine is associated with its basicity, pK_a. As the pK_a of the protonated amine enhances, the rate constant for imidization process also increases, that is, an amine with a greater basicity will react faster. The pK_a values for some common diamines are given in Table 2.2. The diamine bridged by an electron-withdrawing group, for example, diaminobenzophenone, shows reduced nucleophilicity. If the dianhydride structure is fixed, the coloration of PIs is dominated by the electron-donating property of diamine. According to the charge transfer complex formation theory, increase in electron-donating properties of diamine and a larger E_a correspond to an increase in electron-accepting properties of dianhydrides that lead to a deep color in polyirnides.[39]

A lower molecular weight would be expected with the diamines and dian-hydrides of low reactivity, when compared to a highly reactive diamine and/ or dianhydride system. During the initial stages of the reaction, highly basic amines (e.g., aliphatic amines) may form salts, which upsets the stoichiom-etry and arrest the formation of high molecular weight polymer. However, solvents also play a vital role in this reaction. The commonly used solvents in poly(amic acid) preparation are dipolar aprotic solvents such as dimethyl-sulfoxide (DMSO), DMAc, DMF, and *N*-methyl pyrrolidone (NMP), which form strong hydrogen bonds with the carboxyl group that assists in shifting the equilibrium to the forward side of the reaction. As expected the reaction rate is generally rapid in more basic solvents. Several other minor side reac-tions get going with the main reaction. These side reactions may become more remarkable under certain conditions particularly when the main reaction is slower because of low monomer concentration and reactivity.

Besides these, the presence of water also affects the reaction. The presence of water in the reaction system causes lower molecular weight formation of poly(amic acid) due to the hydrolysis of the dianhydride moiety to form the diacid that reacts at a much slower rate with a given amine.[40,41] This is usually done by carefully drying the monomers and distilling the solvents. Water also promotes the reverse reaction. The moisture-free environment is one of the important factors for the preparation of poly(amic) acid. Cyclization of poly(amic) acid will also produce water and may facilitate chain hydrolysis. Three methods have been employed to facilitate the cyclization of poly(amic) acid and the removal of water. Cyclization and the removal of water can be achieved either by heating the poly(amic) acid in solution or as a cast film, or it can be chemically achieved by the addition of dehydrating reagents. During this step, the self-catalyzed cyclization to form PIs cannot take place due to the powerful interconnection between the amic acid and the basic solvent or the larger acylation equilibrium constant.[42] In the next step, cyclization of polyamic acid is taking place with the elimination of water molecule at elevated temperatures (thermal imidization) or in the presence of a cyclizing agent (chemical imidization).

TABLE 2.2 pK_a Values for Some Common Diamines.

Structure of the diamine	Diamine name	pK_a
	m-PDA	4.80
	p-PDA	6.08
	p-PTA	6.39
	4,4'-ODA	5.20

TABLE 2.2   *(Continued)*

Structure of the diamine	Diamine name	pK_a
	MDA	4.8
	Bz	4.60
	DAB	3.10

The advantages of two-step polymerization method over the one-step polymerization are the use of less toxic solvents and straight processing of the soluble polyamic acids to form the final PI products in the form of films or fibers by thermal imidization. However, the storage instability of polyamic acid intermediate and the control of thermal imidization are still crucial issues.[43] A detailed explanation of thermal imidization and chemical imidization of poly(amic acid) is discussed below.

2.2.2.1 THERMAL IMIDIZATION OF POLY(AMIC ACID)

The most common method for the transformation of the poly(amic acid) to the PI is the bulk (or melt) imidization process.[44–46] At elevated temperature, irreversible cyclodehydration reaction occurs that leads to the development of high molecular weight PI. The thermal imidization method is generally used in industry where poly(amic acid) is heated at ~200–300 °C for a given amount of time to form the imide ring by removing the solvent and the water. Films of the poly(amic acid)s are often cast from polar aprotic solvents such as *N,N*-DMF, *N,N*-DMAc, etc. then subsequently dried and imidized. This method is applicable for the formation of PIs in the form of films, powder, fibers, and also useful for coating in order to permit the diffusion of by-product and solvent without the formation of voids and bristles in the final PI products. The trouble of film cracking as a consequence of shrinking can be avoided by carefully controlling the curing profile. The thermal imidization includes a typical heating process below 150°C for overnight, followed by a relative rapid temperature rise to the second stage above the glass transition temperature T_g of the resulting PIs, resulting in the conversion of poly(amic acid) to PI.[47] The cast films are dried and heated cautiously up to 250–350°C depending on the stability and T_g of the PIs. Most of the solvents are slowly evaporate off in the first stage and imidization occurs in the second stage, where curing and shrinkage are reliable.[48] Two probable pathways for the imidization are possible during thermal imidization, and they are shown in Figure 2.4.[49] At the beginning stage of the imidization, a small amount of the poly(amic acid) undergoes a reversible reaction with the anhydride and amine in lieu of forming the imide ring resulting in a lower molecular weight development.[42,50] The heating cycle allows about 92–99% of transformation of poly(amic acid)s into corresponding imides. Further heating at above 300°C cannot lead to 100% conversion because of the so called "kinetic interruption" effect.[42] This results in the formation of defect sites and therefore complete imidization is more difficult to achieve.

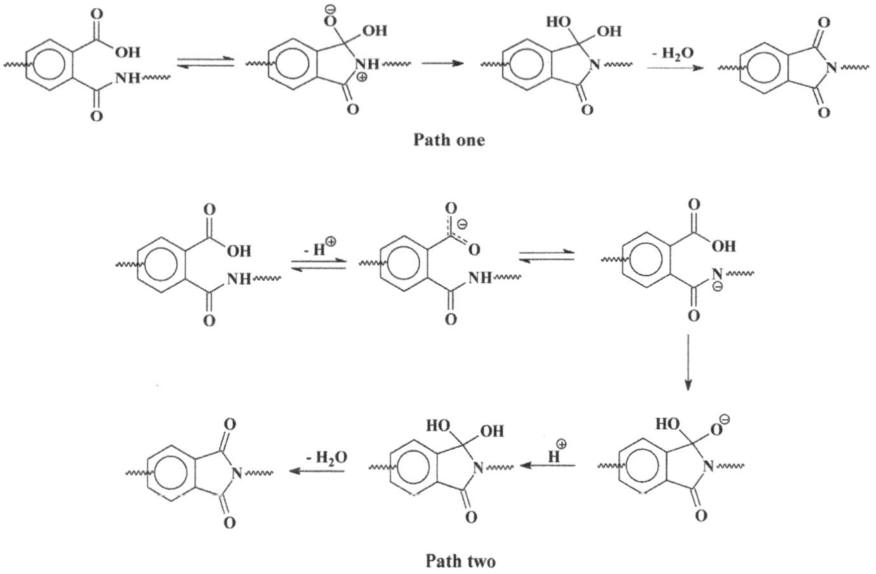

Path one

Path two

FIGURE 2.4   Two possible pathways in thermal imidization.

2.2.2.2   *CHEMICAL IMIDIZATION OF POLY(AMIC ACID)*

Poly(amic acid)s can also be chemically imidized. The chemical imidization of amic acid to imide needed the use of a chemical dehydrating agent in combination with basic catalysts.[51,52] These reagents may be used in low range of temperature (20–80°C). The commonly used chemical reagents are acetic anhydride, propionic anhydride, and *n*-butyric anhydride as dehydrating agents and pyridine, methylpyridine, isoquinoline, and triethylamine as basic chemical catalysts. The formation of PIs relies on the type of dehydrating agent, monomer, and the reaction temperature. For example, high molecular weight PIs can be achieved in the presence of trialkylamines with high pK_a values (>10.65), and in the presence of tertiary amine low molecular weight PIs may be obtained.[52] The time period of complete imidization decreases with the increase in imidization temperature. In the chemical polymerization process of polyamic acid to PI, the molecular mass of the resultant polymer remains the same because; in this case, depolymerization process does not occur like in thermal polymerization.[53] However, chemical imidization process is not much convenient for the chemical application, because it is an expensive method and also it involves complex steps. A reaction pathway for chemical imidization is shown in Figure 2.5.

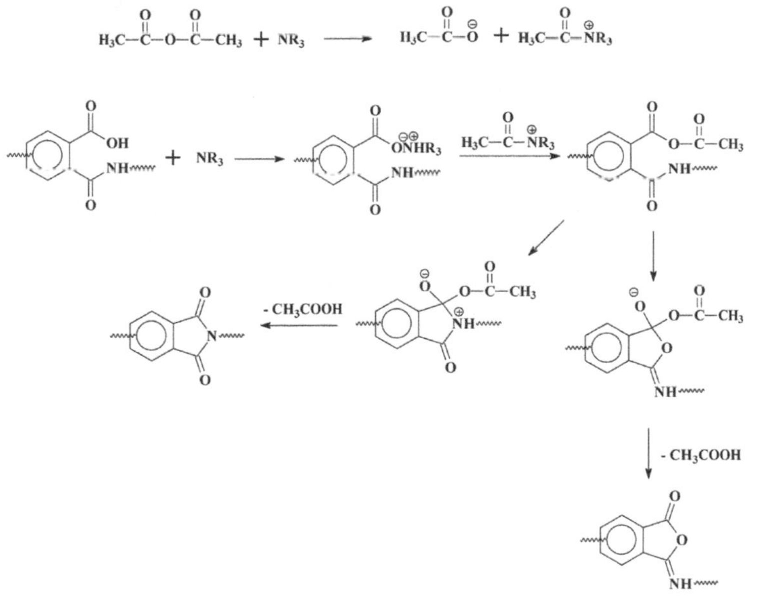

FIGURE 2.5   Reaction pathway in chemical imidization.

The chemistry of PIs itself covers an extensive area. The properties of PIs can be dramatically adjusted by slight variations in their structure. This includes a huge variation in available monomers and different synthetic methodologies.

Various numbers of monomers can be used for PI synthesis. The properties of PI can be changed with a minor modification in its backbone structure.

Therefore, monomer selection is a major concern in tailoring segment engineered PI with desirable properties. Depending upon the need, different structural modifications have been attempted, such as introduction of noncoplanar, kink, spiro, cardo, bulky or pendant, branched substituents, flexible, or unsymmetrical linkages in the backbone.[4,7,54–61]

Examples of monomers with noncoplanar conformation are 2,2`-substituted biphenylenes or binaphthyl derivatives (Fig. 2.6). Substitution in the 2,2' position makes the phenyl units to be in a noncoplanar conformation. This results from steric hindrance in this position. That decreases the intermolecular interactions and induces chain stiffness.[62]

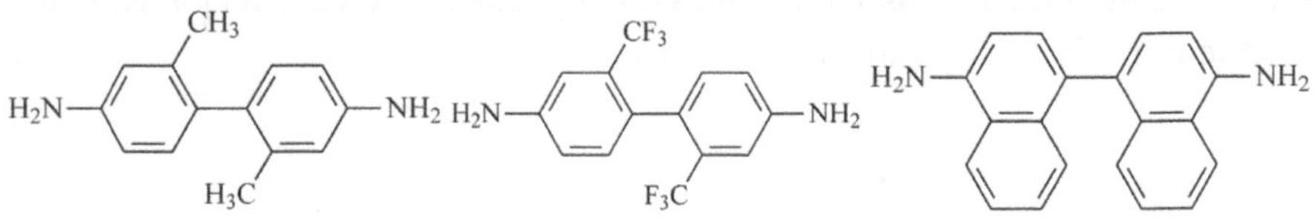

FIGURE 2.6   Structure of noncoplanar monomers.

A kink moiety has a twisted, crank-like, and non-coplanar structure. The kink structure in polymer chains prohibits chain alignments and inhibits the formation of charge transfer complexes. Using a kink structure in diamines (Fig. 2.7) decreases chain packing, enhances the gas solubility, and induces a high degree of crystallinity for PIs.[5,63,64]

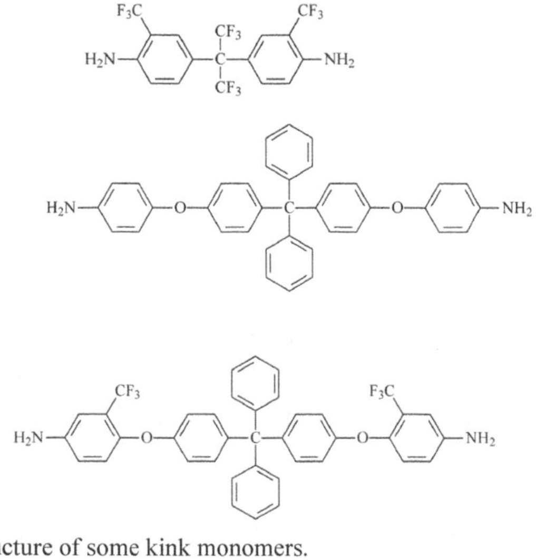

FIGURE 2.7   Structure of some kink monomers.

A spiro center is made up of two rings connected orthogonally through a tetrahedral bonding atom. Generally, a carbon atom acts as the spiro center. In the spiro structures containing PIs, generally, a polymer backbone is periodically twisted at 90° angles at each spiro center. This conformational feature resists the efficient packing of the polymer chains and decreases the probability of interchain interactions, that is, lower crystallinity. These polymers generally are more soluble and have a remarkably increased T_g and thermal stability. Different spiro diammines and dianhydrides were synthesized in different methods; some of their structures are shown in Figure 2.8.[65–68]

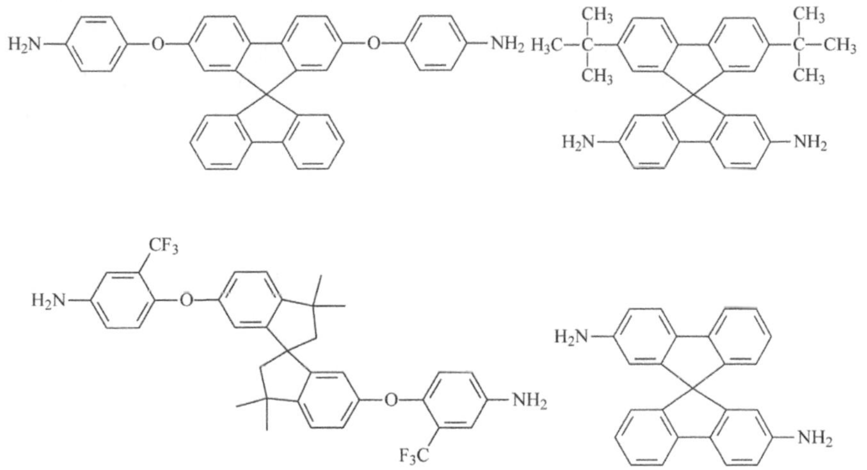

FIGURE 2.8   Structure of some spiro monomers.

Cardo is a Latin word that means "hinge" or "loop". Cardo polymers contain at least one repeating unit with a cyclic (loop shaped) side group are denoted as "cardo" polymers.[69,70] The spiro and cardo structures are very similar. The difference between the spiro and cardo structures is the number of rings at the center of their structure, that is, one ring connects to the center of cardo while two rings attach to a spiro center (Fig. 2.9).

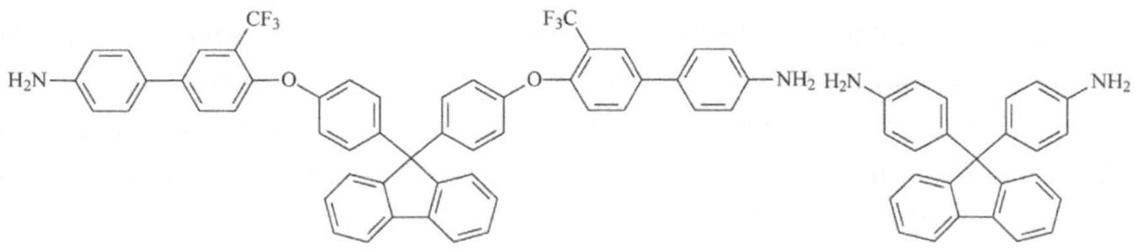

FIGURE 2.9   Structure of some cardo monomers.

The incorporation of asymmetrical diamines and dianhydrides (Fig. 2.10) in the main chain of a polymer can increase the free volume of the polymer. As interchain packing and intermolecular interactions of the polymer chains decrease, the solubility of the polymer increases. But the polymer chain rigidity does not change; therefore, thermal properties are maintained.[71–74]

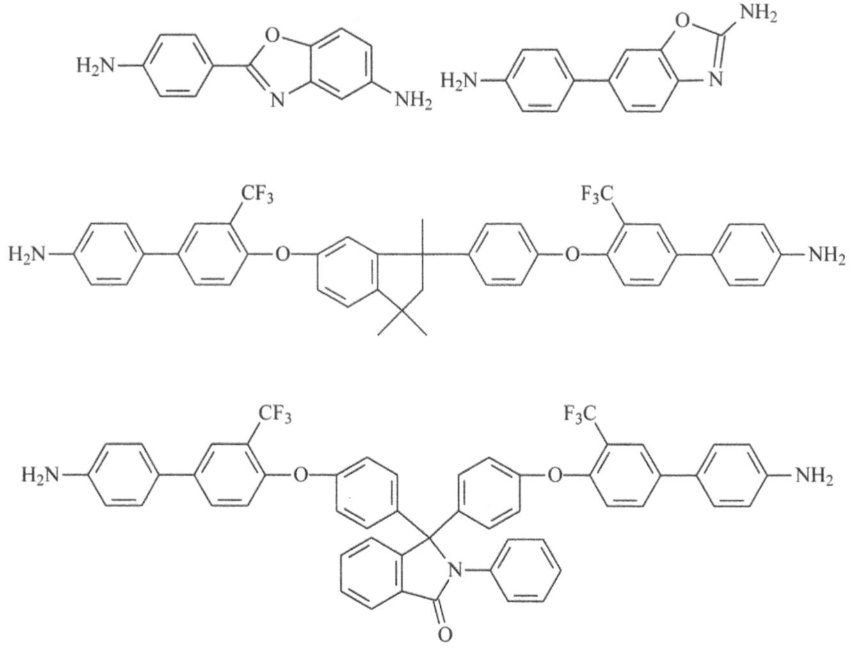

FIGURE 2.10   Structure of asymmetric monomers.

Incorporation of bulky side group or pendent groups (Fig. 2.11) improves solubility, process ability, and the thermal and gas transport properties. For example, introduction of tert-butyl side groups[75,76] increases the solubility of the PIs.

Kasashima et al.[77] showed that incorporation of *o*-terphenyldiyl structures provides thermally stable and soluble PIs. Incorporation of bulky, propeller-shaped packing disruptive triphenylamine (TPA) groups[78] into the polymer backbone enhances solubility of PIs without sacrificing thermal stability. Akutsu et al.[79] reported the improvement of solubility and the other properties of the aromatic PIs by introduction of rigid and zigzag structures, namely, 2,3-quinoxalinediyl.

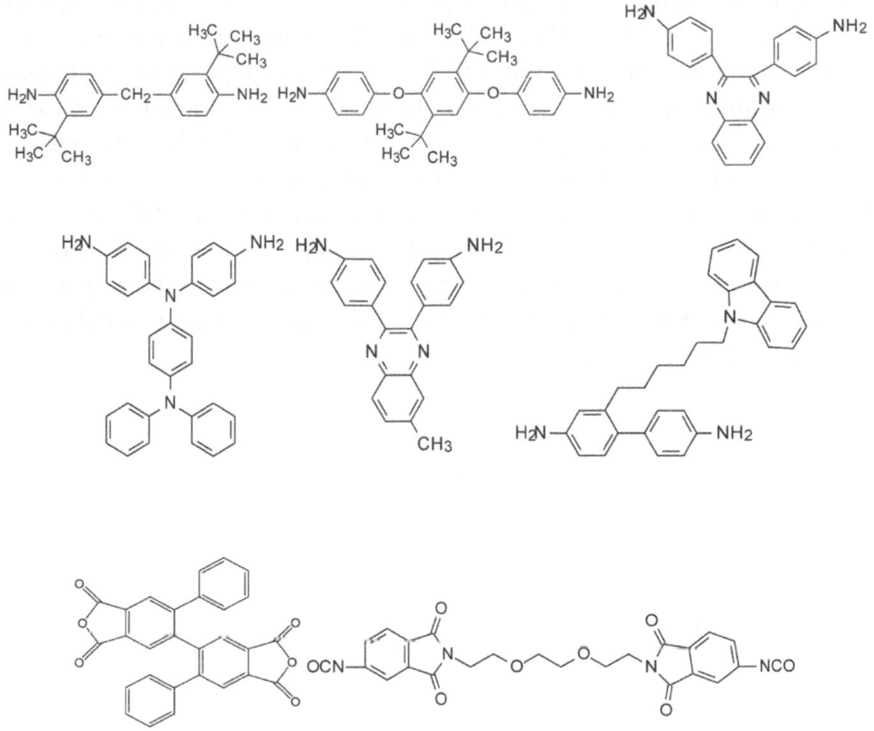

FIGURE 2.11 Structure of monomers with bulky pendent group and zigzag structure.

Mathews et al.[80] reported carbazole containing diamines, namely, 4″-carbazole-9-yl-[1,1′;2′,1″] terphenyl-4-4′-diamine (CzTPDA) and 2-(6-carbazol-9-yl-hexyl)-biphenyl-4,4′-diamine (CzHBPDA). They prepared the PI with the synthesized diamine and PMDA and 6FDA, which showed improved solubility. They noticed that the presence of aliphatic moieties imparted more flexibility than the aromatic groups, while the side chain units disturb the close chain packing of polymer chains leading to increased solubility.

Introduction of twisted biphenyl moiety in PI backbones was reported by the synthesized 2,2′-dibromo-4,4′-5,5′-biphenyltetracarboxylic dianhydride (DBBPDA) and 2,2′-diphenyl-4,4′-5,5′-biphenyltetracarboxylic dianhydride (DPBPDA) and their polymerization with 4,4′-diamino 2,2′-disubstituted biphenyls in refluxing *m*-cresol containing isoquinoline.[81] Presence of the biphenyl group in the 2- and 2′-positions, the solubility of the resulting PIs was improved. Yeganeh et al.[82] prepared a diisocyanate containing aliphatic oxyethylene moieties and with that they synthesized new soluble PIs with high thermal stability. The polymers were soluble in common aprotic solvents such as *NMP*, DMAc, *DMF*, and DMSO at ambient temperature.

It is found that PIs containing trifluoromethyl groups serve to increase the free volume, decreased intermolecular interactions, and reduced packing density, thereby improving various properties like solubility, gas permeability, electrical insulating properties, flame resistance, environmental stability, and optical transparency without affecting thermal stability to that of nonfluorinated one.[83] Because of all these unique properties of the fluorinated polymers, considerable attention has been devoted to the preparation of new classes of fluorinated aromatic PIs, especially trifluoromethyl (-CF₃) containing PIs (Fig. 2.12).[84–88]

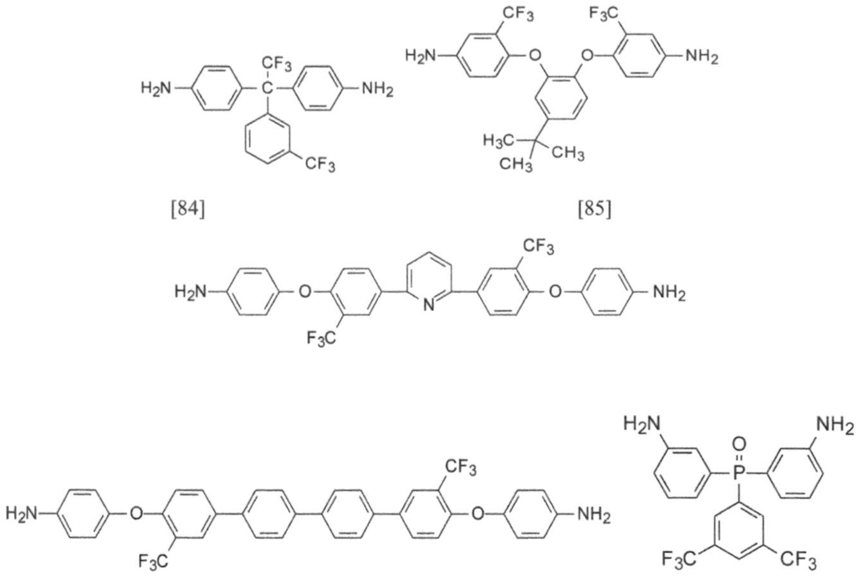

FIGURE 2.12   Monomers containing trifluoromethyl (-CF₃) group.

Ge et al.[84] prepared a fluorinated aromatic diamine and with this diamine, a series of fluorinated PIs with commercial aromatic dianhydrides such as PMDA, 6FDA, BTDA, and ODPA was prepared. The PIs showed good solubility in various solvents such as NMP, DMAc, DMF, *m*-cresol, THF, and CHCl₃.

A novel fluorinated diamine monomer 4-tert-butyl-[1,2-bis(4-amino-2-trifluoromethylphenoxy) phenyl]benzene with a *tert*-butyl group was synthesized by Yang et al.[85], and the PIs were prepared by reacting the monomer with various aromatic dianhydrides by thermal imidization method. The PIs were soluble in various organic solvents. This was due to the presence of tert-butyl group and the -CF₃ group in the molecular chain leading to increased solubility. The PIs were soluble in polar aprotic solvents such as NMP, DMF, DMAc, and DMSO and were partially soluble in less polar solvent like m-cresol, pyridine, THF, dioxane, CH_2Cl_2, and acetone.

Banerjee et al.[86,87] had introduced trifluoromethyl groups (-CF₃) along with the ether linkages in the PI backbone by designing new diamine monomers. The incorporation of ether linkage improved the mobility that ultimately led to improvement in solubility not only in aprotic solvents but also, to some extent, in protic solvents. Furthermore, a reduced dielectric constant and low moisture uptake, high optical transparency, and high gas permeability were also achieved with these types of fluorinated PIs.

Diamine monomer containing phosphine oxide and fluorine, bis(3-aminophenyl) 3,5-bis(trifluoroethyl)phenyl phosphine oxide (mDA6FPPO) was synthesized by Jeong et al..[88] This diamine monomer on reaction with different dianhydrides produced different PIs. The solubility of the PIs was greatly influenced by the phosphine oxide moiety and further by trifluoromethyl phenyl group. The excellent solubility of the mDA6FPPO-based PIs could be attributed to the presence of bulky -CF₃ groups. These bulk groups leading to enhanced free volume. The phosphine oxide segments provided strong intermolecular forces with solvent molecules.

2.3 EFFECT OF BULKY GROUPS ON GAS TRANSPORT PROPERTIES

The effect of bulky substituents on the gas transport properties of polyheteroarylenes (and in particular PIs) were discussed in the literature ever since the 90s.[89] Thus, the introduction of alkyl groups into the aromatic rings of the main chain leads to an increase in gas permeability due to decrease of the chain packing density and, as a rule, to a certain decrease in the selectivity of gas separation. The effect of such substitution on gas transport characteristics was examined for polyesters,[90–92] polycarbonates,[89,93] polysulphones,[89,94] polyamidoimides,[95] and aromatic polyamides.[96,97] Apparently, in this case, hindrance rotation of aromatic rings increases[98] and accordingly the chain rigidity increases,[99] which contributes to the decrease of the chain packing density.

Similar phenomena are observed for PIs (Table 2.3).

TABLE 2.3 Gas Separation Properties and Glass Transition Temperatures of Some Polyimides Based on 6FDA Dianhydride and Diamines with Side Alkyl Substituents.

PI No.	Diamine moiety	$P(O_2)$, Barrer[a]	$\alpha(O_2/N_2)$	T_g, °C	Reference
1	H_2N—C₆H₄—CH_2—C₆H₄—NH_2	3.15	5.6	296	[13]
		3.26	5.1		[100]
2	H_2N—C₆H₄—C(CH₃)₂—C₆H₄—NH_2 (central carbon bearing two CH_3)	5.13	5.9	310	[101]
3	H_2N—C₆H₃(H_3C)—CH_2—C₆H₃(CH_3)—NH_2	2.51	6.2	288	[13]
4	H_2N—C₆H₃(HC(CH₃)₂)—CH_2—C₆H₃(CH(CH₃)₂)—NH_2	9.22	5.4		[100]
5	H_2N—C₆H₃(H_3C—C(CH₃)₂)—CH_2—C₆H₃(C(CH₃)₂—CH_3)—NH_2	20.8	4.4		[100]
6	H_2N—C₆H₂(H_3C, H_3C)—CH_2—C₆H₂(CH_3, CH_3)—NH_2	10.3	4.8	306	[13]
		12.2	3.9		[100]
7	H_2N—C₆H₂(H₂C—CH₃, H₂C—CH₃)—CH_2—C₆H₂(H_3C—CH₂, CH₂—H_3C)—NH_2	20.7	4.0		[100]

TABLE 2.3   *(Continued)*

PI No.	Diamine moiety	P(O$_2$), Barrer[a]	α(O$_2$/N$_2$)	T$_g$, °C	Reference
8		50	3.5		[100]
9		4.22	5.3	351	[102]
		7.9	5.1	342	[103]
10		122	3.4	420	[102][104]
		125	3.6	424	
11		3.01	6.7	298	[102]
		2.61	7.3	285	[105]
		2.37	6.6	301	[106]
12		11	5.2	372	[102][105]
		11.3	5.0	335	
13		7.44	5.7	342	[105][107]
		115	3.6	383	
14		109	3.5	377	[102]
		140	3.9	394	[108]
		159	3.2	–	[95]

[a]1 Barrer = 10^{-10} cm^3(STP) cm/cm^2/s/cmHg

Thus, if in the case of methylation of the bridging carbon (PI 2) permeability increases insignificantly compared to (PI 1), the asymmetric methylation of aromatic rings (PI 3) even leads to a decrease in permeability and

T_g. Further increase in the volume of alkyl groups (PI 4, PI 5) leads to an increase in permeability. However, a more significant increase in permeability is observed in the case of symmetrical alkylation (PI 6–PI 8). The dramatic increase in permeability and T_g occurs in the case of the most rigid main chain (PI 9, PI 10). So, if for a pair of polymers PI 1/PI 5 oxygen permeability increases by a factor 3–4, then for a pair PI 9/PI 10 permeability increases by 30 times. The effect of the position of the substituent in the aromatic ring is also interesting. Thus, the example of PI based on methyl substituted derivatives of m-phenylenediamine (PI 11–PI 14) shows that the introduction of the methyl group to position 2 (PI 12) leads to an increase in T_g and to an increase in permeability 3.5- to 5-folds, which indicates an increase in chain rigidity. The introduction of the methyl group in position 6 (PI 13) does not lead to a noticeable increase in T_g compared with methylation in position 2 (PI 12). However, in this case, one may see an increase in the permeability by a factor 40 compared to the unsubstituted diamine (PI 11). The introduction of three methyl groups in positions 2, 4, and 6 (PI 14) leads to a significant increase in T_g (and chain rigidity, respectively) and to a 70-fold increase in permeability compared to unsubstituted diamine (PI 11). At the same time, for all PIs, the increase in permeability is accompanied by a decrease in selectivity.

In recent decades, some significant synthetic efforts were focused on improving the solubility and process ability as well as gas transport properties without sacrificing their thermal and mechanical properties.

Xu et al. synthesized a series of PIs containing laterally attached p-terphenyls and biphenyls moiety. Their effects on gas transport properties were also studied.[109] Laterally attached phenyl ring in the polymer backbone act as a bulky substituent group (Fig. 2.13).[109] These groups inhibit the polymer chain packing and increase the free volume of the PIs. The introduction of the rigid terphenyl unit (6FDA-terphenyl) caused a significant enhancement in gas permeability as compared to the unsubstituted analogue (6FDA-phenyl). The increase in gas permeability is due to the increase in gas diffusivity. The changes of gas permeability were according the following order: 6FDA-phenyl < 6FDA-biphenyl < 6FDA-terphenyl, which is in agreement of the free volume of these polymers. 6FDA-terphenyl showed the highest permeability of 21.48 Barrer for CO_2 and 5.26 Barrer for O_2. The PI 6FDA-biphenyl showed the high selectivity value of 36.23 for CO_2/CH_4 gas pair, and 6FDA-phenyl showed the selectivity value of 5.74 for O_2/N_2 gas pair.

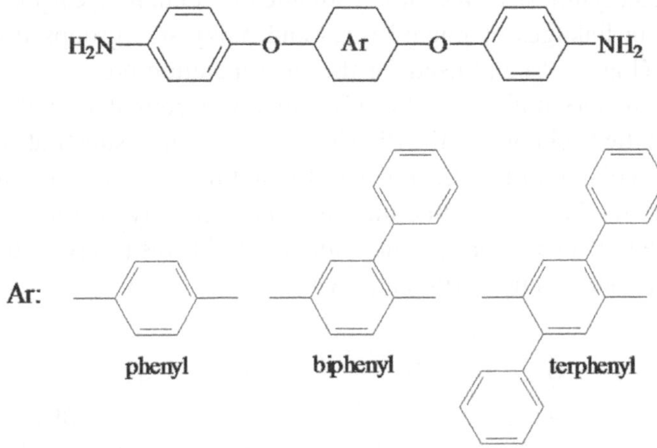

FIGURE 2.13 Structure of diamine moeties containing laterally attached phenyl groups [109].

Chang et al. synthesized a series of PIs substituted with TPA to investigate the effect of the bulky substituents like methyl, N,N bisphenyl amine, and methoxy substituted N,N bisphenyl amine (Fig. 2.14) on their properties.[79] All the PIs had good thermal stability associated with high softening temperatures (279–300°C), 10% weight loss temperatures in excess of 505°C in nitrogen. The substituted triphenyl amine containing PI can be used for gas separation as it reveals very good combination of gas permeability and permselectivity. The 4-methoxy substituted TPA-containing PIs (PI IV) showed better gas permeability ($P_{CO2} = 12.97$; $P_{O2} = 2.94$ Barrer) than that of the unsubstituted PIs (PI III) ($P_{CO2} = 11.77$; $P_{O2} = 1.97$ Barrer). Also, PI IV exhibited higher gas selectivity value of 32.43 for CO_2/CH_4 gas pair (Table 2.4).

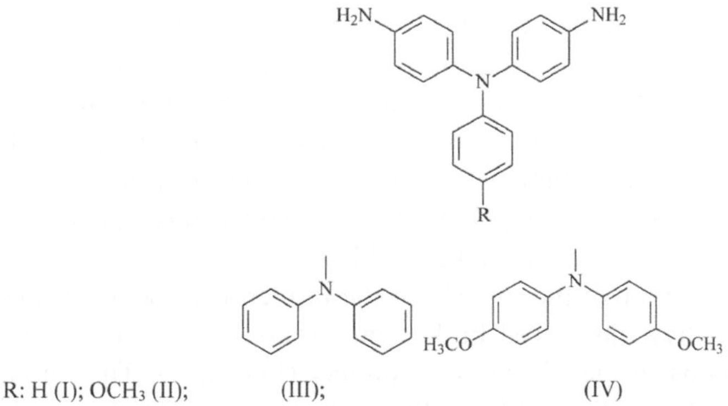

R: H (I); OCH₃ (II);   (III);   (IV)

FIGURE 2.14 Structure of different triphenylaminediamines.

Ayala et al. synthesized a series of soluble PIs containing carbonyl groups as connecting linkages of phenyl rings and bulky side groups like phenyl and t-butyl (Fig. 2.15) and used for the gas separation process.[110] The T_g of these PIs were very high (257–265°C), which was consistent with the rigid, highly aromatic backbone of the PI. The PIs with t-butyl substitution had the highest free volume (Table 2.4). Unsubstituted polymer showed the lowest free volume. As the free volume increases, permeability also increases. The order of free volume and gas permeability of the PI was t-butyl substituent > phenyl substituent > unsubstituted polymer.

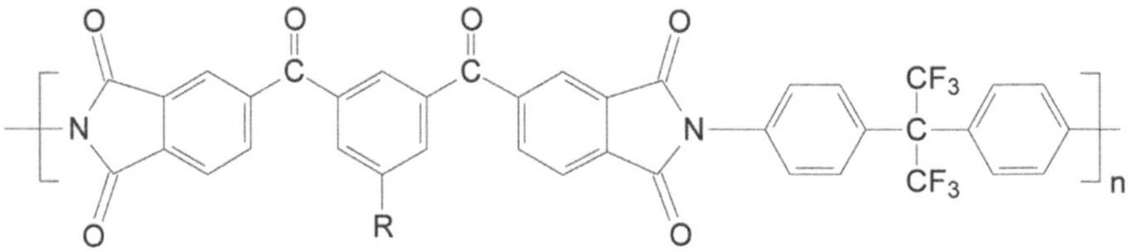

R: H (HDCDA); phenyl (PDCDA); t-butyl (BDCDA)

FIGURE 2.15   Structure of different polyimides containing carbonyl groups.

Calle et al. had designed and synthesized different PIs incorporating different bulky side groups.[111] They synthesized PIs (Fig. 2.16) from bulky tert-butyl side groups containing diamine, 1,4-bis(4-aminophenoxy)2,5-di-tert-butylbenzene (TBAPB). All the PIs showed very good thermal stability and gas permeability. T_g values were in the range of 310–366°C. The data are given in (Table 2.4). The TBAPB monomer that preferably adopts a twisted, rotation-restricted conformation enhanced the fractional free volume of PIs and maintained the rigidity of the molecular chain. Interestingly, the PI prepared from TBAPB and nonfluorinated dianhydride PMDA showed high permeability with slightly lower selectivities in comparison to the fluorinated PIs with higher FV (6FDA—6FpDA). This can be explained on the basis of the difference in distribution of free volume caused by the combination of a bulky and twisted diamine (TBAPB) with a very rigid dianhydride (PMDA), which inhibits the polymer chain packing.

In continuation of their work for the further increase in gas permeability of the PIs, they introduced bulky side groups like tert-butyl and pivaloylimino groups in the dianhydride moiety (Fig. 2.17). That monomer was used it for the synthesis of the PIs with the commercially available dianhydrides. They evaluated the effect of these bulky side groups on the gas permeation performance of the PI. These groups disturbed the polymer chain packing. The *d*-spacing value obtained from wide angle X-ray diffraction also supported the data. These bulky side groups effectively enhances the free volume, consequently the gas permeability of the PIs was also increased (Table 2.4). The enhancement in gas permeability was mainly due to the increase in the gas diffusivity. As free volume increases, gas diffusivity also increases. Interestingly, the permeability in BTPDA was around three times higher than that of PBTPDA; despite the latter has a slightly larger free volume.[112] The three PIs showed high thermal stability with degradation temperature above 500°C. They had very high T_g also. In fact, the T_g of PI PBTPDA was very high (420°C), which was the highest values ever reported for lineal, soluble PIs. In BTPDA, the higher average size of free volume elements helps increase the diffusivity and the permeability. Whereas, in PBTPDA a narrower, more homogeneous, size distribution of free volume elements due to the higher regularity in the chain packing is observed from the X-ray diffraction pattern, accordingly the gas diffusivity and permeability decreases but the overall selectivity increases.

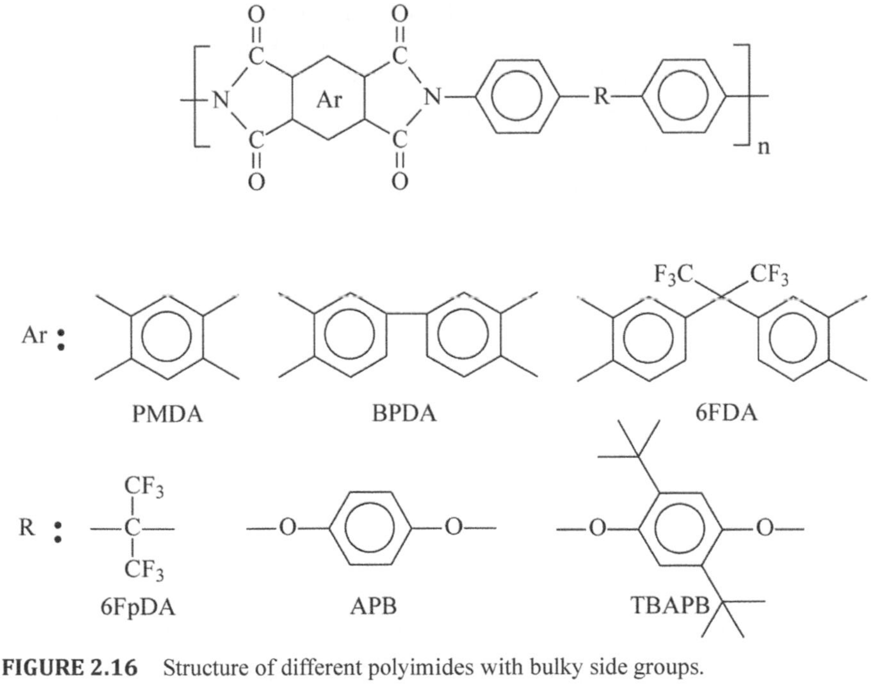

FIGURE 2.16 Structure of different polyimides with bulky side groups.

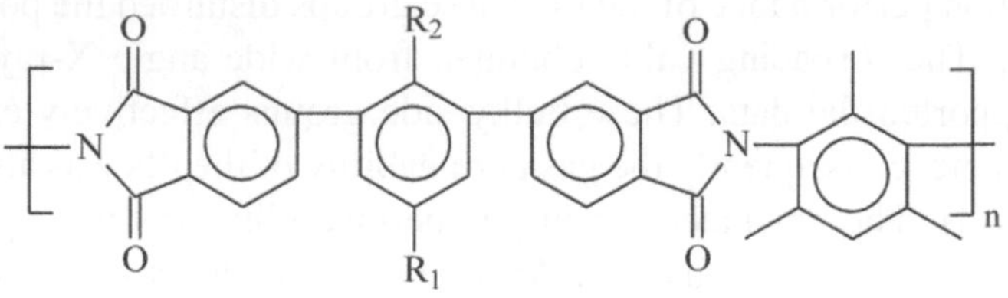

R_1 = H; R_2 = H TDPA

R_1 = *t*-butyl; R_2 = H BTDPA

R_1 = *t*-butyl; R_2 = $NHCOC(CH_3)_2$ PBTDPA

FIGURE 2.17 Structure of different polyimides based on dianhydrides with bulky side groups.

Maya et al. developed a new synthetic method for the incorporation of the bulky groups on the PI structure by post modification of the polymer.[113] They have introduced bulky adamantane moiety by esterification of a functionalized copolyimide (Fig. 2.18), having free carboxylic acid (size: 43.08 Å^3) as pendant groups. This carboxylic acid containing copolyimide reacted with different adamantyl alcohols such as 1-adamantane methanol (ADA-1) (size: 200 Å^3), 2-adamantanol (ADA-2) (size: 179.61 Å^3), and 1-adamantanol (ADA-3) (size: 179.78 Å^3) to give adamantly ester copolyimides (PI-ADAs). This route is more suitable as it replaced the difficult synthetic procedure to synthesize adamantane containing diamine.

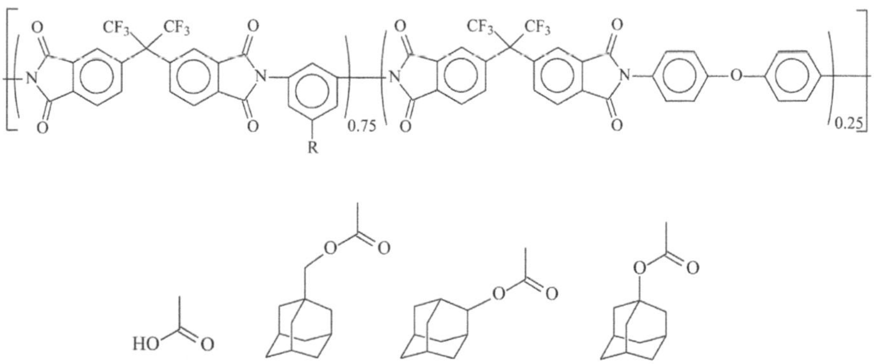

R: PI-A; PI-ADA1; PI-ADA2; PI-ADA3

FIGURE 2.18 Structure of different copolyimides with bulky adamantane moieties.

The effect of these bulky adamantane moieties on the gas transport properties was evaluated. It was found that the larger size of the adamantane moiety inhibited the polymer chain packing and produced higher free volume. All the PIs are thermally very stable. It was noticed that PI-ADA-1 showed two degradation steps, like the starting copolyimide PI-A, while PI-ADA-2 and PI-ADA-3 showed three decomposition steps. The first degradation of PI-ADA-1 occurred at around 420°C followed by the second degradation at 525°C, and it can be ascribed to the generalized polymer decomposition. The adamantanyl ester groups of PI-ADA-2 and PI-ADA-3 degrade in two stages, at 410 and 480 °C and at slightly lower temperatures in PI-ADA-3, that is, 390 and 475 °C. The T_g values were in the range of 268–295°C. Also, the rigidity generated due to the introduction of the bulky rigid adamantane moiety helps maintain the high permselectivity. Higher permeability (P_{CO2} up to 16 and P_{O2} up to 4.7 Barrer) without significant loss of permselectivity was found when the precursor copolyimide (PI-A) was modified with pendant adamantane group (PI-ADA). The gas diffusivity of these PI-ADAs was also greater than the PI-A. However, the solubility coefficients of all the gases except CH_4 were lower for the PI-ADAs. The increased high solubility coefficient of CH_4 was due to its greater affinity to the aliphatic adamantane moiety.

Zhang et al. synthesized three spirobichroman-based diamines, FSBC, SBC, and MSBC (Fig. 2.19), with different substituted groups (-CF_3, -H, and -CH_3).[114] The diamines were reacted 6FDA to form the corresponding PIs. Such organosoluble PIs had high T_g, excellent thermal stabilities. SBC-6FDA showed highest thermal stability. About 5% weight loss value was 448 °C. T_gs of the PIs are in the range of 261–281°C. The value decreased in the order: 6FDA-MSBC > 6FDA-SBC > 6FDA-FSBC. Among the three spirobichroman-based PIs, 6FDA-FSBC exhibited the highest gas permeability that could be attributed to its insufficient chain packing due to the presence of spriobichroman backbone with the noncoplaner and spiro nature, and the bulky CF_3 substituting group that enhances the polymer free volume and decreased the polymer chain flexibility.

Li et al. reported a new family of microporous PIs with bulky tetra-*o*-isopropyl and naphthalene by microwave-assisted polymerization process.[115] 4,4'-(naphthalen-1-ylmethylene)bis(2,6-diisopropylaniline)) (BAN-3) was synthesized via a one-step electrophilic substitution reaction and used for the preparation of PI with the reaction of three commercial dianhydrides. The incorporation of naphthalene and tetra-*o*-isopropyl groups effectively increased rigidity of the polymeric backbone and disrupted chain packing, leading to high gas permeabilities in all the PI films. All the PIs showed very good solubility in common solvents like NMP, DMF, THF, chloroform, and dichloromethane (DCM) at room temperature. They possessed good mechanical properties (tensile strength > 64 MPa and tensile modulus > 1.7 GPa.[116] BAN-3-6FDA showed high CO_2 permeability values of 849 Barrer with maintaining comparable selectivities for CO_2/CH_4 gas pairs. Also, the PI BAN-3-BPDA exhibits excellent gas separation performance for O_2/N_2 gas pair.

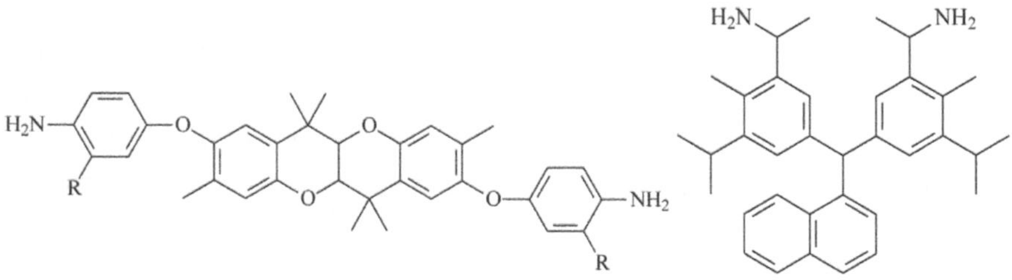

R: CF$_3$ (FSBC); H (SBC); CH$_3$ (MSBC)   BAN-3

FIGURE 2.19   Structure of bulky diamines FSBC, SBC, MSBC, and BAN-3.

Zhang et al. synthesized two carboxylic acid containing diamines, CADA1 and CADA2 (Fig. 2.20), with or without bulky CF$_3$ group.[117] The PIs were shown to have good thermal stability. T_g values were in the range of 270–285°C. PIs with 100 times higher CO_2 permeabilities, similar CO_2/CH_4 selectivities and much better plasticization resistant properties were obtained after the thermal-induced decarboxylation crosslink reaction, compared with the un-crosslinked PIs. The PIs were crosslinked at different temperatures. It was observed that permeabilities gradually increased with the crosslinking temperature maintaining almost the same selectivities. The permeability values of the PIs crosslinked at 200°C are given in the Table 2.4.

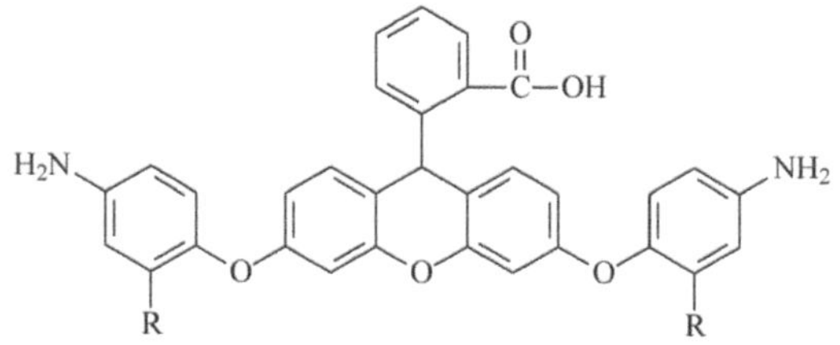

R: CF$_3$ (CADA1); H (CADA2)

FIGURE 2.20   Structure of carboxylic acid containing diamines.

TABLE 2.4   Gas Transport Properties of Some Polyimides.

Polyimide	P_{CO_2} (Barrer)	P_{O_2} (Barrer)	α (CO_2/CH_4)	α (O_2/N_2)	FFV	T_g(°C)	Reference
I-6FDA	4.73	0.69	59.13	8.63	–	299	[79]
II-6FDA	16.82	4.28	33.64	4.81	–	308	[79]
III-6FDA	11.77	1.97	9.49	5.05	–	290	[79]
IV-6FDA	12.97	2.94	32.43	4.98	–	285	[79]
HDCDA-6F	4.1	1.06	41	8.8	0.156	257	[110]
PDCDA-6F	5.0	1.2	31.3	6.7	0.159	260	[110]
BDCDA-6F	15.6	3.8	39	5.9	0.164	265	[110]
APB-PMDA	3.17	0.57	24.4	4.38	0.083	366	[111]
6FpDA-PMDA	70.4	16.6	27.1	4.22	0.186	359	[111]
TBAPA-PMDA	141.8	22.9	18.1	3.95	0.183	329	[111]
6FpDA-6FDA	70.0	17.8	35.7	4.94	0.208	310	[111]
TPDA	140	27.4	19.7	4.5	0.144	414	[112]
BTPDA	465	111	11.7	3.7	0.182	367	[112]
PBTPDA	210	35.1	17.9	4.2	0.193	420	[112]
PI A	5.41	1.52	41.6	5.84	0.163	335	[113]
PI ADA1	14.39	3.79	36.8	5.26	0.170	268	[113]
PI ADA 2	16.06	4.79	35.7	5.77	0.177	294	[113]
PI ADA 3	12.93	3.23	34.9	4.82	0.166	278	[113]
FSBC- 6FDA	66.0	17.0	25.8	4.60	0.160	261	[114]
SBC-6FDA	32.1	8.19	25.1	4.38	0.142	267	[114]
MSBC-6FDA	21.2	5.55	27.6	3.92	0.133	281	[114]
BAN3-ODPA	229	47.1	16.5	3.4	0.102	296	[115]
BAN3-BPDA	391	81.3	15.2	5.0	0.099	339	[115]
BAN3-6FDA	845	99.8	15.4	1.9	0.144	294	[115]
CADA1-6FDA	18.2	4.55	31.3	4.10		270	[117]
CADA2-6FDA	11.0	2.51	33.3	4.11		285	[117]
CADA1-BTDA	6.54	1.58	32.2	5.55		273	[117]
CADA2-DSDA	8.07	1.66	32.9	3.71		270	[117]

Banerjee et al. synthesized a number of new diamines containing bulky groups. Structures of these diamines are given in the Table 2.5.

TABLE 2.5   Chemical Structures of Diamines Containing Bulky Groups.

Structure of diamine	Reference
	[87]

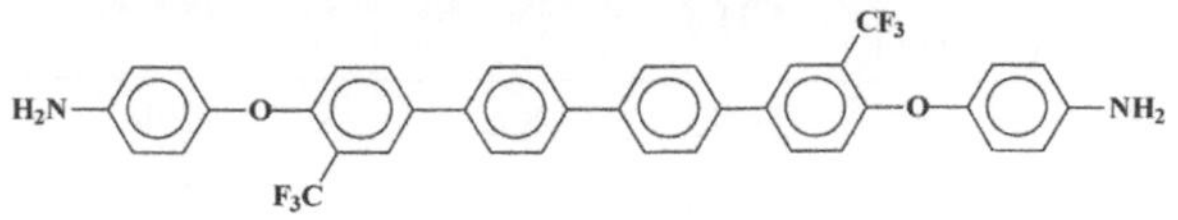

BAQP

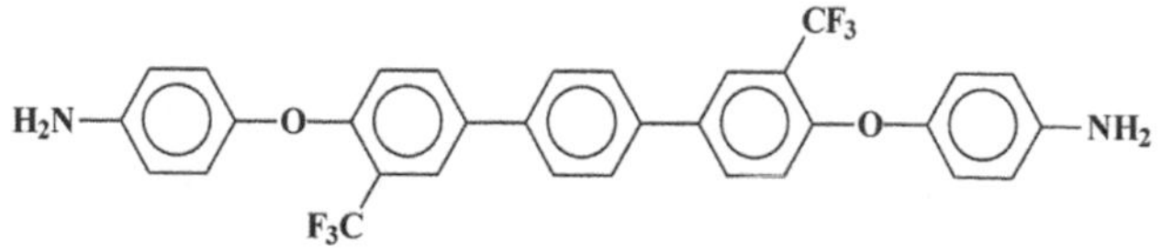

| | [118] |

BATP

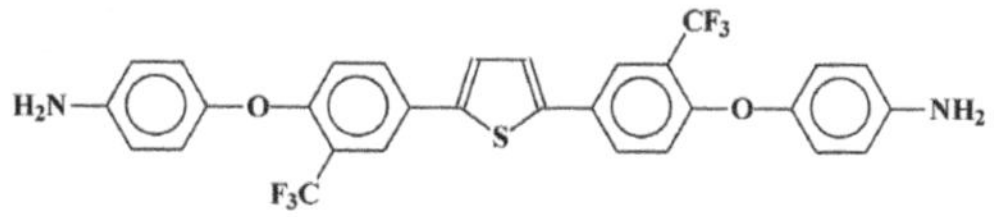

| | [86] |

BATh

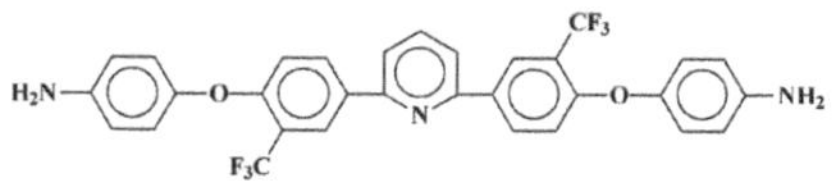

| | [86] |

BAPy

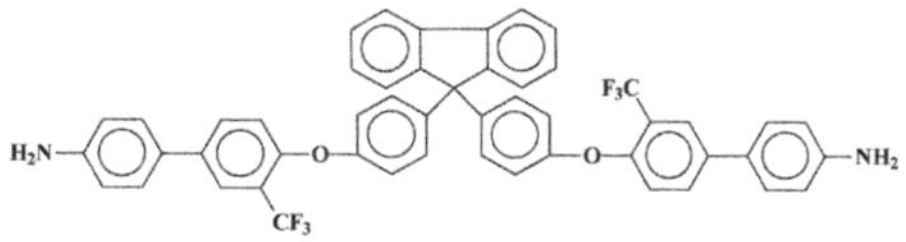

| | [69] |

FBP

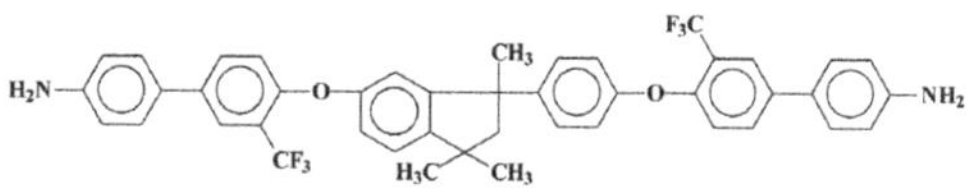

| | [73] |

BPI

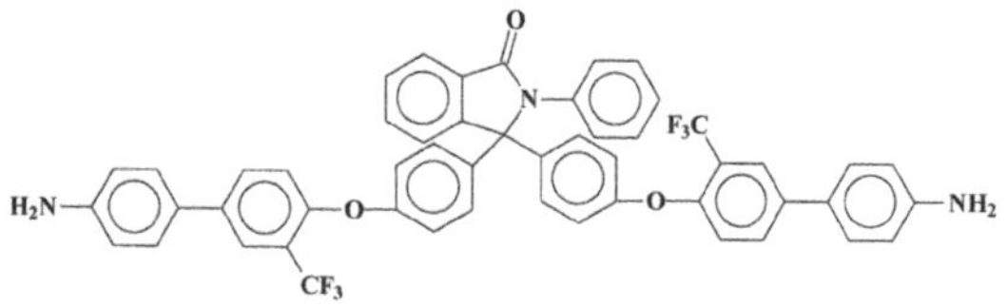

| | [74] |

BAPA

TABLE 2.5 *(Continued)*

Structure of diamine	Reference
SBPDA	[67]
BIDA	[119]
EATPF	[9]
Diamine A	[120]
TPA	[121]

TABLE 2.5   *(Continued)*

Structure of diamine	Reference
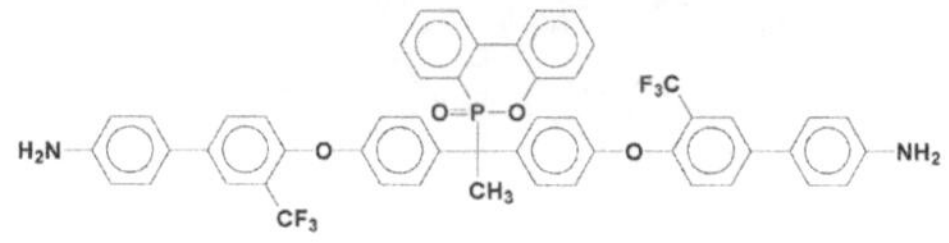 Diamine B Diamine B	[122]
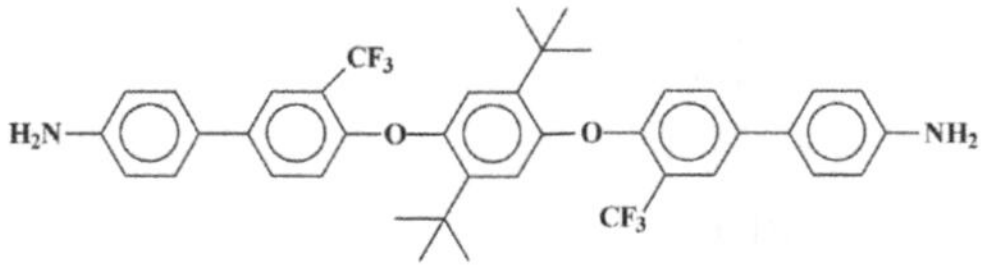 FMTBDA FMTBDA	[123]

The gas transport properties of PIs based on these diamines were measured for CO_2, O_2, N_2, and CH_4. The permeability and permselectivity values are reported in Table 2.6.

TABLE 2.6   Gas Permeability Coefficients of the Polyimides Containing Bulky Groups.

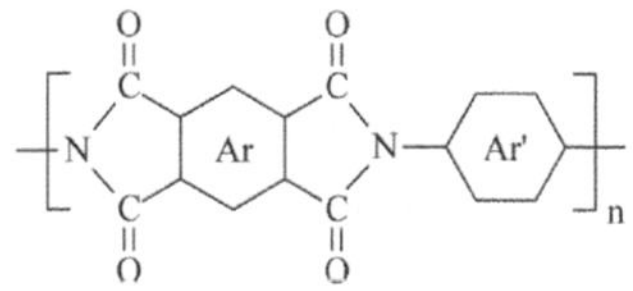

Ar:

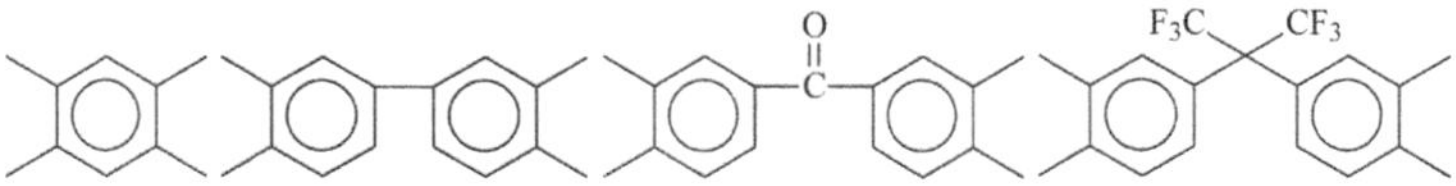

PMDA;   BPDA;   BTDA;   6FDA;

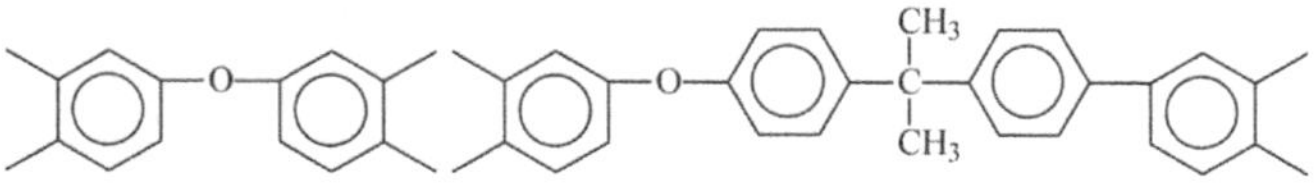

ODPA;   BPADA

Ar': (See Table 2.5)

TABLE 2.6   *(Continued)*

Polymer	P_{CO_2} (Barrer)	P_{O_2} (Barrer)	α (CO$_2$/CH$_4$)	α (O$_2$/N$_2$)	T_g (°C)	Reference
BAQP-6FDA	36.61	17.08	24.3	5.5	273	[124]
BATP-6FDA	33.12	15.17	28.2	5.3	278	[85]
BAPy-6FDA	51.92	12.15	26.6	6.4	264	[124]
BATh-6FDA	45.31	11.65	27.0	6.6	257	[124]
BPI-BPADA	39.45	10.95	28.59	5.26	241	[126]
BPI-6FDA	57.45	14.98	35.46	6.14	262	[126]
BPI-BTDA	34.20	7.98	38.86	6.05	252	[126]
BPI-ODPA	35.78	8.95	36.51	6.30	249	[126]
BPI-PMDA	44.68	12.22	29.39	5.18	265	[126]
BAPA-BPADA	16.61	4.25	19.77	4.47	243	[127]
BAPA-6FDA	53.85	10.23	53.32	5.98	292	[127]
BAPA-BTDA	17.09	4.33	27.57	5.22	280	[127]
BAPA-ODPA	14.59	4.22	17.37	4.40	276	[127]
BAPA-PMDA	39.57	7.62	50.09	6.35	327	[127]
FBP-BPADA	22.52	7.01	22.30	5.89	247	[128]
FBP-6FDA	53.09	13.46	39.62	6.53	297	[128]
FBP-BTDA	36.07	6.24	38.37	6.57	284	[128]
FBP-ODPA	25.91	7.86	24.91	6.44	254	[128]
BIDA-BPADA	25.65	10.32	37.17	8.39	274	[119]
BIDA-6FDA	71.32	25.37	35.84	6.01	315	[119]
BIDA-BTDA	16.06	6.99	29.20	7.13	310	[119]
BIDA-ODPA	16.99	7.74	25.36	5.34	301	[119]
SBPDA-BPADA	23.87	9.92	68.2	9.73	232	[67]
SBPDA-6FDA	52.98	36.08	43.79	10.77	269	[67]
SBPDA-ODPA	22.24	13.07	55.22	11.67	253	[67]

TABLE 2.6   *(Continued)*

Polymer	P_{CO_2} (Barrer)	P_{O_2} (Barrer)	α (CO_2/CH_4)	α (O_2/N_2)	T_g (°C)	Reference
EATPF -6FDA	73.91	17.12	67.20	17.47	248	[9]
EATPF -BPADA	59.3	13.11	61.77	15.05	171	[9]
EATPF -ODPA	51.41	11.90	63.47	15.26	183	[9]
EATPF -BTDA	45.82	10.80	65.46	15.88	195	[9]
EATPF -PMDA	35.21	8.50	78.24	20.24	214	[9]
Diamine A-BPDA	58.6	12.4	68.9	10.3	250	[120]
Diamine A-6FDA	70.3	16.7	63.9	9.2	288	[120]
Diamine A-ODPA	41.2	7.5	73.5	13.3	258	[120]
TPA-BPDA	76.2	27.3	48.2	7.1	230	[121]
TPA-6FDA	100.8	40.4	50.9	7.6	270	[121]
TPA-ODPA	64.9	20.5	47.5	7.3	260	[121]
Diamine B-6FDA	175	64	51	7.1	261	[122]
Diamine B-BPADA	113	57	45	6.1	235	[122]
Diamine B-ODPA	97	40	49	6.2	251	[122]
Diamine B-BTDA	94	34	48	6.4	258	[122]
FMTBDA-BPADA	23	3.9	23	6.6	215	[123]
FMTBDA-6FDA	73	12	21	4.5	253	[123]
FMTBDA-ODPA	22	3.4	24	6.7	246	[123]
FMTBDA-BTDA	16	3.0	25	5.3	252	[123]

It could be noted that the PIs synthesized from 6FDA showed drastically improved properties. They possesses very good solubility in several organic solvents such as *N*-methylpyrrolidone (NMP), *N,N*-DMF, *N,N*-DMAc, tetrahydrofuran (THF), chloroform ($CHCl_3$), and DCM at room temperature. They also showed better thermal and mechanical response than the other analogous PIs. These PIs had the highest permeability due to the presence of bulky $>C(CF_3)_2$ moieties.[9,67,119–128] Presence of two trifluoromethyl groups in 6FDA makes the PI chains bulky, which increase the rigidity and fractional free volume. Accordingly the permeability along with the selectivity of different gas pairs increases.

The effects of incorporation of heterocyclic moieties into fluorinated PI membranes were evaluated. These polymers also showed very good solubility and processability. The PIs were thermally stable. T_g values were in the range of 257–278°C. Highest tensile strength was observed for BAQP-6FDA: 109 MPa. The interaction of the gas molecules with the polymer chain enhances with the increase in the group polarity as observed for BAPy and BATh diamines.[124] This enhances the solubility of the gas molecules in the PIs responsible for the high gas permeability, for example, in case of high permeability for CO_2.

The effect of polyhedral oligomeric silsesquioxanes (POSS) on the thermal, mechanical, and gas transport properties of the PI was also evaluated. A series of PI-POSS nanocomposite membranes was prepared with POSS as nanofiller. Incorporation of POSS increases the thermal and gas transport properties of the composites without much sacrificing their mechanical properties. The permeability of all the gases was increased significantly with keeping almost similar selectivities. For the membrane BAPy-6FDA, the value of P_{CO2} increases from 51.92 to 62.14 Barrer.[125]

One of the PIs used commercially is Matrimid®, which has a bulky indan-based polymeric structure. Consequently, a series of PIs from indan-based diamine, 3-(4'-amino-3-trifluoromethyl-biphenyl-4-yloxy-phenyl)-5-(4'-amino-3-trifluoromethyl-biphenyl-4-yloxy)-1,1,3-trimehylindane (BPI), was synthesized. All the polyimides showed excellent solubility in different solvents. Very low water absorption (0.19–0.30%) with good optical transparency was noticed for the light yellow polymers. The polymers exhibited high tensile strength up to 85 MPa and modulus up to 2.5 GPa. The polymers showed high thermal stability up to 526°C in N_2 atmosphere with 5% weight loss, and high T_g up to 265°C. They showed good mechanical properties such as high tensile strength up to 85 MPa, modulus up to 2.5 GPa, and elongation at break up to 38%, depending on the exact polymer structure.[68] Their gas transport properties were also investigated.[126] The PIs showed remarkably high permselectivity ($\alpha(CO_2/CH_4)$ = 38.86, $\alpha(O_2/N_2)$ = 6.05)) with high CO_2 permeability (up to 57.45 Barrer) due to the presence of rigid indan unit.

Phthalimidine is another vital bulky group that was incorporated in PI membranes. This moiety is very helpful in achieving the high solubility along with the enhanced thermal, mechanical, and gas separation properties. The diamine 3,3-bis-[4-{2'-trifluoromethyl-4'-(4"-aminophenyl)phenoxy} phenyl]-2-phenyl-2,3-dihydroisoindole-1-one (BAPA) was used with the different anhydrides.[74] The polymers provided high thermal stability up to 500 °C in air with 5% weight loss, and high T_g up to 327°C. They showed good mechanical properties such as high tensile strength up to 98.4 MPa and elongation at break up to 9.3%, depending on the exact polymer structure. The high permselectivity was due to the presence of phenyl substituted bulky phthalimidine unit along with comparable permeability. The BAPA-6FDA and BAPA-PMDA showed high CO_2 permselectivity 53.32 and 50.09, respectively.[127]

The effect of the bulky diphenylfluorene moiety on their thermal, mechanical, and gas transport properties was also evaluated. A diamine containing diphenylfluorene moiety (FBP) was synthesized and used for the preparation of PIs.[69] The bulky fluorine moiety in a diphenylfluorene-based PI projected vertically from the polymer main chain, and increased the rigidity and the free volume by reducing the polymer chain packing, which in turn is responsible for the high permeability. The phenylfluorene containing cardo PIs also had high thermal stability, low refractive index, high optical transparency, and low dielectric constant, because of the relatively high free volume. The T_g values were in the range of 247–297 °C. The PIs exhibited very good mechanical properties. Highest tensile strength (122 MPa) with good elongation value was achieved for FBP-BTDA. All the polymers showed good solubility in different solvents. All the FBP-containing PIs showed high gas permeability and permselectivity. The FBP-6FDA showed the highest permeability (P_{CO2} = 53.09, P_{O2} = 13.46 Barrer) with good permselectivity $\alpha(CO_2/CH_4)$ = 39.62, $\alpha(O_2/N_2)$ = 6.53).[128] Another series of polyimide was synthesized with bulky cardo phenyl fluorene moiety with BPADA, 6FDA, and ODPA. The T_g values of the PIs were in the range of 250–288°C. PI-containing 6FDA showed the highest 10% weight loss temperature (521°C) measured by under synthetic air. The tensile strength of the PIs was in the range of 61–76 MPa and percentage of elongation of break at around 4–9%. The PI membranes showed high gas permeability (P_{CO2} = 70.3 and P_{O2} = 16.7 Barrer) with high permselectivity up to 73.6 for the CO_2/CH_4 gas pair and up to 13.4 for the O_2/N_2 gas pair.[120]

A series of PI was synthesized from diamine containing cycloapliphatic moiety, like spiro-biindane namely 6,6'-bis-[2"-trifluoromethyl4"-(4'''-aminophenyl)phenoxy]-3,3,3',3'-tetramethyl-1,1'-spirobiindane (SBPDA). All the PIs were well characterized and their gas transport properties were measured along with the thermal and mechanical properties.[67] As the bulky spiro-biindane moiety present in the backbone, the polymer showed very good thermal and mechanical properties. The polymer SBPDA-6FDA showed high T_g value (269°C). The PEIs exhibited excellent thermooxidative stability (409–491°C for 5% weight loss) and high tensile strengths (up to 60 MPa). BTDA and PMDA containing polymers were brittle due to their high rigidity of the main chain. PI prepared from SBPDA having the rigid spiro-biindane linkage is characterized by high permselectivity ($\alpha(CO_2/CH_4)$ = 68.2, $\alpha(O_2/N_2)$ = 11.7) and high permeability [P_{O2} (Barrer) = 36.08].

Another important bulky group incorporated in the PI backbone is benzoisoindoledione group. A series of PIs was synthesized by reacting the diamine, 4,9-Bis-(4-hydroxy-phenyl)-2-phenyl-benzo[f]isoindole-1,3-dione (BIDA) with different dianhydrides.[119] It was found that the incorporation of benzoisoindoledione moiety helps restrict mutual rotation of the phenyl rings due to its bulkiness, and formation of intermolecular charge transfer complex between the diimide moiety and the planar aromatic moiety. The PIs showed extremely high T_g up to 335°C, outstanding thermal stability up to 559°C, and high tensile strength up to 102 MPa. The polymer BIDA-6FDA showed the highest permeability coefficient for all the gases ($P(CO_2)$ = 71.3 Barrer, $P(O_2)$ = 25.4 Barrer). Whereas BIDA-BPADA based polymer exhibited the highest permselectivity (CO_2/CH_4 37.2 and O_2/N_2 8.4), as the flexible anhydride moiety was attached.

PIs with long branched aliphatic chain from a diamine monomer namely 9,9-bis(2-ethylhexyl)-2,7-bis[4-aminophenoxy-3-trifluoromethylphenyl]-9H-fluorene (EATPF) were also synthesized.[9] The polymers were soluble in many organic solvents, such as $CHCl_3$, DCM, DMF, NMP, DMAc, and THF. The high solubility of the polymers was attributed to the long-branched aliphatic chains at the 9-position of fluorene moiety present in the polymers backbone that inhibited the interchain packing. This resulted in an enhanced free volume and improved solubility. The polymers showed high molecular weight, T_g up to 248°C, 10% degradation temperature up to 409 and 437°C under air and nitrogen, respectively, tensile strength up to 100 MPa, and dielectric constant (ε) as low as 2.35. All the PIs showed very good gas transport properties. The larger free volume and chain rigidity were responsible for such gas selectivity. The PI synthesized with 6FDA showed highest CO_2 and O_2 permeability (P_{CO2} = 73.9 and P_{O2} = 17.1 Barrer) and EATPF-PMDA exhibited highest permselectivity for different gas pairs (P_{O2}/P_{N2} = 20.2 and P_{CO2}/P_{CH4} = 78.2).

Another new aromatic PIs containing bulky tert-butyl group, propeller-shaped TPA unit in its structure[121] were synthesized. The polymers are highly soluble. With an increase in inter-segmental distance, the FFV of the polymers also increased.

The PIs showed high glass transition temperature (T_g up to 270°C) and thermal stability (10% weight loss temperature up to 475°C). The PI membranes showed good mechanical properties with tensile strength up to 70 MPa. The strong affinity between CO_2 and nitrogen atoms of tertiary amine in TPA, made the polymer extremely high solubility selectivity for the CO_2/CH_4 gas pair. Excellent separation performance [$P(CO_2)$ = 100.8, $P(O_2)$ = 40.4 Barrer] and good permselectivity [$P(CO_2)/P(CH_4)$= 50.9, $P(O_2)/P(N_2)$ = 7.6] was achieved.

A series of new PIs with phosphaphenanthrene unit was synthesized. The effect of bulky spiral phosphaphenanthrene skeleton, its spatial arrange-ment, and size-distribution function of the free-volume were also studied using molecular dynamics simulation and correlated with the experimental data obtained.[122] The polymers showed a good combination of thermal and mechanical properties. About 10% weight loss temperature was up to 416°C under synthetic air and tensile strength up to 91 MPa was observed with low dielectric constant (2.10–2.55 at 1 MHz). The T_g values of the PIs were ranged from 235 to 261°C. All the PI films showed high gas permeability with high permselectivity (P_{CO2}/P_{CH4} up to 51 and P_{O2}/P_{N2} up to 7.1). The highest permeability obtained for CO_2 was up to 175 and for O_2 was up to 64 Barrer.

Polymers with bulky substituents reveal very pronounced effect on their thermal, mechanical, and gas transport application. A series pendant di-tert-butyl and trifluorometyl groups containing new PIs was synthesized. The PIs were prepared with the new diamine monomer, namely 1,4-bis-[{2′-trifluoromethyl   4′-(4″-aminophenyl)phenoxy}]   2,5-di-t-butylbenzene (FMTBDA) with four different commercially available aromatic dianhy-drides. The polymers were soluble in many organic solvents.[123] The polymers were thermally stable up to 474 °C with 10% weight loss temperature in air and had a glass transition temperature as high as 253°C. The PI membranes showed good mechanical properties with tensile strength up to 95 MPa. All the PI films showed very good separation performance (P_{CO2}/P_{CH4} up to 25 and P_{O2}/P_{N2} up to 6.7). The highest permeability obtained for CO_2 was up to 73 and for O_2 was up to 12 Barrer.

For better understanding the structure—properties relations and the effect of some different bulky groups on the gas separation properties, the

$P(CO_2)$ vs $\alpha(CO_2/CH_4)$ (Fig. 2.21) and $P(O_2)$ vs $\alpha(O_2/N_2)$ (Fig. 2.22) have been drawn. The polymers reported in the Table 2.6 were close to the upper boundary of the Robeson plot[129,130] justifying their superiority as gas separation membrane.

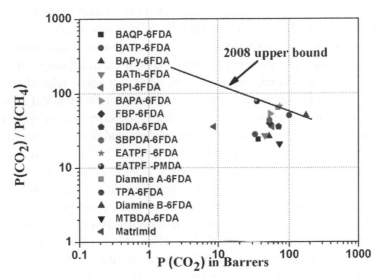

FIGURE 2.21   Robeson plot for comparison of CO_2/CH_4 selectivity vs CO_2 permeability.

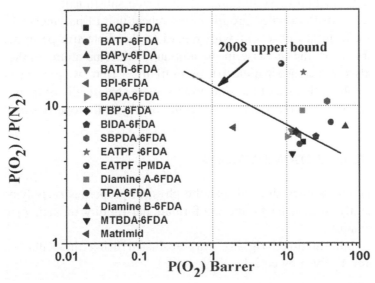

FIGURE 2.22   Robeson plot for comparison of O_2/N_2 selectivity vs O_2 permeability.

2.4   EFFECT OF BULKY GROUPS ON PHYSICAL PROPERTIES

Many physical properties of amorphous polymers, such as glass transition temperature, density, free volume, dielectric permittivity, gas transport properties, etc. depend on the basic characteristics of macromolecules like the conformational rigidity of their chains and energy of interchain interactions or cohesion energy. Due to the complexity of polymer structures and their polydispersity, direct correlations between these basic characteristics and physical properties of polymers generally do not exist, although in certain cases such correlations can be evidenced. Thus, for some related series of polymers, where the energy of interchain interactions is identical or it only slightly varies from one polymer to another, good correlations were found in glass transition temperature, gas permeability, and dielectric permittivity vs conformational rigidity of their polymer chains.[131–143] For a large number of PIs, such correlations are significantly less successful.[13,14] However, for those physical properties of polymers that depend on free volume, like gas permeability, dielectric permittivity, and glass transition temperature, good correlations exist between a large number of data of polymers having various chemical structures.[131,136–139,144,145] Thus, the increase of free volume generally leads to the decrease of dielectric permittivity and glass transition temperature of those polymers. The increase of free volume is determined by the change of chemical structure of the repeat unit of polymer, for example, by the introduction of voluminous (bulky) side substituents[143,144] or by the introduction of fluorinated groups in the repeat unit of polymers.[131,136,141,144] Similar effects are observed with regard to the dielectric permittivity of polymers. Thus, the increase in the amount of fluorine atoms in the repeat unit leads to a decrease in dielectric permittivity.[136] Here, we present a review regarding the influence of the side groups, more specifically of their van der Waals volume and conformational rigidity on physical properties of PIs.

2.4.1   CALCULATION METHODS

Our previous studies showed that the physical properties of polymers are significantly influenced by the conformational rigidity of their macromolecular chains.[132]

The conformational rigidity was evaluated using the Kuhn segment parameter A_{fr} given by eq 2.1.

$$V_f = \frac{1}{\rho} - \frac{N_A \bullet V_w}{M_o} \qquad (2.1)$$

where $<R^2>$ is mean square distance between the ends of the chain calculated by assuming the free rotation; n is the number of repeat units; l_o is the contour length of the repeat unit; and $L = nl_o$ is the contour length of the chain. These calculations were performed by using the Monte Carlo method.[135] The model of a repeat unit was constructed in molecular editor and refined by using quantum-mechanical method AM1.[146]

Most of these polymers did not have any hindrance of rotation, or their hindrance was too low and it was neglected. As shown previously, the values of conformational parameters calculated under the assumption of free rotation in the absence of bulky substituents are practically equal to the values found experimentally from the hydrodynamic data.[99]

The van der Waals volume (V_w), free volume (V_f), occupied volume (V_{occ}), accessible volume (V_{acs}), and fractional accessible volume (FAV) were calculated by using the method previously described.[147] The values of van der Waals radii of atoms were taken from literature.[148] The calculations were done by using the Monte Carlo method[149] so that, without making any assumptions about packing of the polymer chains in glassy state, we could quickly calculate the van der Waals volume and free volume.

The *free volume* (V_f) was calculated with the eq 2.2.

$$V_f = \frac{1}{\rho} - \frac{N_A \bullet V_w}{M_o} \qquad (2.2)$$

where ρ is the density of polymer; N_A is the Avogadro number, and M_o is the molecular weight of the repeat unit. The value V_f shows the volume that is not occupied by the macromolecules in 1 cm^3 of polymer film.

The occupied volume (V_{occ}) of a repeat unit is given by eq 2.3 and it presents the sum of van der Waals volume (V_w) of the repeat unit and the space around this unit that is not accessible for a given type of gas molecule that is named "dead volume" (V_{dead}).[147] The dead volume was calculated with the method described in Ref. [149]. It is evident that the dead volume and therefore the occupied volume of a repeat unit depend on the size of gas molecule.

$$V_{occ} = (V_w + V_{dead}) \qquad (2.3)$$

The accessible volume of a polymer (V_{acs}) is given by eq 2.4.

$$V_{acs} = \frac{1}{\rho} - \frac{N_A \bullet V_{occ}}{M_o} \qquad (2.4)$$

The correlation of permeability and diffusion coefficients with accessible volume is clearly reflected by the so-called "fractional accessible volume" (*FAV*), without any dimensions, that is given by eq 2.5.[149]

$$FAV = V_{acs} \bullet \rho \qquad (2.5)$$

The gas separation characteristics of polymers depend on the packing of polymeric chains into polymer matrix, which is on free volume (V_{fr}). Thus, it is considered that permeability (*P*) and diffusion (*D*) coefficients of polymers are linearly correlated with the free volume in semilogarithmic coordinates,[13,14,144,145] which means that for a system one gas–different polymers at isothermic conditions we have the following equations:

$$\ln P = a_p - b_p/V_{fr} \qquad (2.6)$$

$$\ln D = a_D - b_D/V_{fr} \qquad (2.7)$$

where a_p and b_p are inclination and slope of permeability coefficients, and a_D and b_D are similar parameters for diffusion coefficients.

The value of V_{fr} for polymers depends on the method of calculation. In our previous works, we have used analog dependences for the systems one polymer–different gases, according to eqs 2.3 and 2.4, and we calculated the volume of polymer that is accessible for molecules of gas of different dimensions and forms V_{acs} and the specific accessible volume *FAV*.[147,149]

$$\ln P = A_p - B_p/FAV \qquad (2.8)$$

$$\ln D = A_D - B_D/FAV \qquad (2.9)$$

where A_p and B_p are inclination and slope for coefficient of permeability, and A_D and B_D are also for coefficient diffusion.

2.4.2 EFFECT OF SIDE BULKY GROUPS ON PHYSICAL PROPERTIES

The influence of the side groups on the conformational flexibility of the polymer depends on the position of this substituent in the repeat unit of the PI. So, for the polymers of Table 2.7, this is significant. If a small volume substituent (CH_3, $V_w = 28.34$ Å³) is located on the phenyl ring of the amine components in the ortho-position, this leads to hindered rotation around the imide cycle—phenyl ring. The value of the Kuhn segment calculated taking into account the retardation of rotation increases by about 3%.

TABLE 2.7 Repeat Unit, Conformational Parameters (Kuhn Segment A_{fr}, van der Waals Volume V_w, Free Volume V_f), and Dielectric Constant ε_0 of First Series of Polyimides.

Ar:BPADA; BTDA; 6FDA (see Table 2.6)

Ar	X	ρ (g/cm³)	A_{fr} (Å)	V_w (Å³)	V_f (cm³/g)	ε_0
BPADA	H	1.200	25.43	625.4	0.2915	2.54
	CH_3	1.169	26.20	656.1	0.3056	2.11
					4.6%	−16.9%
BTDA	H	1.398	28.03	426.7	0.1889	3.28
	CH_3	1.337	28.87	460.6	0.2085	2.84
					10.3%	−13.4%
6FDA	H	1.546	28.50	488.6	0.1648	2.31
	CH_3	1.488	29.51	522.7	0.1793	1.99
					8%	−13.8%

In the same cases, when the introduction of a more bulky substituent does not lead to inhibition of rotation, its effect mainly affects the loosening of the polymer packaging in a glassy state. And this leads to an increase in free volume. The increase in free volume largely depends on the van der Waals volume of the side group and to a lesser extent on the conformational rigidity of the chain (Table 2.8).[143]

TABLE 2.8 Increase of the Free Volume of Polymer (ΔV_f,%) by Introduction of Bulky Substituent.

Ar: BPADA; BTDA; 6FDA; ODPA (see Table 2.6)

Ar	X = CH$_3$ $V_w = 28.34$ Å^3 A_{fr} (Å)	Y= ⬡–OH $V_w = 82.00$ Å^3 A_{fr} (Å)	Y = ⬡–O–⬡(CN)(CN) $V_w = 208.34$ Å^3, A_{fr} (Å)
BTDA	10.3; 28.87	22.9; 28.03	31.3: 28.03
6FDA	8; 29.51	54.8; 28.50	–
ODPA	–	15.7; 27.54	16.7; 27.54
BPADA	4.6; 26.20	29.1; 25.30	–

Tables 2.9–2.11 show the groups of polymers, each of which has its own substituent. Side groups have different van der Waals volumes and, therefore, increase the free volume to different extend. The smaller the increase in free volume, the lower is the glass transition temperature decrease.

Table 2.9 shows the data obtained for the second series of PIs.[150,151] Here, the nitrogen atom connecting two phenylene rings in the diamine segment of the PI has a phenyl substituent that contains a cyano (CN) group in para-position. The van der Waals volume of CN group is 30.71 Å^3. As seen in Table 2.9, the increase in free volume is significantly lower compared with the first series of PIs (the maximum increase of free volume of the first pair of polymers in this second series is 0.67%, and in the second pair is 2.9%). This is why, there is no significant influence on the decrease of glass transition temperature; only in case of the first polymer, the T_g is slightly lower when CN group is introduced, while in the case of the other two polymers T_g is slightly higher, but in all cases it is close to the accuracy of determining the glass transition temperature.

TABLE 2.9 Repeat Unit, van der Waals Volume V_w, Free Volume V_f, Glass Transition Temperature T_g, and Density ρ of Second Series of PIs.

Ar: 6FDA; BTDA; ODPA (see Table 2.6)
R: H; CN

Ar	R	T_g (°C)	ρ (g/cm³)	V_w (Å³)	V_f (cm³/g) ΔV_f (%)
	H	307[150]	1.365[150]	561.0	0.2384
6FDA	CN	299[151]	1.08[151]	579.6	0.2400
					0.67%
	H	292[150]	1.311[150]	497.7	0.1242
BTDA	CN	296[151]	1.52[151]	516.6	0.1278
					2.9%
	H	288[150]	1.311[150]	487.9	0.2230
ODPA	CN	295[151]	1.32[151]	506.7	0.2265
					1.6%

In the third series of PIs, we have examined four polymers having identical amine segment and different bridges between imide rings, and four related PIs containing hydroxyphenyl substituent connected to the methylene bridge between phenylene rings in the diamine segment (Table 2.10).[143,145,152–154]

TABLE 2.10 Repeat Unit, Kuhn Segment A_{fr}, van der Waals Volume V_w, Free Volume V_f, Density ρ, and Glass Transition Temperature T_g of the Third Series of Polyimides.

Ar: 6FDA; BTDA; ODPA; BPADA (see Table 2.6)

R: H;   (V_w = 82.00 Å³)

TABLE 2.10   *(Continued)*

Ar	R	A_{fr} (Å)	V_w (Å³)	T_g (°C)	ρ (g/cm³)	V_f (cm³/g) ΔV_f (%)
6FDA	H	28.50	488.6	297[145]	1.546	0.1616
	⬡—OH	28.50	572.3	286*	1.345	0.2502
						54.8%
BTDA	H	28.03	426.7	280[153]	1.398	0.1880
	⬡—OH	28.03	509.0	284*	1.311	0.2311
						22.9%
ODPA	H	27.54	417.2	262[154]	1.339	0.2151
	⬡—OH	27.54	499.2	278[152]	1.280	0.2488
						15.7%
BPADA	H	25.43	625.4	200[153]	1.200	0.2818
	⬡—OH	25.30	703.9	260[154]	1.098	0.3637
						29.1%

*T_g values are calculated from the dependence of the glass transition temperature on the Kuhn segment.[143]

The most significant increase in free volume, 54.8%, was observed for the first pair of polymers, followed by the fourth pair, 29.1%, then the second pair, 22.9%, and the less significant increase in free volume was observed in case of the third pair, 15.7%. This is explained in the case of the first pair of polymers by the presence of electronegative fluorine atoms of hexafluoroisopropylidene groups, and in case of the fourth pair by the highest conformational rigidity.[155] Between two hexafluoroisopropylidene groups of different chains, a rejection appears, the packing of polymer is less dense and the free volume increases. The introduction of voluminous substituents decreases the packing of polymer even more, and as a consequence the probability of formation of interchain hydrogen bonds decreases. Only in the first pair of polymers, by introduction of substituents and increase in free volume, the glass transition temperature decreases. In the next three pairs of polymers, the glass transition temperature increases. This behavior is explained by possible interchain hydrogen bonds between oxygen of imide carbonyl groups and hydrogen of hydroxyl groups in side substituent of polymers with hydroxyl groups.[143]

Table 2.11 presents the structures and properties of the fourth series of polymers that contains side substituents. The first pair of 6FDA PIs shows insignificant increase in free volume by the introduction of substituents. Thus, the glass transition temperature of 6FDA PI with side group decreased by 25% and the dielectric permittivity decreased by 44%. In the BPADA PIs, the influence of side groups on glass transition temperature is weak, but substituents significantly influence the free volume: the free volume of polymer matrix increased by 64% and the dielectric permittivity decreased by 24%. However, the free volume of polymers in 6FDA PIs increased less than in the case of BPADA PIs.

TABLE 2.11 Repeat Unit, Kuhn Segment A_{fr}, van der Waals Volume V_w, Free Volume V_f, Glass Transition Temperature T_g, Density ρ, and Dielectric Permittivity ε of the Fourth Series of Polyimides.

Ar: 6FDA; BPADA (see Table 2.6)

R: H; ($V_w = 117.42$ Å^3)

Ar	R	A_{fr} (Å)	V_w (Å^3)	T_g (°C) ΔT_g (%)	ρ (g/cm^3)	V_f (cm^3/g) ΔV_f (%)	ε
6FDA	H	22.58	397.1	298 [156]	1.472[156]	0.2163	3.44 [143]
				301 [106]	1.474 [106]	0.2154	
		23.95	502.8	225 [157] 24.5% 25.2%	1.375[158]	0.2509 16% 16.5%	1.92[157]
BPADA	H	19.75	533.5	225[159]	1.379 [159]	0.1860	3.19 [159]
		21.13	639.0	230[157]	1.289[159]	0.3065 64.6%	2.42 [157]

In Table 2.12, we examine fifth series of PIs with substituents, whose van der Waals volume is 208.34 Å^3

TABLE 2.12   Repeat Unit, Conformational Parameters (Kuhn Segment A_{fr}, van der Waals Volume V_w, and Free Volume V_f), Glass Transition Temperature T_g, and Density ρ of the Fifth Series of Polyimides.

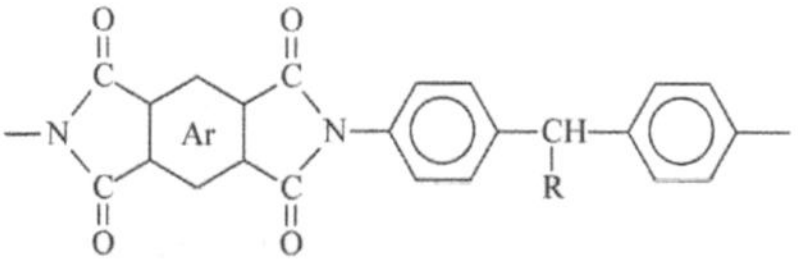

Ar: 6FDA; BTDA; ODPA; BPADA (see Table 2.6)

R: H; 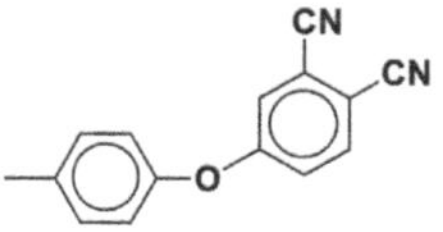   ($V_w = 208.343$ Å^3)

Ar	R	A_{fr} (Å)	V_w (Å^3)	T_g (°C), ΔT_g (%)	ρ (g/ cm^3)	V_{fr} (cm^3/g)
ODPA	H	27.54	417.2	262 [153]	1.339	0.2151
	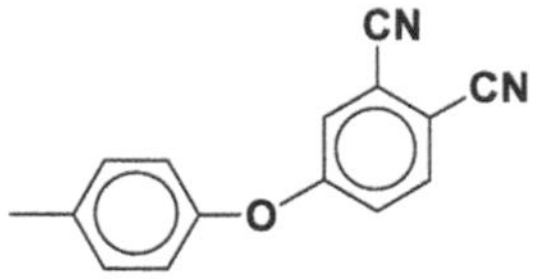	27.54	613.3	211 [140] 20%	1.299	0.2351 16.7%
BTDA	H	28.03	426.7	280 [153]	1.346	0.2126
	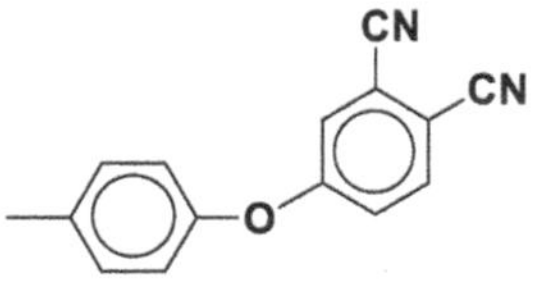	28.03	623.2	197 [140] 30%	1.230	0.2792 31.3%

The data in Table 2.12 again show that the introduction of substituents increases the free volume of the polymer and, consequently, in the absence of specific interactions, lowers the glass transition temperature. Thus, for example, different series of PIs with various structures shows that the glass transition temperature and dielectric constant reduce due to the introduction of various side groups. The effect of bridging groups in the amine component of PIs was studied in Bruma et al.[160] It was shown that even small bridging groups increase the free volume in the polymer.

2.4.3 RELATIONSHIPS BETWEEN PHYSICAL AND GAS TRANSPORT PROPERTIES

In a number of studies, it was shown that an increase in the free volume, for example, by swelling in supercritical CO_2, affect the transport parameters of gas permeation for polymers of various structures increase.[161–163] The main regularities of the effect of an increase in free volume on gas permeability can be traced using as examples of a number of PIs containing bulky groups synthesized by Banerjee et al. (see Tables 2.5 and 2.6).

Eighteen PIs, divided in four series by diamine moieties, were studied with regard to the dependence of permeability, diffusion, and solubility coefficients of gases on the reverse value of fractional accessible volume. The calculation of FAV according to the eq 2.5 given above can be performed only in the case when we know the density of polymers; therefore to make the correlation with transport parameters we selected a series of PIs containing voluminous bridging groups reported in publications[9,67,126,127] (Table 2.13). Such polymers were selected due to the reported data on permeability and diffusion coefficients. For each polymer, we built the models of the repeating unit and we calculated the Van der Waals volume (V_w) and the free volume (V_f). All these data and the values of glass transition temperatures are shown in Table 2.13.

TABLE 2.13   Repeat unit, Glass Transition Temperature, Density, van der Waals Volume, and Free Volume of Polyimides Containing Bulky Groups Synthesized by Banerjee et al.

No.	Polyimide	T_g (°C)	ρ (g/cm³)	V_w (Å³)	V_{fr} (cm³/g)	V_w (Å³)	Reference
1	EATPF-6FDA	248	1.21	1140.416	0.2988	441.116	[9]
2	EATPF-BPADA	183	1.16	1276.936	0.3039		[9]
3	EATPF-ODPA	171	1.22	1068.286	0.2686		[9]
4	EATPF-BTDA	195	1.23	1078.188	0.2625		[9]
5	EATPF-PMDA	214	1.26	982.860	0.2432		[9]
1a	SBPDA-6FDA	269	1.28	1006.468	0.2831	296.406	[67]
2a	SBPDA-BPADA	232	1.34	1142.989	0.2015		[67]
3a	SBPDA-ODPA	253	1.39	934.339	0.1852		[67]

TABLE 2.13   *(Continued)*

No.	Polyimide	T_g (°C)	ρ (g/cm³)	V_w (A³)	V_{fr} (cm³/g)	V_w (A³)	Reference
1b	BPI-6FDA	262	1.17	963.473	0.3489	255.588	[126]
2b	BPI-BPADA	241	1.12	1099.993	0.3515		[126]
3b	BPI-ODPA	249	1.14	891.343	0.3474		[126]
4b	BPI-BTDA	252	1.15	901.240	0.3401		[126]
5b	BPI-PMDA	265	1.27	805.917	0.2604		[126]
1c	BAPA-6FDA	292	1.25	1058.135	0.2991	362.773	[127]
2c	BAPA-BPADA	243	1,29	1194.655	0.2417		[127]
3c	BAPA-ODPA	276	1.319	986.005	0.2396		[127]
4c	BAPA-BTDA	280	1.332	995.907	0.2275		[127]
5c	BAPA-PMDA	327	1.323	900.579	0.2457		[127]

T_g = glass transition temperature; ρ = density; V_{fr} = free volume; V_w = Van der Waals volume.

For four gases, CO_2, O_2, N_2, and CH_4, we calculated the values of the occupied volume (V_{occ}), the accessible volume (V_{acs}) and the specific accessible volume **FAV**. Then, we built the dependence of permeability and diffusion coefficients on the reverse value of **FAV** according to the eqs 2.8 and 2.9. These dependences are demonstrated for the polymer 1 (EATPF-6FDA) as typical example on Figure 2.23A and B.

At low pressure of gases, the permeability and diffusion coefficients are correlated by eq 2.10.

$$P = D \cdot S \qquad (2.10)$$

where S is the solubility coefficient.

All these polymers show linear correlations, with high accuracy, in eqs 2.8 and 2.9. Therefore, according to eqs 2.8, 2.9, and 2.10, it can be concluded that solubility coefficients should linearly depend on the specific accessible volume, as shown in eq 2.11.

$$\ln S = A_S + B_S/\text{FAV} \qquad (2.11)$$

where A_S and B_S are the inclination and overall selectivity of solubility. This dependence is demonstrated for the polymer 1 (EATPF-6FDA) as typical example on Figure 23C. The results of calculations according to the eqs 2.8, 2.9, and 2.11 are shown in Table 2.14.

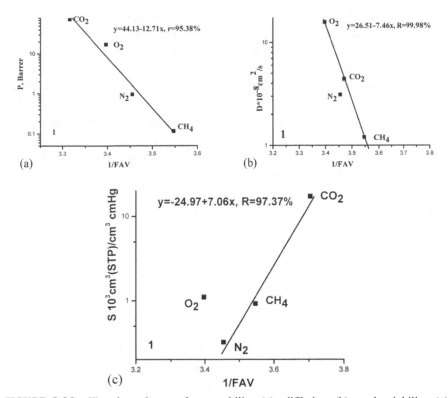

FIGURE 2.23 The dependence of permeability (a), diffusion (b), and solubility (c) coefficients on the specific accessible volume (FAV) for the EATPF-6FDA PI.

As it was shown in our previous publications,[143,149,164,165] for all gases except for CO_2, the accessible volume for the permeability coefficient P is calculated by taking into consideration the dead volume for the maximum axis. On the dependence **P(1/FAV)**, the point corresponding to CO_2 is situated on the overall line when the accessible volume is calculated by taking into consideration the dead volume with the minimum radius of gas molecule. As seen in Table 2.14, all the dependences **P(1/FAV)** calculated according to eq 2.8 have high correlation coefficients.

The B_p value represents the overall selectivity of gas separation of polymer. The increase in the slope of **P(1/FAV)** dependence and in the B_p value indicates the increase in the selectivity of this polymer.

As seen in Table 2.14, the highest overall selectivity in each series of polymers corresponds to the first polymer, which contains a hexafluoroiso-propylidene bridge in the dianhydride fragment. The free volume (Table

2.13) corresponding to the first polymer in each series is maximum, in the limits of the accuracy of density measurement. This is explained by the presence of those 12 fluorine atoms in the repeating unit of these polymers that leads to the increase in free volume due to the repulsion between these negative atoms of different chains and as a consequence to the loosening of the packing in glassy state.[166] The polymer 5 from the first and the third series, containing the pyrromellitimide fragment, being the most rigid of all these polymers, shows the lowest overall selectivity. An exception is made by polymer 5 in the fourth series. The overall selectivity of polymers 3 and 4 in all series is practically identical.[9,67,126,127]

TABLE 2.14 The Parameters of eqs 2.8, 2.9, and 2.11 for the Permeability, Diffusion, and Solubility Coefficients.

No.	P			D			S		
	A	-B	R%	A	-B	R%	-A	B	R%
1	44.13	12.71	95.38	26.61	7.48	99.98	24.97	7.06	97.37
2	44.73	12.59	96.14	24.76	6.82	89.78	25.40	7.00	97.42
3	36.59	9.24	96.92	19.89	4.89	93.22	21.14	5.19	97.24
4	36.60	9.09	96.30	18.54	4.48	95.08	20.89	5.03	93.76
5	26.72	6.21	88.50	19.31	4.36	96.88	17.21	3.84	95.45
1a	26.71	7.45	95.39	17.93	4.87	87.02	27.14	9.02	92.90
2a	22.76	4.36	93.53	12.67	2.34	86.10	14.88	2.74	99.89
3a	16.64	2.89	88.18	10.83	1.79	86.31	12.58	2.08	95.24
1b	57.84	20.05	98.48	42.63	14.43	95.57	29.15	10.04	90.65
2b	55.69	18.56	98.54	38.10	12.55	93.29	26.52	8.77	88.78
3b	55.95	18.92	98.84	38.41	12.77	97.01	27.14	9.02	92.97
4b	56.92	18.91	98.56	37.14	12.22	97.14	27.00	8.88	96.16
5b	36.45	9.66	99.63	27.45	7.17	94.01	19.08	5.02	90.11
1c	74.26	23.64	99.99	70.22	22.38	99.97	21.71	6.04	94.81
2c	45.20	11.44	99.89	53.97	13.65	98.32	14.39	3.62	97.02
3c	42.28	10.67	99.57	47.03	11.84	99.99	17.76	4.41	72.86
4c	45.95	11.32	99.67	42.93	10.53	99.80	24.18	5.83	82.36
5c	59.22	16.08	99.88	49.49	13.40	99.52	33.64	9.01	80.40

The overall selectivity B_P depends on the free volume inside of each series of polymers. The dependence of overall selectivity on the free volume for all four series of polymers is shown in Figure 2.24. Each of them is well described by the line $B_p = E + F/V_{fr}$. As seen in Figure 2.24, the most visible dependence on free volume is shown by polymers belonging to the fourth series. In this series, the value of the slope F is maximum. The bridging groups in the diamine component of all four series of polymers can be described in the following way: in the first and the second series, the bridges are totally rigid, in the third series the bridge is semirigid (the rotation is possible around one bond phenyl–carbon), and finally in the fourth series the rotation is possible around both bonds phenyl–carbon.

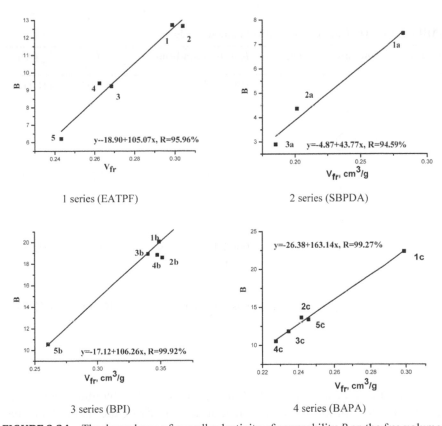

FIGURE 2.24 The dependence of overall selectivity of permeability B on the free volume.

Table 2.15 presents the fragments of diamine component, rotation conditions around the bridge in these fragments, the value of slope F, and the Van der Waals volume of fragments of diamine components of all four series. On one hand, the rotation conditions around the bridge in the main chain of polymer show the possibility of conformational transitions in polymer matrix during passing of gas molecules through membrane. On another hand, the increase in the volume of bridging group leads to the increase in the free volume of polymer matrix. This can be seen in the case of the first polymer in the first series: the value of F is high enough and the coefficient of permeability for CO_2 is 73.91 Barrer.[127] The polymers of the fourth series have the highest value of F, the van der Waals volume of the bridge is the second regarding the value, and the highest overall selectivity is shown by the first polymer in this series. But the coefficient of permeability for CO_2 is not too high, being only 17.04 Barrer.[127]

TABLE 2.15   The Characteristics of Voluminous Bridges.

Series	Voluminous bridge	Rotation conditions	F	V_w (A^3)
1 EATPF		Hindered	105.07	441.116
2 SBPDA		Hindered	43.77	296.406
3 BPI		Possible around one bond	106.21	255.588
4 BAPA		Possible around both bonds	163.14	362.773

Thus, the introduction of bridging groups in the diamine component having the possibility of rotation around the bonds of the main chain that are close to these bridges leads to the increase in the overall selectivity of permeability of the polymer at a high enough permeability.

As we have shown in previous publications[143,149,164,165] for all gases, except CO_2, the accessible volume for diffusion coefficient D is calculated taking into account the dead volume for the maximum radius. The accessible volume for CO_2 is calculated with formula $0.4 R_{max} + 0.6 R_{min}$. Only in this case, the point of CO_2 is on the overall direct line. This can also be seen for other polymers.[133,166] As can be seen in Table 2.14, all the dependences **D(1/FAV)** calculated with eq 2.9 have a high correlation coefficient.

The coefficient B_D in this case represents also the overall selectivity of diffusion. In all four series, the first polymers having 12 fluorine atoms in the main chain show the highest overall selectivity. But the highest selectivity of diffusion, as in the case of permeability coefficients, is shown by the first polymer of the fourth series.

Figure 2.23C presents the dependence of solubility coefficients S on the specific accessible volume according to eq 2.11. Here, we can see the increase in solubility with the increase in 1/FAV, which means that the solubility increases with the decrease in specific accessible volume. In all these diagrams showing the dependence $S(1/FAV)$, the points corresponding to oxygen go out of the general dependence. Here, the behavior of CO_2 is unusual: the point corresponding to the solubility coefficient of this gas is situated on the overall dependence if the calculation of accessible volume takes into consideration the dead volume calculated for the maximum radius of the gas molecule.

There are articles that report the dependence of solubility coefficients of gases on the accessible contact surface of gas molecule and on the surface of the repeating unit of the polymer,[167,168] based on the van der Waals interactions. Here it is considered that higher the surface of the gas molecule, higher the solubility and therefore the solubility coefficient. Indeed, the surface of CO_2 molecule is substantially higher than that of methane, for example, and as a consequence the solubility coefficient is higher.[167,168] This is why the use of the maximum radius of CO_2 for the calculation of FAV in correlation with the solubility coefficient is totally logic.

The correlations according to eqs 2.6–2.9 have been intensively discussed,[143,144,149,159,162,164–166] but the linear aspect of the dependence of solubility coefficients on the specific accessible volume has been investigated just once.[169] However, the increase in solubility coefficients with the increase in free volume, or the dimensions of elements of free volume have been studied in the literature.[167,170–173] Usually, the dimensions of the elements of free volume were studied in correlation with the value of Langmuir sorption capacity according to the dual sorption model,[172,173] and then the correlations one gas—different polymers were discussed. For the case one polymer–different gases, a linear dependence of solubility coefficients on the accessible volume was found for the first time.[169] This dependence can be explained by the sorption of gas molecules of different sizes in the free volume elements.

Thus, the study of the system one–polymer different gases allows to differentiate the polymers in function of the overall selectivity of permeability, diffusion, and solubility. The overall selectivity depends on the free volume of the polymer. Detailed analysis of the overall selectivity of the polymers showed that the van der Waals volume of the bridge in the diamine component of the PI monomer unit along with the conditions of the rotation around the bridge bonds with the main chain has a great influence on the transport characteristics of the polymer. Therefore, PIs with bulky groups are very perspective as the membrane materials.

ACKNOWLEDGMENT

S. Banerjee and Y. Yampolskii acknowledge the research grant received from DST and RFBR against Indo-Russian (DST-RFBR) joint project (INT/RUS/RFBR/P-303 and #17-58-45045) that allows in framing the proposal of writing this book chapter.

I. Ronova and A. Alentiev acknowledge the financial support provided by the Ministry of Science and Higher Education of the Russian Federation.

KEYWORDS

- **polyimide**
- **synthesis**
- **bulky groups**
- **free volume**
- **membranes**
- **gas separation**
- **conformational rigidity**

REFERENCES

1. Bogert, T. M.; Renshaw, R. R. 4-amino-o-phthalic Acid and Some of Its Derivatives. *J. Am. Chem. Soc.* **1908,** *30* (7), 1135–1144.
2. Edwards, W. M.; Robinson, I. M. Polyimides of Pyromellitic Acid. U.S. Patent 2,710,853, June 14, 1955.
3. Ghosh, M. K.; Mittal, K. L. (Eds.). *Polyimides: Fundamentals and Applications*; Marcel Dekker: New York, 1996.

4. Bessonov, M. I.; Koton, M. M.; Kudryavtsev, V. V.; Laius, L. A. *Polyimides: Thermally Stable Polymers*; Plenum Press: New York, 1987.
5. Liaw, D.-J.; Wang, K.-L.; Huang, Y.-C.; Lee, K.-R.; Lai, J.-Y.; Ha, C.-S. Advanced Polyimide Materials: Syntheses, Physical Properties and Applications. *Prog. Polym. Sci.* **2012**, *37*, 907–974.
6. Wang, K.-L.; Liu, Y.-L.; Shih, I.-H.; Neoh, K.-G.; Kang, E.-T. Synthesis of Polyimides Containing Triphenylamine-Substituted Triazole Moieties for Polymer Memory Applications. *J. Polym. Sci. Part A: Polym. Chem.* **2010**, *48* (24), 5790–5800.
7. Ohya, H.; Kudryavtsev, V. V.; Semenova, S. I. *Polyimide Membranes—Applications, Fabrications and Properties*; Gordon and Breach Pbs.: Kodansha, Amsterdam, Tokyo, 1996.
8. Tanaka, K.; Okamoto, K.-I. Structure and Transport Properties of Polyimides as Materials of Gas and Vapor Membrane Separation. In *Material Science of Membranes for Gas and Vapor Separation*; Yampolskii, Yu, Pinnau, I., Freeman, B. D., Eds.; Wiley: Chichester, 2006; pp 271–291.
9. Ghosh, S.; Banerjee, S. Synthesis of 9-alkylated Fluorene-based Poly(ether imide) and their Gas Transport Properties. *J. Membr. Sci.* **2016**, *497* (1), 172–182.
10. Kim, J.-H.; Chang, B.-J.; Lee, S.-B.; Kim, S. Y. Incorporation Effect of Fluorinated Side Groups Into Polyimide Membranes on their Pervaporation Properties. *J. Membr. Sci.* **2000**, *169* (2), 185–196.
11. Gas Permeation Parameters of Glassy Polymers. Database, Inform-registr RF, No. 3585, 1998.
12. Alentiev, A.; Yampolskii, Yu.; Ryzhikh, V.; Tsarev, D. The Database "Gas Separation Properties of Glassy Polymers" (Topchiev Institute): Capabilities and Prospects. *Petrol. Chem.* **2013**, *53* (8), 554–558.
13. Hirayama, Y.; Yoshinaga, T.; Kusuki, Y.; Ninomiya, K.; Sakakibara, T.; Tamari, T.Relation of Gas Permeability with Structure of Aromatic Polyimides I, II. *J. Membr. Sci.* **1996**, *111*, 169–192.
14. Hirayama, Y.; Yoshinaga, T.; Nakanishi, S.; Kusuki, Y. Relation between Gas Permeabilities and Structure of Polyimides. In *Polymer Membranes for Gas and Vapor Separation: Chemistry and Material Science*; Freeman, B. D., Pinnau, I., Eds.; ASC Symposium Series 733; American Chemical Society: Washington, DC, 1999; pp 194–214.
15. Xiao, Y.; Low, B. T.; Hosseini, S. S.; Chung, T. S.; Paul, D. R. The Strategies of Molecular Architecture and Modification of Polyimide-based Membranes for CO_2 Removal from Natural Gas—A Review. *Prog. Polym. Sci.* **2009**, *34*, 561–580.
16. Heuchel, M.; Hofmann, D.; Pullumbi, P. Molecular Modeling of Small-Molecule Permeation in Polyimides and Its Correlation to Free-Volume Distributions. *Macromolecules* **2004**, *37* (1), 201–214.
17. Pinel, E.; Brown, D.; Bas, C.; Mercier, R.; Albérola, N. D.; Neyertz, S. Chemical Influence of the Dianhydride and the Diamine Structure on a Series of Copolyimides Studied by Molecular Dynamics Simulations. *Macromolecules* **2002**, *35* (27), 10198–10209.
18. Yampolskii, Y.; Shishatskii, S.; Alentiev, A.; Loza, K. Group Contribution Method for Transport Property Predictions of Glassy Polymers: Focus on Polyimides and Polynorbornenes. *J. Membr. Sci.* **1998**, *149*, 203–220.

19. Alentiev, A. Y.; Loza, K. A.; Yampolskii, Y. P. Development of the Methods for Prediction of Gas Permeation Parameters of Glassy Polymers: Polyimides as Alternating Copolymers. *J. Membr. Sci.* **2000**, *167*, 91–106.

20. Velioglu, S.; Tantekin-Ersolmaz, S. B. Prediction of Gas Permeability Coefficients of Copolyimides by Group Contribution Methods. *J. Membr. Sci.* **2015**, *480*, 47–63.

21. Velioglu, S.; Tantekin-Ersolmaz, S. B.; Chew, J. W. Towards the Generalization of Membrane Structure-property Relationship of Polyimides and Copolyimides: A Group Contribution Study. *J. Membr. Sci.* **2017**, *543*, 233–254.

22. Low, Z.-X.; Budd, P. M.; McKeown, N. B.; Patterson, D. A. Gas Permeation Properties, Physical Aging, and Its Mitigation in High Free Volume Glassy Polymers. *Chem. Rev.* **2018**, *118*, 5871–5911.

23. Banerjee, S.; Bera, D. Polycondensation Materials Containing Bulky Side Groups: Synthesis and Transport Properties. In *Membrane Materials for Gas and Vapor Separation. Synthesis and Application of Silicon-containing Polymers*; Yampolskii, Yu. P., Finkelshtein, E., Eds.; Wiley: Chichester, 2017; pp 223–269.

24. Hsiao, S.-H.; Yang, C.-P.; Yang, C.-Y. Synthesis and Properties of Polyimides, Polyamides and Poly(amide-imide)s from Ether Diamine Having the Spirobichroman Structure. *J. Polym. Sci. Part A: Polym. Chem.* **1997**, *35* (8), 1487–1497.

25. Sonnett, J. M.; McColloung, R. L.; Beeler, A. J.; Gannett, T. P. The Kinetics of One-step Linear Polyimide Formation: Reaction of p-phenylenediamine and 4,4'-(hexafluoroisopropylidene) Bisphthalic Acid. In *Advances in Polyimide Science and Technology*; Feger, C., Khojasteh, M. M., Eds.; M/S/Htoo Technomic Publishing Co: Lancaster, 1993, pp 313–325.

26. *Progress in Polyimide Chemistry II*; Kriecheldorf, H. R., Ed.; Advances in Polymer Science 141; Springer-Verlag: Berlin Heidelberg, New York, 1999.

27. *Polyimides and Other High Temperature Polymers*; Mittal, K. L., Ed.; VSP/Brill: Leiden, The Netherlands, 2009, vol. 5.

28. Mittal, V. High Performance Polymers: An Overview. In *High Performance Polymers and Engineering Plastics*; Mittal, V., Ed.; Wiley: Hoboken, 2011, pp 1–20.

29. *High Performance Polymers—Polyimides based—From Chemistry to Application*; Abadie, M. J. M., Ed.; Intech Prepress: Novi Sad, 2012.

30. Kuznetsov, A. A. One-pot Polyimide Synthesis in Carboxylic Acid Medium. *High Perform. Polym.* **2000**, *12*, 445–460.

31. Kuznetsov, A. A.; Yablokova, M.; Buzin, P. V.; Tsegelskaya, A. Y. New Alternating Copolyimides by High Temperature Synthesis in Benzoic Acid Medium. *High Perform. Polym.* **2004**, *16*, 89–100.

32. Kuznetsov, A. A.; Tsegelskaya, A. Yu.; Buzin, P. V. One-Pot High-Temperature Synthesis of Polyimides in Molten Benzoic Acid: Kinetics of Reactions Modeling Stages of Polycondensation and Cyclization. *Polym. Sci., Ser. A.* **2007**, *49* (11), 1157–1164.

33. Batuashvili, M. R.; Tsegelskaya, A. Yu.; Perov, N. S.; Semenova, G. K.; Abramov, I. G.; Kuznetsov, A. A. Chain Microstructure of Soluble Copolyimides Containing Moieties of Aliphatic and Aromatic Diamines and Aromatic Dianhydrides Prepared in Molten Benzoic Acid. *High Perform. Polym.* **2014**, *26* (4), 470–476.

34. Hendrix, W. R. Process for Preparing Polyimides by Treating Polyamide-acids with Aromatic Monocarboxylic Acid Anhydrides. U.S. Patent 3,179,632A, April 20, 1965.

35. Endrey, A. L. Aromatic Polyimides from Meta-phenylene Diamine and Para-phenylene Diamine. U.S. Patent 3,179,633A, April 20, 1965.

36. Frost, L. W.; Kesse, J. Spontaneous Degradation of Aromatic Polypromellitamic Acids. *J. Appl. Polym. Sci.* **1964,** *8* (3), 1039–1051.
37. *Polyimides: Materials, Chemistry and Characterization*; Feger, C., Khojasteh, M. M., McGrath, J. E., Eds.; Elsevier: Amsterdam, 1989.
38. *Advanced Polyimide Materials: Synthesis, Characterization, and Applications*; Yang, S.-Y., Ed.; Elsevier: Amsterdam, 2018.
39. Ando, S.; Matsuura, T.; Sasaki, S. Coloration of Aromatic Polyimides and Electronic Properties of their Source Materials. *Polym. J.* **1997,** *29* (1), 69–76.
40. Hsu, L.-C., Ph.D. Dissertation, University of Akron., 1991.
41. Kim, Y.; Glass, T.; Lyle, G.; McGrath, J. Kinetic and Mechanistic Investigations of the Formation of Polyimides Under Homogeneous Conditions. *Macromolecules* **1993,** *26,* 1344–1358.
42. Harris, F. Synthesis of Aromatic Polyimides from Dianhydrides and Diamines. In *Polyimides*; Wilson, D., Stenzenberger, H. D., Hergenrother, P. M., Eds.; Chapman and Hall: New York, 1990, pp. 1–37.
43. Dhara, M. G.; Banerjee, S. Fluorinated High-performance Polymers: Poly(arylene ether)s and Aromatic Polyimides Containing Trifluoromethyl Groups. *Prog. Polym. Sci.* **2010,** *35* (8), 1022–1077.
44. Synder, R. W.; Thomson, B.; Bartges, B.; Czeriwski, D.; Painter, P. C. FTIR Studies of Polyimides: Thermal Curing. *Macromolecules* **1989,** *22* (11), 4166–4172.
45. Scroog, C. E. Polyimides. In *Encyclopedia of Polymer Science and Technology*; Bikales, N. M., Ed.; Interscience: New York, 1969, vol. 11, p 247.
46. Cassidy, P. C.; Fawcett, N. C. Polyimides. In *Kirk-Othmer Encyclopedia of Chemical Technology*; Wiley: New York, 1982, p 704.
47. Volksen, W. Condensation Polyimides: Synthesis, Solution Behavior, and Imidization Characteristics. In *High Performance Polymers*; Hergenrother, P. M., Ed. Adv. Polym. Sci. *117*; Springer: Berlin, Heidelberg, 1994.
48. Angelo, R. J.; Golike, R. C.; Tatum, W. E.; Kreuz, J. A. In *Recent Advances in Polyimide Science and Technology*; Weber, W. D., Gupta, M. R., Eds.; Soc. Plas. Eng.: Brookfield, CT, 1985, p 67.
49. Sroog, C. E. Polyimides. *J. Polym. Sci.: Macromolecular Reviews* **1976,** *11*(1), 161–208.
50. Johnston, J. C.; Mcador, M. A. B.; Alston, W. B. A Mechanistic Study of Polyimide Formation from Diester-diacids. *J. Polym. Sci. Part A: Polym. Chem.* **1987,** *25* (8), 2175–2183.
51. Koton, M.; Meleshko, T.; Kudryavtsev, V.; Nechayev, P.; Kamzolkina, Y. V.; Bogorad, N. Investigation of the Kinetics of Chemical Imidization. *Polym. Sci. USSR* **1982,** *24,* 791–800.
52. Vinogradova, S.; Vygodskii, Y. S.; Vorob'ev, V.; Churochkina, N.; Chudina, L.; Spirina, T.; Korshak, V. Chemical Cyclization of Poly (Amido-acids) in Solution. *Polym. Sci. USSR* **1974,** *16,* 584–589.
53. Wallach, M. Polyimide Solution Properties. *J. Polym. Sci. Part B: Polym. Phys.* **1969,** *7,* 1995–2004.
54. Buhler, K. U. *Spezialplaste*; Academie-Verlag: Berlin; 1978 [in German, Chap. 7.1.11.1].
55. Liaw, D. J. Synthesis and Characterization of New Highly Soluble Organic Polyimdes. In *Macromolecular Nanostructured Materials*; Ueyama, N., Harada, A., Eds.; Springer: Osaka; 2004, pp. 80–100.

56. Imai, Y.; Yokota, R. *New Polyimide: Basic and Application*; Saishin: Tokyo, 2002 [in Japanese].
57. Hasegawa, M.; Horie, K. Photophysics, Photochemistry, and Optical Properties of Polyimides. *Prog. Polym. Sci.* **2001**, *26*, 259–335.
58. Hrdlovic, P. Photochemical Reactions and Photophysical Processes. *Polym. News* **2004**, *29* (2), 50–53.
59. Negi, Y. S.; Damkale, S. R.; Ansari, S. Photosensitive Polyimides. *J. Macromol. Sci. Rev. Macromol. Chem. Phys.* **2001**, *41*, 119–138.
60. Ding, M. X. Isomeric Polyimides. *Prog. Polym. Sci.* **2007**, *32*, 623–668.
61. Mittal, K. L. *Polyimides and Other High Temperature Polymers: Synthesis, Characterization*, vol. 2. VSP: Utrecht, 2003.
62. Greiner, A.; Schmidt, H.-W. Synthesis, Structure and Properties: Aromatic Main Chain Liquid Crystalline Polymers. In *Handbook of Liquid Crystals: High Molecular Weight Liquid Crystals*; Demus, D., Goodby, J., Gray, G. W., Spiess, H.-W., Vill, V., Eds.; New York: John Wiley & Son, Inc, 1998; pp. 1–25.
63. Liaw, D.-J.; Liaw, B.-Y.; Yang, C.-M. Synthesis and Properties of New Polyamides Based on bis[4-(4-aminophenoxy)phenyl]diphenylmethane. *Macromolecules* **1999**, *32*, 7248–7250.
64. Liaw, D.-J., Hsu, P.-N.; Chen, W.-H., Liaw, B.-Y. Novel Organosoluble Poly(amide-imide)s Derived from Kink Diamine bis [4-(4-trimellitimidophenoxy)phenyl]-diphenylmethane. Synthesis and Characterization. *Macromol. Chem. Phys.* **2001**, *202*, 1483–1487.
65. Zhang, S. J.; Li, Y. F.; Ma, T.; Zhao, J. J.; Xu, X. Y., Yang, F. C.; Xiang, X. Y. Organosolubility and Optical Transparency of Novel Polyimides Derived from 2′,7′-bis(4-aminophenoxy)-spiro(fluorene-9,9′-xanthene). *Polym. Chem.* **2010**, *1* (4), 485–493.
66. Jiang, G.-M.; Jiang, X.; Zhu, Y.-F.; Huang, D.; Jing, X.-H.; Gao, W.-D. Synthesis and Characterization of Organo-soluble Polyimides Derived from a New Spirobifluorene Diamine. *Polym. Int.* **2010**, *59* (7), 896–900.
67. Sen, S. K.; Banerjee, S. *Spiro*-biindane Containing Fluorinated Poly(ether imide)s: Synthesis, Characterization and Gas Separation Properties. *J. Membr. Sci.* **2010**, *365* (1–2), 329–340.
68. Ma, X.; Salinas, O.; Litwiller, E.; Pinnau, I. Novel Spirobifluorene- and Dibromospirobifluorene-based Polyimides of Intrinsic Microporosity for Gas Separation Applications. *Macromolecules* **2013**, *46*, 9618–9624.
69. Kute, V.; Banerjee, S. Polyimides 7: Synthesis, Characterization, and Properties of Novel Soluble Semifluorinated Poly(ether imide)s. *J. Appl. Polym. Sci.* **2007**, *103*, 3025–3044.
70. Kazama, S.; Teramoto, T.; Haraya, K. Carbon Dioxide and Nitrogen Transport Properties of bis(phenyl)fluorene-based Cardo Polymer Membranes. *J. Membr. Sci.* **2002**, *207*, 91–104.
71. Zhuang, Y.; Seong, J. G.; Lee, W. H.; Do, Y. S., Lee, M. J.; Wang, G.; Guiver, M. D.; Lee, Y. M. Mechanically Tough, Thermally Rearranged (TR) Random/Block Poly(benzoxazole-co-imide) Gas Separation Membranes. *Macromolecules* **2015**, *48*, 5286–5299.
72. Luo, L.; Zhang, J.; Huang, J.; Feng, Y.; Peng, C.; Wang, X.; Liu, X. The Dominant Factor for Mechanical Property of Polyimide Films Containing Heterocyclicmoieties:

In-plane Orientation, Crystallization, or Hydrogen Bonding. *J. Appl. Polym. Sci.* **2016**, *133*, 44000. DOI: 10.1002/app.44000.

73. Dasgupta, B.; Sen, S. K.; Maji, S.; Chatterjee, S.; Banerjee, S. Synthesis and Characterization of Highly Soluble Poly(ether imide)s Containing Indane Moieties in the Main Chain. *J. Appl. Polym. Sci.* **2009**, *112* (6), 3640–3651.

74. Maji, S.; Sen, S. K.; Dasgupta, B.; Chatterjee, S.; Banerjee, S. Synthesis and Characterization of New Poly(ether amide)s Based on a New Cardo Monomer. *Polym. Adv. Technol.* **2009**, *20* (4), 384–392.

75. Huang, W.; Yan, D.; Lu, Q.; Tao, P. Preparation of Aromatic Polyimides Highly Soluble in Conventional Solvents. *J. Polym. Sci.: Part A: Polym. Chem.* **2002**, *40*, 229–234.

76. Yagci, H.; Mathias, L. J. Synthesis and Characterization of Aromatic Polyamides and Polyimides from Trimethyl- and di-t-butylhydroquinone-based Ether-linked Diamines. *Polymer* **1998**, *39*, 3779–3786.

77. Kasashima, Y.; Kumada, H.; Yamamoto, K.; Akutsu, F.; Naruchi, K.; Miura, M. Preparation and Properties of Polyamides and Pfrom 4,4″-diamino-o-terphenyl. *Polymer* **1995**, *36*, 645–650.

78. Chang, C.-W.; Yen, H.-J.; Huang, K.-Y.; Yeh, J.-M.; Liou, G.-S. Novel Organo Soluble Aromatic Polyimides Bearing Pendant Methoxy-substituted Triphenylamine Moieties: Synthesis, Electrochromic, and Gas Separation Properties. *J. Polym. Sci. Part A: Polym. Chem.* **2008**, *46* (24), 7937–7949.

79. Akutsu, F.; Kuze, S.; Matsuo, K.; Naruchi, K.; Miura, M. *Makromol. Chem., Rapid Commun.* **1990**, *11* (12), 673–677.

80. Mathews, A. S.; Kim, D.; Kim, Y.; Kim, I. L. Ha, C.-S. Synthesis and Characterization of Soluble Polyimides Functionalized with Carbazole Moieties. *J. Polym. Sci. Part A: Polym. Chem.*, **2008**, *46* (24), 8117–8130.

81. Harris, F. W.; Lin, S. H.; Li, F.; Cheng, S. Z. D. Organo-soluble Polyimides: Synthesis and Polymerization of 2,2′-disubstituted-4,4′,5,5′-biphenyltetracarboxylic Dianhydrides. *Polymer* **1996**, *37* (22), 5049–5057.

82. Yeganeh, H.; Mehdipour-Ataei, S. Preparation and Properties of Novel Processable Polyimides Derived from a New Diisocyanate. *J. Polym. Sci. Part A: Polym. Chem.* **2000**, *38* (9), 1528–1532.

83. Qiu, Z.; Wang, J.; Zhang, Q.; Zhang, S.; Ding, M.; Gao, L. Synthesis and Properties of Soluble Polyimides Based on Isomeric Ditrifluoromethyl Substituted 1,4-bis(4-aminophenoxy)benzene. *Polymer* **2006**, *47* (26), 8444–8452.

84. Ge, Z.; Fan, L.; Yang, S. Synthesis and Characterization of Novel Fluorinated Polyimides Derived from 1,1′-bis(4-aminophenyl)-1-(3-trifluoromethylphenyl)-2,2,2-trifluoroethane and Aromatic Dianhydrides. *Eur. Polym. J.* **2008**, *44* (4), 1252–1260.

85. Yang, C.-P.; Su, Y.-Y.; Chiang, H.-C. Organosoluble and Light-colored Fluorinated Polyimides from 4-tert-butyl-[1,2-bis(4-amino-2-trifluoromethylphenoxy)phenyl] benzene and Aromatic Dianhydrides. *React. Funct. Polym.* **2006**, *66* (7), 689–794.

86. Madhra, M. K.; Salunke, A. K.; Banerjee, S.; Prabha, S. Synthesis and Properties of Fluorinated Polyimides, 2. Derived from Novel 2,6-bis(3′-trifluoromethyl-*p*-aminobiphenyl ether)pyridine and 2,5-bis(3′-trifluoromethyl-*p*-aminobiphenyl ether) thiophene. *Macromol. Chem. Phys.* **2002**, *203* (9), 1238–1248.

87. Banerjee, S.; Madhra, M. K.; Salunke, A. K.; Jaiswal, D. K. Synthesis and properties of fluorinated polyimides. 3. Derived from novel 1,3-bis[3′-trifluoromethyl-4′(4″-amino

benzoxy) benzyl] benzene and 4,4-bis[3'-trifluoromethyl-4'(4-amino benzoxy) benzyl] biphenyl. _Polymer_ **2003**, _44_ (3), 613–622.

88. Jeong, K. U.; Kim, J.-J.; Yoon, T.-H. Synthesis and Characterization of Novel Aromatic Polyimides from Bis(3-aminophenyl)3.5-bis(trifluoromethyl)phenyl Phosphine Oxide. _Korea Polym. J.,_ **2000**, _8_ (5), 215–223.
89. Pixton, M. R.; Paul, D. R. Relationships between Structure and Transport Properties for Polymers with Aromatic Backbones. In Polymeric Gas Separation Membranes; Paul, D. R., Yampolskii, Yu. P., Eds.; CRC Press: Boca Raton, 1994; pp. 83–154.
90. Pixton, M. R.; Paul, D. R. Gas Transport Properties of Polyarylates Part I: Connector and Pendant Group Effects. _J. Polym. Sci. Part B: Polym. Phys._ **1995**, _33_, 1135–1149.
91. Pixton, M. R.; Paul, D. R. Gas Transport Properties of Polyarylates: Substituent Size and Symmetry Effects. _Macromolecules_ **1995**, _28_ (24), 8277–8286.
92. Pixton, M. R.; Paul, D. R. Gas Transport Properties of Polyarylates Based on 9,9-bis(4-hydroxyphenyl)anthrone. Polymer **1995**, _36_, 2745–2751.
93. Costello, L. M.; Koros, W. J. Effects of Structure on the Temperature Dependence of Gas Transport and Sorption in a Series of Polycarbonates. _J. Polym. Sci. Part B: Polym. Phys._ **1994**, _32_, 701–713.
94. McHattie, J. S.; Koros, W. J.; Paul, D. R. Gas Transport Properties of Polysulfones: 1. Role of Symmetry of Methyl Group Placement on Bisphenole Rings. _Polymer_ **1991**, _32_, 840–850.
95. Fritsch, D.; Peinemann, K.-V. Novel Highly Permselective 6F-poly(amide-imide)s as Membrane Host for Nano-sized Catalysts. _J. Membr. Sci._ **1995**, _99, 29–38._
96. Ghosal, K.; Morisato, A.; Freeman, B. D.; Chern, R. T.; Alvarez, J. C.; de la Campa, J. G.; de Abajo, J. Synthesis and Gas Separation Properties of a Family of New Aromatic Polyamides for Petrochemical Application. _Polym. Prepr._ **1994**, _35_ (1), 731–735.
97. Ghosal, K.; Freeman, B. D., Chern, R. T., Alvarez, J. C., de la Campa, J. G., Lozano, A. E., de Abajo, J. Gas Separation Properties of Aromatic Polyamides with Sulfone Groups. _Polymer_ **1995**, _36,_ 793–800.
98. Ronova, I. A.; Dubrovina, L. V.; Kovalevsky, A. Yu.; Hamchuk, C.; Bruma, M. The Effect of Side Substituents on Rotation Hinderance in Polyheteroarylenes. Russian Bulletin **1998**, _47_ (7), 1248–1256.
99. Hamchuk, C.; Ronova, I. A.; Hamchuc, E.; Bruma, M. The Effect of Rotation Hinderance on Physical Properties of Some Heterocyclic Polyamides Containing Pendent Imide Groups. _Angew. Makromol. Chem._ **1998**, _254,_ 67–74.
100. Langsam, M.; Burgoyne, W. F. Effects of Diamine Monomer Structure on the Gas Permeability of Polyimides. I. Bridged Diamines. _J. Polym. Sci. Part A: Polym. Chem._ **1993**, _31_, 909–921.
101. Kim, T. H.; Koros, W. J.; Husk, G. R. Temperature Effects on Gas Permselection Properties in Hexafluoro Aromatic Polyimides. _J. Membr. Sci._ **1989**, _46, 43–56._
102. Tanaka, K.; Okano, M.; Toshino, H.; Kita, H.; Okamoto, K. Effect of Methyl Substituents on Permeability and Permselectivity of Gases in Polyimides Prepared from Methyl-substituted Phenylenediamines. _J. Polym. Sci. Part B: Polym. Phys._ **1992**, _30_ (8), 907–914.
103. Matsumoto, K.; Xu, P. Gas Permeation Properties of Hexafluoro Aromatic Polyimides. _J. Appl. Polym. Sci._ **1993**, _47_ (11), 1961–1972.

104. Lin, W.-H.; Vora, R. H.; Chung, T. S. Gas Transport Properties of 6FDA-durene/1,4-phenylenediamine (pPDA) Copolyimides. *J. Polym. Sci. Part B: Polym. Phys.* **2000**, *38* (21), 2703–2713.

105. Yamamoto, H.; Mi, Y.; Stern, S. A.; St. Clair, A. K. Structure/Permeability Relationships of Polyimide Membranes. II. *J. Polym. Sci. Part B: Polym. Phys.* **1990**, 28, 2291–2304.

106. Chung, T. S.; Lin, W.-H.; Vora, R. H. Gas transport Properties of 6FDA-durene/1,3-phenylenediamine (mPDA) Copolyimides. *J. Appl. Polym. Sci.* **2001**, *81, 3552–3564*

107. Yeom, C. X.; Lee J. M.; Hong, Y. T.; Choi, K. Y.; Kim, S. C. Analysis of Permeation Transients of Pure Gases through Dense Polymeric Membranes Measured by a New Permeation Apparatus. *J. Membr. Sci.* **2000**, *166, 71–83.*

108. Matsui, S.; Sato, H.; Nakagawa, T. Effects of Low Molecular Weight Photosensitizer and UV Irradiation on Gas Permeability and Selectivity of Polyimide Membrane. *J. Membr. Sci.* **1998**, *141, 31–43.*

109. Xu, J. W.; Chng, M. L.; Chung, T. S.; He, C. B.; Wang, R. Permeability of Polyimides Derived from Non-coplanar Diamines and 4,4′-(hexafluoroisopropylidene) Diphthalic Anhydride. *Polymer* **2003**, *44* (16), 4715–4721.

110. Ayala, D.; Lozano, A. E.; de Abajo, J.; García-Perez, C.; de la Campa, J. G.; Peinemann, K. V.; Freeman, B. D.; Prabhakar, R. Gas Separation Properties of Aromatic Polyimides. *J. Membr. Sci.* **2003**, *215* (1–2), 61–73.

111. Calle, M.; Lozano, A. E.; de Abajo, J. Design of Gas Separation Membranes Derived of Rigid Aromatic Polyimides. 1. Polymers from Diamines Containing Di-tert-butyl Side Groups. *J. Membr. Sci.* **2010**, *365* (1–2), 145–153.

112. Calle, M.; García, C.; Lozano, A. E.; de la Campa, J. G.; de Abajo, J.; Álvarez, C. Local Chain Mobility Dependence on Molecular Structure in Polyimides with Bulky Side Groups: Correlation with Gas Separation Properties. *J. Membr. Sci.* **2013**, *434*, 121–129.

113. Maya, E. M.; Yoldi, I. G.; Lozano, A. E.; de la Campa, J. G.; de Abajo, J. Synthesis, Characterization, and Gas Separation Properties of Novel Copolyimides Containing Adamantyl Ester Pendant Groups. *Macromolecules* **2011**, *44* (8), 2780–2790.

114. Zhang, C.; Li, P.; Cao, B. Effects of the Side Groups of the Spirobichroman-based Diamines on the Chain Packing and Gas Separation Properties of the Polyimides. *J. Membr. Sci.* **2017**, *530*, 176–184.

115. Li, T.; Liu, J.; Zhao, S.; Chen, Z.; Huang, H.; Guo, R.; Chen, Y. Microporous olyimides Containing Bulky Tetra-o-isopropyl and Naphthalene Groups for Gas Separation Membranes. *J. Membr. Sci.* **2019**, *585*, 282–288.

116. Li, T.; Huang, H.; Wang, L.; Chen, Y. High Performance Polyimides with Good Solubility and Optical Transparency Formed by the Introduction of Alkyl and Naphthalene Groups Into Diamine Monomers. *RSC Adv.* **2017**, *7*, 40996–41003.

117. Zhang, C.; Li, P.; Cao, B. Decarboxylation Crosslinking of Polyimides with High CO2/CH4 Separation Performance and Plasticization Resistance. *J. Membr. Sci.* **2017**, *528*, 206–216.

118. Banerjee, S.; Madhra, M. K.; Salunke, A. K.; Maier, G. Synthesis and Properties of Fluorinated Polyimides. 1. Derived from Novel 4,4″-bis(aminophenoxy)-3,3″-trifluoromethyl Terphenyl. *J. Polym. Sci. Part A: Polym. Chem.* **2002**, *40* (8), 1016–1027.

119. Sen, S. K.; Banerjee, S. High T_g, Processable Fluorinated Polyimides Containing Benzoisoindoledione Unit and Evaluation of their Gas Transport Properties. RSC Adv. **2012**, 2 **(15)**, 6274–6289.

120. Chatterjee, R.; Ghosh, S.; Bisoi, S.; Banerjee, S. Synthesis, Characterization, and Gas Transport Properties of New Semifluorinated Poly(ether imide)s Containing Cardo Moiety. *J. Appl. Polym. Sci.* **2017,** *134* (34), 45213. DOI: 10.1002/app.45213.

121. Dutta, A.; Bisoi, A.; Mukherjee, R.; Chatterjee, R.; Das, R. K.; Banerjee, S. Soluble Polyimides with Propeller Shape Triphenyl Core for Membrane Based Gas Separation. *J. Appl. Polym. Sci.* **2018,** *135*, 46658. DOI: 10.1002/app.46658.

122. Chatterjee, R.; Bisoi, S.; Kumar, A. G.; Padmanabhan, V.; Banerjee, S. Polyimides Containing Phosphaphenanthrene Skeleton: Gas-Transport Properties and Molecular Dynamics Simulations. *ACS Omega* **2018,** *3*, 13510–13523.

123. Belov, N.; Chatterjee, R.; Nikiforov, R.; Ryzhikh, V.; Bisoi, S.; Kumar, A. G.; Banerjee, S.; Yampolskii, Yu. New Poly(ether imide)s with Pendant Di-*tert*-butyl Groups: Synthesis, Characterization and Gas Transport Properties. *Separat. Purif. Technol.* **2019,** *217*, 183–194.

124. Sen, S. K.; Dasgupta, B.; Banerjee, S. Effect of Introduction of Heterocyclic Moieties Into Polymer Backbone on Gas Transport Properties of Fluorinated Poly (ether imide) Membranes. *J. Membr. Sci.* **2009,** *343* (1), 97–103.

125. Dasgupta, B.; Sen, S. K.; Banerjee, S. Aminoethylaminopropylisobutyl POSS—Polyimide Nanocomposite Membranes and their Gas Transport Properties. *Mater. Sci. Eng. B* **2010,** *168*, 30–35.

126. Dasgupta, B.; Sen, S. K.; Banerjee, S. Gas Transport Properties of Fluorinated Poly (ether imide) Membranes Containing Indan Moiety in the Main Chain. *J. Membr. Sci.* **2009,** *345* (1), 249–256.

127. Sen, S. K.; Banerjee, S. Gas Transport Properties of Fluorinated Poly (ether imide) Films Containing Phthalimidine Moiety in the Main Chain. *J. Membr. Sci.* **2010,** *350* (1), 53–61.

128. Dasgupta, B.; Banerjee, S. A Study of Gas Transport Properties of Semifluorinated Poly(ether imide) Membranes Containing Cardo Diphenylfluorene Moieties. *J. Membr. Sci.* **2010,** *362* (1), 58–67.

129. Robeson, L. M. Correlation of Separation Factor Versus Permeability for Polymeric Membranes. *J. Membr. Sci.* **1991,** *62*, 165–185.

130. Robeson, L. M. The Upper Bound Revisited. *J. Membr. Sci.* **2008,** *320*, 390–400.

131. Ronova, I. A. On the Flexibility of Polyheteroarylenes and the Effect on Several Physical Properties of these Polymers. In *Polymers Chains Structure, Physical Properties and Industrial Uses*; Penzkofer, R., Wu, Y., Eds.; Nova Science Publisher: New York, 2012; pp 1–97.

132. Ronova, I. A.; Pavlova, S. S. A. The Effect of the Conformational Rigidity on Several Physical Properties of Polymers. *High Perform. Polym.* **1998,** *10* (3), 309–329.

133. Alentiev, A. Yu.; Ronova, I. A.; Schukin, B. V.; Yampolskii, Yu. P. Correlation between Gas Permeability of Amorphous Polymers and Conformational Rigidity of their Chains. *Polym. Sci. A* **2007,** *49* (2), 217–226.

134. Ronova, I.; Sokolova, E. A.; Bruma, M. Influence of Chemical Structure of the Repeating Unit on Physical Properties of Aromatic Polymers Containing Phenylquinoxaline Rings. *J. Polym. Sci., Part B: Polym. Phys.* **2008,** *46* (17), 1868–1877.

135. Ronova, I. Structural Aspects in Polymers. Interconnections between Conformational Parameters of the Polymers with their Physical Properties. *Struct. Chem.* **2010,** *21* (3), 541–553.

136. Chisca, S.; Ronova, I. A.; Sava, I.; Medvedeva, V; Bruma, M. Influence of Conformational Parameters on Physical Properties of Polyimides Containing Methylene Bridges. *Mater. Plast.* **2011**, *48* (1), 38–44.

137. Bruma, M.; Damaceanu, M. D.; Ronova, I. A. Correlation between Conformational Rigidity and Physical Properties of Some Poly(oxadiazole-imide)s. *Rev. Roum. Chim.* **2012**, *57* (4–5), 383–391.

138. Ronova, I. A.; Bruma, M. Influence of Conformational Rigidity on Membrane Properties of Polyimides. *Struct. Chem.* **2012**, *23* (1), 47–54.

139. Ronova, I.; Bruma, M.; Schmidt, H. W. Conformational Rigidity and Dielectric Properties of Polyimides. *Struct. Chem.* **2012**, *23* (1), 219–226.

140. Carja, I.D; Hamciuc, C.; Vlad-Bubulac, T., Bruma, M.; Ronova, I. A. Effect of Conformational Parameters on Physical Properties of Polymers Containing Pendant Phenoxyphtalonitrile Substituents. *Struct. Chem.* **2013**, *24* (5), 1693–1703.

141. Ipate, A. M.; Hamciuc, C.; Bruma, M.; Ronova, I. A.; Buzin, M. I. Influence of Conformational Rigidity on Physical Properties of Some Poly(1,3,4 oxadiazole ether)s Containing Trifluoromethyl Groups. *Rev. Roum. Chim.* **2014**, *59* (6-7), 475–483.

142. Sava, I.; Bruma, M.; Ronova, I. A. The Influence of Conformational Parameters on Some Physical Properties of Polyimides Containing Naphthalene Units. *High Perform. Polym.* **2015**, *27* (5), 583–589.

143. Ronova, I. A.; Alentiev, A. Yu.; Bruma, M. Influence of Voluminous Substituents in Polyimides on Their Physical Properties. *Polym. Rev.* **2018**, *58* (2), 376–402.

144. Matteucci, S.; Yampolskii, Yu. P.; Freeman, B. D.; Pinnau, I. Transport of Gases and Vapors in Glassy and Rubbery Polymers. In *Materials Science of Membranes for Gas and Vapor Separation*; Yampolskii, Yu. P., Pinnau, I., Freeman, B. D., Eds.; Wiley: New York, 2006; pp 1–47.

145. Matsumoto, K.; Xu, P.; Nishikimi, T. Gas Permeation of Aromatic Polyimides. I. Relationship between Gas Permeabilities and Dielectric Constants. *J. Membr. Sci.* **1993**, *81* (1), 15–22.

146. Dewar, M. J. S.; Zoebisch, E. G.; Healy, E. F.; Stewart, J. J. P. Development and Use of Quantum Mechanical Molecular Models. 76. AM1: A New General Purpose Quantum Mechanical Molecular Model. *J. Am. Chem. Soc.* **1985**, *107* (13), 3902–3909.

147. Rozhkov, E.M; Schukin, B. V.; Ronova, I. A. Methods for Calculations of Occupied Volumes in Glassy Polymers. The Lattice Integration and the Monte Carlo Methods. *Cent. Eur. J. Chem.* **2003**, *1* (4), 402–426.

148. Askadskii, A. A. *Computational Materials Science of Polymers*. Cambridge International Science Publishing: Cambridge, 2003.

149. Ronova, I. A.; Rozhkov, E. M.; Alentiev, A. Yu.; Yampolskii, Yu. P. Occupied and Accessible Volumes in Glassy Polymers and their Relation with Gas Permeation Parameters. *Macromol. Theor. Simul.* **2003**, *12* (6), 425–439.

150. Okamoto, K.; Tanaka, K.; Kita, H.; Ishida, M.; Kakimoto, M.; Imai, Y. Gas Permeability and Permselectivity of Polyimides Prepared from 4,4'-diaminotriphenylamine. *Polym. J.* **1992**, *24* (5), 451–457.

151. Dinari, M.; Ahmadizadegan, H. Novel and Processable Polyimides with a *N*-benzonitrile Side Chain: Thermal, Mechanical and Gas Separation Properties. *RSC Adv.* **2015**, *5* (33), 26040–26050.

152. Serbezeanu, D.; Carja, I. D.; Bruma, M.; Ronova, I. A. Correlation between Physical Properties and Conformational Rigidity of Some Aromatic Polyimides Having Pendant Phenolic Groups. *Struct. Chem.* **2016,** *27* (3), 973–981.
153. Li, Y.; Wang, X.; Ding, M.; Xu, J. Effects of Molecular Structure on the Permeability and Permselectivity of Aromatic Polyimides. *J. Appl. Polym. Sci.* **1996,** *61* (5), 741–748.
154. Li, Y.; Ding, M.; Xu, J. Relationship between Structure and Gas Permeation Properties of Polyimides Prepared from Oxydiphtalic Dianhydride. *Macromol. Chem. Phys.* **1997,** *198* (9), 2769–2778.
155. Pauling, L. *General Chemistry*; W.H. Freeman and Company: San Francisco, 1957.
156. Tanaka, K.; Osada, Y.; Kita, H.; Okamoto, K. Gas Permeability and Permselectivity of Polyimides with Large Aromatic Rings. *J. Polym. Sci. Part B: Polym. Phys.* **1995,** *33* (13), 1907–1915.
157. Ronova, I. A.; Ryvkina, N. G.; Jablokova, M. Yu.; Zhukova, E. S.; Sinitsina, O. V.; Alentiev, A. Yu.; Busin, M. I. Influence of Side Substituents on the Dielectric Constant in Polyimides. *XX International Scientific-technical Conference "High technologies in Russian industry, thin films in electronics"* October 9, 2015; Moscow. Devices Processing, pp 77–79.
158. Zhukova, E. K.; Kuznetsov, A. A.; Yablokova, M. Yu.; Alentiev, A. Yu. Gas Separation Properties of New Thermoplastic Polyimides with Phenylamide Groups in Diamine Moiety. Effect of Polymer Structure. *Pet. Chem.* **2014,** *54* (7), 544–550.
159. Ronova, I. A.; Bruma, M.; Nikolaev, A. Yu.; Kuznetsov, A. A. Lowering the Dielectric Constant of Polyimides by Swelling in Supercritical Carbon Dioxide. *Polym. Adv. Technol.* **2013,** *24* (7), 615–622.
160. Bruma, M.; Alentiev, A. Yu.; Ronova, I. A. Influence of Bridging Groups on Gas Separation Properties of Aromatic Polyimedes: A Comparative Analysis. *Rev. Roum. Chim.* **2018,** *63* (7–8), 697–710.
161. Ronova, I. A.; Alentiev, A. Yu.; Chisca, S.; Sava, I.; Bruma, M.; Nikolaev, A. Yu.; Belov, N. A.; Buzin, M. I. Change of Microstructure of Polyimide Thin Films Under the Action of Supercritical Carbon Dioxide and Its Influence on the Transport Properties. *Struct. Chem.* **2014,** *25* (1), 301–310.
162. Ronova, I. A.; Belov, N. A.; Alentiev, A. Yu.; Nikolaev, A. Yu.; Chirkov, S. V. Influence of Swelling in Supercritical Carbon Dioxide of Ultem and Polyhexafluoropropylene Thin Films on their Gas Separation Properties: Comparative Analysis. *Struct. Chem.* **2018,** *29* (2), 457–466.
163. Belov, N. A.; Alentiev, A. Yu.; Ronova, I. A.; Sinitsyna, O. V.; Nikolaev, A. Yu.; Zharov, A. A. The Relaxation Process of the Microstructure in Polyhexafluoropropylene Afterswelling in Supercritical Carbon Dioxide. *J. Appl. Polym. Sci.* **2016,** *133* (14), 43105. DOI:10.1002/app.43105.
164. Ronova, I. A.; Alentiev, A. Yu.; Bruma, M. Trends in Polymeric Gas Separation Membranes. In *Physical Chemistry for the Chemical and Biochemical Sciences*; Lopez-Bonilla, J. L., Abdullin, M. I., Zaikov, G. E., Eds.; Apple Academic Press: New York, 2016, pp 31–98.
165. Ronova, I. A.; Khokhlov, A. R.; Shchukin, B. V. Transport Parameters of Glassy Polymers: Effect of Occupied and Accessible Volumes. *Polym. Sci. A* **2007,** *49,* 517–531.
166. Ronova, I. A.; Khokhlov, A. R.; Alentiev, A. Yu.; Bruma, M. Correlation between Conformational Rigidity and Membrane Properties of Polyheteroarylenes. In *New*

Steps in Chemical and Biochemical Physics. Pure and Applied Science; Pearce, E. M., Kirshenbaum, G., Zaikov, G. E., Eds.; Nova Science: New York, 2010, pp 161–202.

167. Goubko, M.; Miloserdov, O.; Yampolskii, Yu. P.; Alentiev, A. Yu.; Ryzhikh, V. A Novel Model to Predict Gas Solubility in Glassy Polymers. *J. Polym. Sci. Part B: Polym. Phys.* **2017,** *55,* 228–244.

168. Yampolskii, Yu. P.; Wiley, D.; Maher, C. Novel Correlation for Solubility of Gases in Polymers: Effect of Molecular Surface Area of Gases. *J. Appl. Polym. Sci.* **2000,** *76,* 552–560.

169. Ronova, I.; Alentiev, A.; Bruma, M. Correlation of Accessible Volume with Selectivity in Polyimides Having Voluminous Bridge in Diamine Component. *Int. J. Eng. Res. App..* **2018,** *8* (12 part III), 42–50.

170. Yampolskii, Yu. P. Methods for Investigation of the Free Volume in Polymers. *Russ. Chem. Rev.* **2007,** *76,* 59–78.

171. Yampolskii, Yu. P. Polymeric Gas Separation Membranes. *Macromolecules* **2012,** *45,* 3298–3311.

172. Kanehashi, S.; Nagai, K. Analysis of Dual-mode Model Parameters for Gas Sorption in Glassy Polymers. *J. Membr. Sci.* **2005,** *253,* 117–138.

173. Ryzhikh, V. E., Alentiev, A. Yu.; Yampolskii, Yu. P. Relation of Gas-transport Parameters of Amorphous Glassy Polymers to their Free Volume: Positron Annihilation Study. *Polym. Sci. A* **2013,** *55,* 244–252.

CHAPTER 3

The Thermal *Cis–Trans* Isomerization of Azopolyimides in the Solid State

JOLANTA KONIECZKOWSKA*

Centre of Polymer and Carbon Materials Polish Academy of Sciences, 34 M. Curie-Sklodowska Str., 41-819 Zabrze, Poland

*E-mail: jkonieczkowska@cmpw-pan.edu.pl

ABSTRACT

Azobenzene-containing polymers, also known as azopolymers, are a wide group of materials interesting from the point of view of their potential utilization in the fields of photonics, optoelectronics, and biology as cell culture substrates and biomaterials. They are utilized as thin films on the substrates or free-standing foils. This class of polymers contains light-sensitive linkages -*N*=*N*- in their chemical structure. Azobenzene derivatives occur in the form of two isomers: an energetically stable *trans*-isomer and a metastable *cis*-isomer. Under the irradiation of azochromophores with a specific wavelength, the reversible *trans–cis–trans* isomerization is observed. The stability of *cis*-isomer in the solid state is crucial in some azopolymers applications, which can be tailored by changing the azochromophore structure, kinds of azomolecules incorporation to the polymer backbone (covalent or non-covalent), and polymer molar masses. The knowledge about the influence of mentioned factors supports designing azopolymers with desired properties adjusted to a particular device. This paper is devoted to showing the impact of azopolyimide structure (main chain, side chain or T-type polymers, and "guest–host" systems) on *trans–cis* and *cis–trans* isomerization process in the solid state based on literature review and investigations published in our previous works.

3.1 INTRODUCTION

The development of technologies in areas such as photonics and optoelectronic increases the interest in new light-sensitive materials. Although many results have been already achieved in this field of research exploration of the new processable materials, especially with a high optical and thermal damage threshold, are still required. Taking into account the current state of knowledge concerning photochromic materials, it can be concluded that the most important ones are materials containing derivatives of azobenzene.[1,2] The possible use of the light-sensitive polymers is the results of generating the photoinduced optical anisotropy (POA) by the polarized light in the material. The irradiation azopolymers by the polarized light cause local changes of absorption coefficient $\Delta\alpha$ (dichroism) or refractive index Δn (birefringence), which are the result of *cis–trans–cis* isomerization reaction and reorientation of *trans*-azochromophore,[2] and also may cause migration of polymer chain that forms the surface relief gratings (SGRs) in the holographic grating recording experiment.[3] Azobenzene derivatives are characterized by the reversible *trans–cis–trans* isomerization from the generally stable *trans*-form to the less stable *cis*-form, upon irradiation with UV or visible light.[4] This phenomenon lets to the structural transformation which in turn leads to macroscopic variation in the chemical and physical properties of polymeric material and the POA is generated. The POA is a key for potential technological applications as materials for high-density optical data storage, optical waveguides, diffractive elements, for the liquid crystal alignment, and preparation of membranes for gas separation.[2,5–7] It has been shown that the photoresponsive behavior, that is, efficiency and stability of the POA, strongly depends on many factors, for example, architecture of the chromophore and polymer main chain, the content and type linkage of photochromic moieties to the polymer chain, the chromophore mobility, the molecular masses of azopolymers, molecular near-neighbor interactions of the photochromic units in the polymer chain, and also stability of *cis–trans* isomerization reaction. Understanding the correlation between the structural features is crucial in searching for the polymers with the required physical and optical properties.

Polymers for potential technological applications should characterize high optical and thermal damage threshold and high POA stability. Polyimides (PIs) meet all these requirements. Their high glass transition temperature, high thermal stability, low dielectric constant, thermo-oxidative stability, outstanding mechanical and electrical properties, superior chemical resistance, optical transparency, and low susceptibility to a laser light damage make them ideal candidates for various uses in the area of photonics, optoelectronics, medicine devices, and membranes for gas separation.[7–9]

The knowledge about the *trans–cis–trans* isomerization of azopolymers in the solid state is important from the point of view of their potential applications in devices. Information about the relationship between the chemical architecture of azochromophore and polymer backbone, azo-dye content in the polymer chain, and their influence in the *trans–cis–trans* isomerization may allow the modeling of polymer chemical structure to obtain materials with defined properties adapted for specific devices.

3.2 PHOTOISOMERIZATION

Photoisomerization is a photoinduced reversible transformation of the structure of a chemical compound, associated with a change in the energy state of the molecule due to excitation through the absorption of electromagnetic waves. The consequence of excitation is a change in both spectral and physicochemical properties. Irradiation of compound A with the light of wavelength λ_1, another structurally different compound B is formed. The return of compound B to A can occur spontaneously, as a result of thermal relaxation (kT) or as a result of irradiation with radiation of wavelength λ_2.[10,11] The isomerization process is schematically shown in Figure 3.1.

$$A \; \underset{\lambda_2,\, kT}{\overset{\lambda_1}{\rightleftharpoons}} \; B$$

FIGURE 3.1   The schemat of *trans–cis* reaction induced by light and *cis–trans* relaxation.

The two spectroscopically different photochromic compounds obtained in this way differ in both the absorption spectrum and a number of properties, that is, dielectric susceptibility, refractive index, dipole moment, polarizability, and oxidation–reduction potential.[10] Photoisomerization is considered to be one of the "purest" photoreactions, due to the lack of formation of by-products, even with countless cycles.[12] Photochromism in

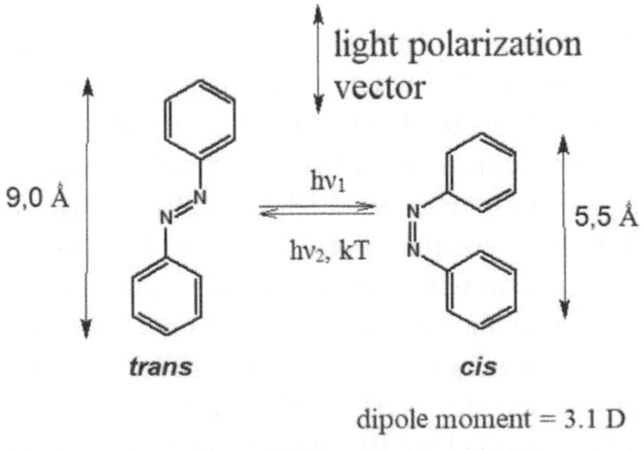

dipole moment = 3.1 D

FIGURE 3.2   *Trans–cis* isomerization of azobenzene.

chemical compounds can occur through various mechanisms: *trans–cis–trans* isomerization, redox reactions, triplet–triplet absorption, and valence tautomerization (opening or closing of a cyclic ring).[10,11,13]

Chromophores containing an azo group in their structure undergo reversible photoinduced *trans–cis–trans* isomerization. As a result of irradiation of the energetic stable *trans*-isomer (*E*) of azobenzene, their less stable *cis*-form (*Z*) is generated.[14] During the photochromic transition, structural and geometric changes occur in the molecule. The distances between the carbon atoms at positions 4 and 4' consist form 9.0 to 5.5 Å in azobenzene molecules (Fig. 3.2). This transition is accompanied by a change in the dipole moment. The *trans*-form does not have a permanent dipole moment, while the *cis*-isomer gains a dipole moment of 3.0 D.[11] In addition, the *trans*-isomer exhibits anisotropic properties, while the *cis*-isomer has isotropic properties. Irradiation induces the changes in electron spectra of azobenzenes. The most *trans*-isomers of azobenzene derivatives are characterized by the high-intensity bandwidth attributed to the π–π* transition in the UV spectral region (ca. 350 nm) and the much less intense band corresponding to the n–π* transition in the visible spectral range (ca. 450 nm).[11,15] After conversion to the *cis*-isomer, the band corresponding to the π–π* transition shift toward shorter waves, while the intensity of the n–π* transition increases.[11] The content of *cis*-isomer in the sample during light irradiation depends on different factors, that is, the chemical structure of azochromophore, the intensity of the light inducing the transformation, the quantum photochemical efficiency of *trans–cis* and *cis–trans* isomerization, and the aggregation ability of chromophore.[12,14–16]

3.3 KINDS OF AZOPOLYIMIDES

Azopolymers are macromolecules containing azobenzene or azopyridine moieties in their structure.[14,16] They are a wide group of polymers differentiated in terms of the structure of the polymer main chain, type and the content of chromophore, as well as the kind of its attachment to the polymer chain. Due to the way the azochromophore is introduced into the macromolecule chain, we distinguish polymers in which the chromophore is connected to the polymer matrix by a covalent bond, so-called functionalized polymers, and guest–host systems, in which the dye is dispersed in the polymer matrix without forming the covalent bonds.[14,16,17] Functionalized PIs can be obtained by pre- or post-polymerization functionalization.[14,18] Pre-polymerization functionalization is a one-step method. The reaction of monomers results in the preparation of the main chain PIs, where two phenyl rings are an integral part of the main chain (Fig. 3.3a) or T-type PIs (Fig. 3.3b), where one of the azochromophore rings is a part of the main chain.[1,14,19]

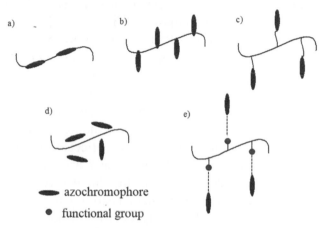

FIGURE 3.3   Kinds of azopolyimides (a) main chain, (b) T-type, (c) side chain, (d) guest–host azosystems, and (e) supramolecular polyimides.

Post-polymerization is a two-step method in which side chain polymers are obtained (Fig. 3.3c).[13,19] At the first step, a polymer containing appropriate functional groups (e.g., OH, COOH, NH_2) is synthesized, which in the second step are used to attach the chromophore.[13,14,18] Post-polymerization allows attachment of the azochromophore to the polymer chain under much milder conditions than pre-polymerization; therefore, it is possible to use azo compounds that are not thermally resistant.[18,20,21] In addition, post-polymerization makes it possible to control the content of azochromophore in the polymer backbone.[18,21] Polymers containing non-covalent bond azochromophores can be divided into two groups: (1) doped systems, where the dye is dispersed molecularly in the polymer matrix (Fig. 3.3d) and (2) supramolecular assembles (polymers) based on interactions such as hydrogen, ionic, coordination bonds, or π–π type interactions (Fig. 3.3e).[22–27]

Regardless of how the azo dye is attached to the polymer backbone, each type of azopolymer has its advantages and disadvantages. The lack of covalent bonds causes the increase in the mobility of azo groups, that is important from the point of view generation of fast *trans–cis–trans* isomerization of azomolecules.[25] In addition, easy control of the azo dye content in the system allows to track its impact on physicochemical properties, as well as to obtain polymers with different molar masses.[25] Despite easy preparation of this kind of azopolymers, the lack of covalent bonds promotes the aggregation of dye molecules, phase separation of chromophore and polymer matrix, evaporation or sublimation of dye from the matrix, and a significant reduction in the glass transition temperature of the azosystem in compared with the polymer matrix.[1,13,28,29] Moreover, in some cases, preparation of guest–host azopolymers may be impossible because of low chromophores solubility in the polymer matrix.[3] Partly, these problems were solved in supramolecular systems, where the non-covalent interactions between the chromophore and the polymer reduce adverse effects. In supramolecular polymers, it is possible to increase the content of azochromophore by over 100% in comparison with doped systems, without observing adverse effects.[25,30,31] In addition, the formation of non-covalent interactions improves the durability of light-induced optical effects, due to the reduction in chromophore mobility.[3,16,29,32] On the other hand, azopolymers containing a covalently attached dye do not exhibit the drawbacks characteristic of guest–host systems. Covalent attachment of a chromophore also contributes to the improvement of thermal stability of azopolymer.[4,25] However, the high-temperature reaction conditions can lead to azo group's damage.[14]

3.4 CIS–TRANS ISOMERIZATION REACTION

In Chapter 4, the *cis–trans* isomerization in the solid state of azopolyimides is discussed. It is well known that the most polymeric applications in light-sensitive devices may be possible only in the solid state, because of that it is important to investigate the isomerization reaction of azopolymers in the form of thin films or free-standing polymer foils. Depending on the specific device or application requirements, the azopolymer should characterize fast or very slow *cis–trans* conversion (Chapter 5).

Azopolyimides discuss in this chapter in the context of *cis–trans* isomerization are different from the point of view their chemical structure, that is, location of the azobenzene moieties, that are in the main chain, side chain of polymer backbone or form T-type polymers; or they can by dispersed in guest–host azosystems. For the chosen polymers, the *cis–trans* reaction occurs spontaneously upon natural visible light or in the dark. For all PIs, the content of *cis*–isomer after UV-light irradiation is discussed.

3.4.1 MAIN CHAIN AZOPOLYIMIDES

The first discussed group of azopolymers are main chain homopolymers, that is, PIs (PI-1–PI-7), polyamide (PA-1), poly(amic acid) (PAA-1), and co-polyimides (PI-8–PI-10).[7,33–37] Polymers PI-1, PI-2, PI-8, PI-9, and PAA-1 contain one azobenzene group, while PI-3–PI-7, PI-10, PI-11, and

PA-1 have two azobenzene moieties in the main chain separated by flexible ether linkages (Fig. 3.4).

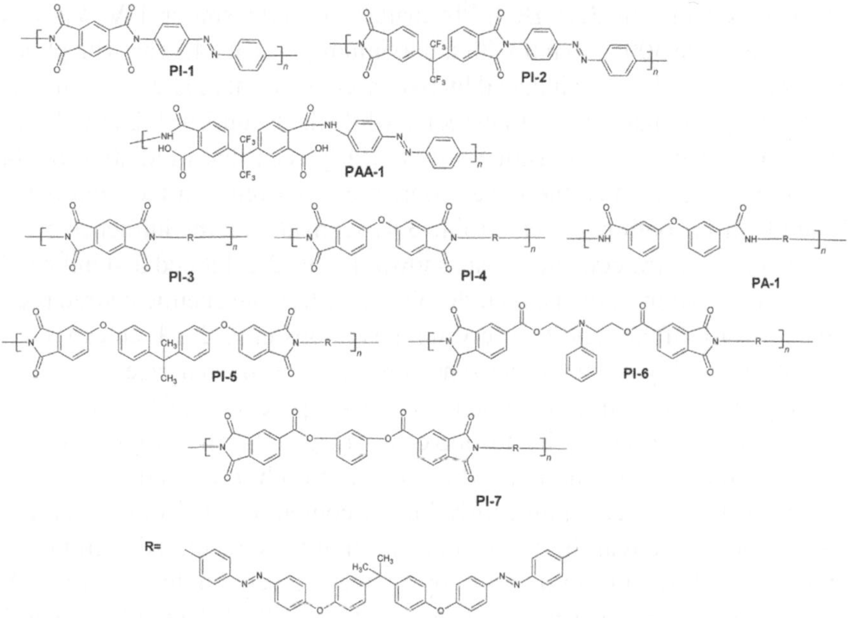

FIGURE 3.4 Chemical structures of main chain homopolymers: azopolyimides (PI-X), azopoly(amic acid) (PAA-1), and azopolyamide (PA-1).

TABLE 3.1 Content of *Cis*-isomer and *Cis-Trans* Relaxation Time in the Dark of Main Chain Homopolymers with Azo Moieties.

Polymer code	Content of *cis*-isomer	Relaxation time (in the dark)	Irradiation conditions
PI-1	50%	–	$I = 50$ mW/cm^2; $\lambda = 375$ nm; $t = 4$ h
	10%		$I = 100$ mW/cm^2; $\lambda = 442$ nm; $t = 16$ h
PI-2	14%	–	$I = 50$ mW/cm^2; $\lambda = 375$ nm; $t = 4$ h
	7%		$I = 100$ mW/cm^2; $\lambda = 442$ nm; $t = 16$ h
PI-3	15%	–	$I = 60$ mW/cm^2; $\lambda = 470$ nm; $t = 1$ h
PI-4	28%	–	$I = 60$ mW/cm^2; $\lambda = 445$ nm; $t = 1$ h
	19%		$I = 50$ mW/cm^2; $\lambda = 445$ nm; $t = 10$ min
PI-5	41%	>7 days	$P = 9$ W; $\lambda = 405$ nm; $t = 5$ min
PI-6	26%	>7 days	$P = 9$ W; $\lambda = 405$ nm; $t = 5$ min
PI-7	33%	>7 days	$P = 9$ W; $\lambda = 405$ nm; $t = 5$ min
PAA-1	70%	>180 min	$I = 29$ mW/cm^2; $\lambda = 360$–470 nm; $t = 1$ min
PA-1	22%	–	$I = 50$ mW/cm^2; $\lambda = 445$ nm; $t = 10$ min

In Lee et al.[33] the *trans–cis* photoisomerization reaction of amorphous PI-1 and semi-crystalline PI-2 polyimides was investigated. Polymer PI-1 contains a rigid phenyl ring between imide moieties, while PI-2 has flexible CF_3 groups in the main chain (Fig. 3.4). Both PIs characterized the similar UV–Vis spectrum in the range 300–550 nm with a maximum (λ_{max}) at 345 nm. The photoisomerization reaction was induced by two laser beams, that is, $\lambda = 375$ nm ($I = 50$ mW/cm²) stimulated π–π* transitions, while the beam $\lambda = 442$ nm ($I = 100$ mW/cm²) induced n–π* transitions. It is clearly seen that irradiation by 442 nm laser beam generated the lower content of *cis*-isomer in the sample than 375 nm light for both polymers (Table 3.4). Considering the influence of 375 nm light on *cis*-form content, it was shown that PI-2 exhibited a significantly lower concentration of *cis*-isomer, despite more flexible chemical structure. It may be a result of the rigidity of crystalline regions in PI-2, which can reduce or negate the ability of the azobenzene chromophore to isomerize.

The poly(amic acid) gel denoted as PAA-1 (precursor of PI-2 with open imide rings) was described in Hosono et al.[34] (Fig. 3.4). This polymer showed λ_{max} at 380 nm. The excitation light generated both π–π* and n–π* transitions. PAA-1 exhibited significantly higher content of *cis*-isomer than their PI analogue. Moreover, high *cis* concentration was generated significantly faster than for PAA-1 (1 min vs 4 h) by a lower-powered light (29 vs 50 mW/cm²). Larger and faster *trans–cis* isomerization of PAA-1 than PI-2 may be a result of different factors: (1) chemical structure—PAA-1 has open imide rings, which increase the flexibility of the polymer chain and increase of azobenzene mobility; (2) using broad absorption light ($\lambda = 360$–470 nm) the π–π* and n–π* transitions are induced; and (3) PAA-1 was prepared as a gel, which can increase the mobility of the azobenzenes. The *cis–trans* relaxation for poly(amic acid) PAA-1 was stable by more than 180 min.

Amorphous PIs, PI-3 and PI-4, with two azobenzene moieties in the main chain were described in Wie et al.[35] PIs are different in rigidity/flexibility of the backbone. PI-3 contains rigid pyromellitic dianhydride part, while PI-4 was synthesized from oxy-4,4'-di(phthalic anhydride) with flexible ether group (Fig. 3.4). Both polymers characterized the same absorption spectra with a maximum at 365 nm. The used 445 nm excitation beam induced the low energetical n–π* transitions. The influence of main chain flexibility was observed. After 1 h of irradiation, the higher *cis*-isomer content was generated for PI-4 with more flexible chemical structure (Table 3.1). Polymer PI-4 was described in two works where two different intensities of 445 nm light were used (Table 3.1).[36,37] Using the 50 mW/cm² light, the photostationary state was reached after 10 min. When the 60 mW/cm² light beam was used, the photostationary state was generated after 30 min. Authors did not give the information about the thickness of films, molar masses, or densities of polymers, because of that it is difficult to consider the possible reason for this result. The key information may be the glass transition temperature (T_g). Higher T_g may indicate the higher molar mass of the polymer. T_gs determined by DMA (dynamic mechanical analysis) measurements were 230°C and 276°C for 50 and 60 mW/cm² light beams, respectively. Longer exposure time (for 60 mW/cm² laser beam) needed to reach the photostationary state may be the result of the higher molar mass of polymer. After 10 min of irradiation, *cis*-isomer contents were 17 and 23% for 50 and 60 mW/cm² of laser beams, respectively. Higher intensity of excitation light generates a slight more *cis*-chromophore in the sample.

It should be noticed that PI-3 is an analogue of PI-2. Both PIs contain the same dianhydride part and different content of azobenzene moieties. PI-2 characterized more rigid structure without flexible ether or methyl groups. The excitation lights at 442 and 445 nm induced n–π* transitions were used for both PIs. After 1 h of irradiation, the content of *cis*-isomer was ca. 4 and 15% for PI-1 and PI-3, respectively. It is difficult to state clearly that flexible ether and methyl groups help in *trans–cis* isomerization in PI-3. The population of *cis*-isomer was higher for PI-3, despite using almost twice more power light in the case PI-2 (60 vs 100 mW/cm²). However, PI-2 has a semicrystalline nature, which can reduce the ability of the azobenzene chromophore to isomerize.

In Wie et al.[35], the influence of the intermolecular H-bonds in polyamide PA-1 and PI PI-4 was discussed in the context of *trans–cis* isomerization. Chemical structures of both polymers were different in the content of the amide or imide groups (Fig. 3.4). The presence of intermolecular H-bonds between amide groups was proved by FTIR spectra for PA-1. The excitation beam at 445 nm induced n–π* transitions for both polymers. Polymer films were irradiated by 30 min but after 10 min the photostationary state was observed. A slight higher content of *cis*-isomer was observed for PA-1 (Table 3.1). Presence of H-bonds did not influence the photoisomerization reaction for PA-1, but changes in λ_{max} before and after 445 nm light were observed. After 15 min of irradiation, the λ_{max} of PA-1 was bathochromically shifted by 6 nm, while the λ_{max} for PI-4 was shifted only by 0.5 nm.

PIs, PI-5–PI-7, were described in Bujak et al.[7] PIs were synthesized from the same azodiamie and different dianhydrides (Fig. 3.4). Polymers have flexible ether (PI-5), ester groups (PI-7), or ester with flexible ethylene groups (PI-6). All materials exhibited similar UV–Vis spectra with λ_{max} localized at ca. 353 nm. Excitation diode-light induced both π–π* and n–π* transitions. After 3 min of UV-light exposure, the photostationary state was observed, where 26–41% of *cis*-isomers was generated (Table 3.1). The lowest content of *cis*-form for PI-6 may be a result of the most flexible structure of the backbone. The excitation beam not only induced both π–π* and n–π* transitions of *trans*-form, but also may generate the back isomerization. In general, *cis*-isomers exhibit absorption in the range of 400–550 nm. Because of that, the 405 nm light can generate *trans–cis* isomerization and *cis–trans* recovery. The most flexible structure of PI-6 can favor the *cis–trans* reaction in comparison to PI-5 and PI-7. For PIs, PI-5, PI-6, and PI-7, the *cis–trans* relaxation in the dark was monitored. The full recovery to *trans*-form was not observed after 7 days for all PIs. Polymers exhibited ca. 8% content of *cis*-isomer after 7 days of relaxation (Table 3.1). It seems that the more rigid structure (PI-7) has an impact on the higher stability of *cis–trans* reaction. PI-7 showed ca. 73% of *cis*-isomer conversion to *trans*-form within 7 days, whereas PI-5 transformed in ca. 83%. The *cis*-isomer in the polymer films was generated also in the linear (PI-8, PI-9) and cross-linked (PI-10, PI-11) co-polyimides with azobenzene moieties. Similar to homo-polyimides, azobenzenes were directly incorporate to the main chain (PI-8, PI-9), or by flexible ether groups (PI-10, PI-11). Chemical structures were shown in Figure 3.5.

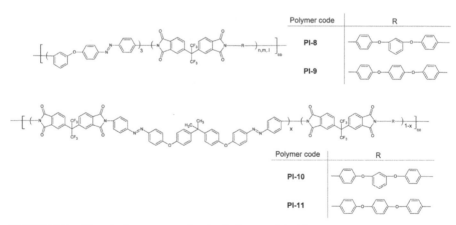

FIGURE 3.5   Chemical structures of main chain co-polyimides.

Trans–cis isomerization was generated by 445 nm light (I = 120 mW/cm^2; t = 1 h), which induced n–π* transitions. After 1 h of irradiation, all co-polyimides showed the same content of *cis*-isomer, which is 25%. Incorporation phenyl rings in ortho or para position did not influence the content of *cis*-isomer in the film. The similar photoresponse can be the result of nearly identical chemical composition and maintain equivalent azobenzene moiety concentrations.

Summarizing, direct incorporation of two phenyl rings of azobenzene to the polymer backbone did not significantly decrease the mobility of the chromophores. PIs characterized relatively high content of *cis*-isomer after light irradiation (5–50%). The content of *cis*-isomers was strongly dependent on the type of transition (π–π* or n–π*) induced by the light beam. When the π–π* transitions were generated the concentration of *cis*-form was 14–50%, while only 5–28% population of *cis*-isomer was observed when n–π* transitions were induced. It should be noticed that when the π–π* transitions are generated, only *trans–cis* isomerization has to take place. In most cases, the *cis*-isomer of azobenzene derivatives absorb the light in the range 400–550 nm. Because of that, the induction of n–π* may generate not only *trans–cis* isomerization but also *cis–trans* recovery. In this case, the content of *cis*-isomer in the photostationary state may be lower than for excitation beams that generated only π–π* transitions. It seems that incorporation of azobenzene moieties to the polymer main chain influence the high stability of the *cis–trans* isomerization. The full recovery to *trans*-form was longer than 7 days for PI-5–PI-7.

3.4.2 T-TYPE AZOPOLYIMIDES

Photoisomerization of azobenzenes was investigated for a series of T-type homo and co-polyimides.[38–41] All homopolyimides contain azobenzenes with CH$_3$ group substituted in para or ortho positions and Cl group attached in para position. Homopolyimides differ in the main chain structure. PIs, PI-12–PI-14, have CF$_3$ groups; PI-15–PI-17 have a rigid structure with C=O group; PI-18–PI-20 contain high volume cardo group; and PI-21–PI-23 contain flexible ether and CH$_3$ groups (Fig. 3.6).

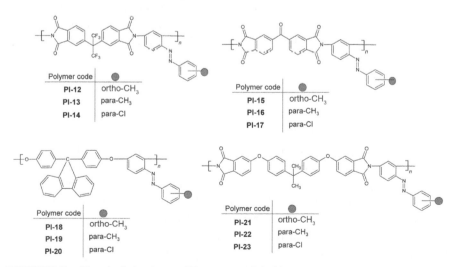

FIGURE 3.6 Chemical structures of T-type azopolyimides.

TABLE 3.2 Content of *Cis*-Isomer and *Cis–Trans* Relaxation Time in the Dark and in Visible Light of T-type Homopolymers with Azo Moieties.

Polymer code	Content of *cis*-isomer	Relaxation time (in visible light)	Relaxation time (in the dark)	Time of irradiation
PI-12	33%[a,b]	6 min[a]	19.45 h[a]	$t = 1200$ s[a] $t = 1100$ s[b]
PI-13	40%[a,b]	3.15 h[a]	25 h[a]	$t = 1200$ s[a] $t = 1100$ s[b]
PI-14	30%[a,b]	3.15 h[a]	39.30 h[a]	$t = 1000$ s[a,b]
PI-15	20%[a,b]	3.15 h[a]	7.35 h[a]	$t = 1400$ s[a] $t = 1100$ s[b]
PI-16	28%[a,b]	5.55 h[a]	24.10 h[a]	$t = 900$[a]; $t = 1100$ s[b]
PI-17	—[a]	–	–	–
PI-18	41%[b]	–	–	$t = 800$ s
PI-19	40%[b]	–	–	$t = 1200$ s
PI-20	55%[b]	–	–	$t = 200$ s
PI-21	23%[a]	>8.30 h	19.45 h	$t = 200$ s
PI-22	28%[a]	>8.30 h	19.45 h	$t = 200$ s
PI-23	23%[a]	>8.30 h	17.05 h	$t = 250$ s

Irradiation conditions:
[a]$P = 100$ W; $\lambda = 350$ nm
[b]$I = 10$–12 mW/cm^2; $\lambda = 365$ nm

All PIs characterized similar UV–Vis spectra with λ_{max} in the range 340–350 nm. For the generation of *cis*-isomer, two different UV lights ($\lambda_{ex} = 350$ or 365 nm) were used (Table 3.2). Irradiation time needed to the generation of photostationary state was different for PIs and was in the range 200–1400 s (Table 3.2). The used excitation light induced π–π^* transitions of all PIs. The used two different irradiation conditions did not influence the values of the *cis* population in the sample. The highest values of *cis*-isomers in the range of 30–55% were observed for PIs, PI-13–PI-15 and PI-18–PI-20 (Table 3.2). Considering the influence of main chain polymer on the content of *cis*-isomer, it is seen that high volume "cardo" group or flexible CF$_3$ moieties improve the *trans–cis* isomerization, which may be the result of increasing the distance between polymer chains and increase in the free volume, needed for the isomerization reaction. PIs with ether or C=O groups exhibited similar *cis*-form content in the range 20–28% (Table 3.2). The influence of kind of substituent and their location in benzene ring was observed. The presence of chlorine atom in para position drastically diminishes

TABLE 3.2   *(Continued)*

the *cis*-isomer content in comparison to PIs with a methyl group (except PI-20). The possible reason may be the character of substituents. Chlorine atom is an electron-withdrawing group, while methyl substituent is an electron-donating group. In our previous work, it was shown that kinetics of *cis–trans* isomerization was slower for azobenzenes with halogen atoms than for compounds with methyl groups.[42] Incorporation of methyl group in para position to the *N=N* linage increases the *cis*-isomer content in comparison with ortho-substituted PIs (Table 3.2). Attachment of substituent in ortho position may cause the steric hindrance that can hamper the *trans–cis* isomerization. The stability of *cis–trans* isomerization was monitored in the dark or upon natural visible light. Considering the stability of *cis–trans* isomerization upon the visible light, PIs with ether groups in the main chain (PI-21, PI-22, and PI-23) characterized the higher time recovery. After 8.5 h of relaxation, the 8–18% of *cis*-isomer was still observed (Table 3.2). For PIs with C=O or CF$_3$ groups, the full recovery to *trans*-form in visible light was observed from 6 min to almost 6 h. The *cis–trans* isomerization was more stable in the dark. The full recovery was observed from 7.5 to 39.5 h (Table 3.2). Attachment of methyl groups in para position increased the time of *cis–trans* relaxation in comparison to ortho-substituted PIs.

Two series of T-type co-polyimides with azobenzene moieties were described in two studies.[40,41] Both kinds of co-polyimides contain methyl-substituted azobenzene moieties (Fig. 3.7).

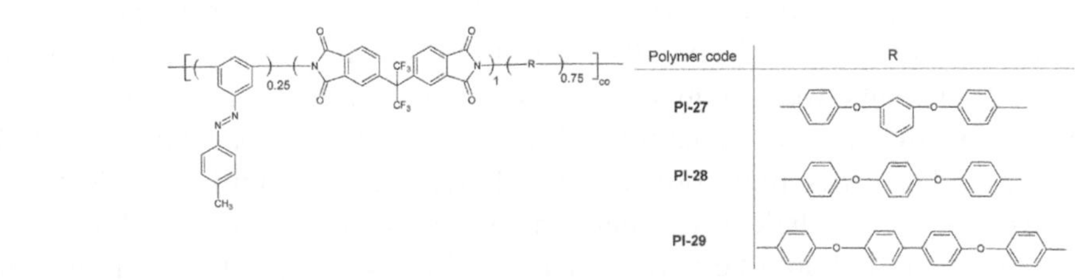

FIGURE 3.7   Chemical structures of T-type azo-co-polyimides.

TABLE 3.3   Content of *Cis*-Isomer of T-type Co-polyimides with Azo Moieties.

Polymer code	Content of *cis*-isomer	Irradiation conditions
PI-24	40%	$P = 100$ W; $\lambda = 350$ nm; $t = 2000$ s
PI-25	44%	$P = 100$ W; $\lambda = 350$ nm; $t = 750$ s
PI-26	39%	$P = 100$ W; $\lambda = 350$ nm; $t = 1400$ s
PI-27	14%	$P = 280$ W; $\lambda = $ UV–Vis broad spectrum; $t = 450$ s
PI-28	8%	$P = 280$ W; $\lambda = $ UV–Vis broad spectrum; $t = 450$ s
PI-29	9%	$P = 280$ W; $\lambda = $ UV–Vis broad spectrum; $t = 450$ s

Both series of co-polyimides exhibited the similar UV–Vis absorption region with λ_{max} at 355 nm. Excitation lights induced the π–π* transitions for polymers PI-24–PI-26 or both π–π* and n–π* for PI-27–PI-29. PIs in series PI-24–PI-26 characterized high content of *cis*-isomer that was ca. 40%. The influence of connection between phenylene rings (denoted as R in Fig. 3.7) was observed in the time of light exposure needed for generation of the photostationary state. Polymers with phenyl ring incorporated between ether groups in para position showed the highest *cis*-isomer population and the fastest *cis–trans* isomerization (Table 3.3). Attachment of phenylene ring in ortho position significantly increased the time needed for the generation of the photostationary state. Polymers in series PI-27–PI-29 exhibited significantly low *cis*-isomer content than previously described co-PIs. The lower content of *cis*-isomer may be a result of shorter light exposition or kind of applied excitation light. Authors irradiated thin films by 450 s (the same irradiation time was applied for isomerization in the solution). It is well known that the isomerization reaction is much slower in the solid-state than in the solution; therefore, the time of irradiation of the film should be longer in order to generate a photostationary state. On the other hand, applying a broad spectrum of UV–Vis light for the generation of *trans–cis* isomerization may induce the *cis–trans* back reaction of generated *cis*-isomer.

T-type PIs characterized larger content of *cis*-isomer than main chain PIs with 8–55% population of *cis*-isomer, but for most of them the *cis*-isomer content was between 23 and 40%. For all polymers, the high energetically π–π* transitions were induced by the excitation light. Incorporation of one of phenyl rings of azochromophore to the polymer backbone resulted in the higher *cis*-isomer content and relatively large stability of *cis–trans* conversion in the dark in comparison to main chain azopolyimides (7–39 h vs 7 days). It should be noticed that despite a significantly lower content of azochromophore in T-type co-polyimides chemical structures, they can exhibit higher content of *cis*-isomers than their homo-polyimide analogues.

3.4.3 SIDE CHAIN AZO-CO-POLYIMIDES

The *cis–trans* isomerization of side chain polyimides was investigated only in one work.[43] Co-polyimides contain 25% of azochromophore attached in the side chain by alkoxy linkage. Aliphatic chain of azochromophore has from 3 to 6 methylene groups (Fig. 3.8).

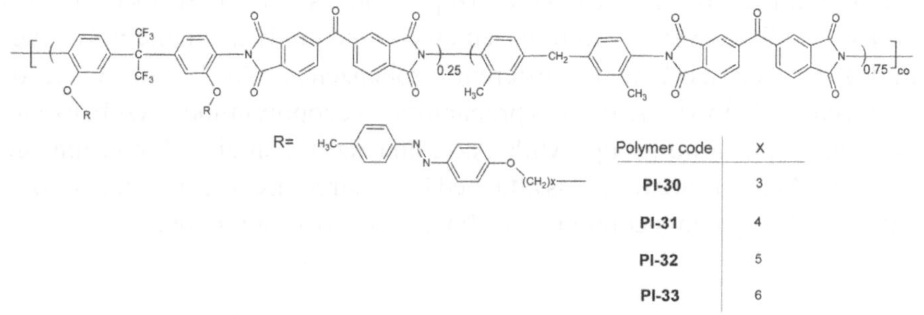

Polymer code	X
PI-30	3
PI-31	4
PI-32	5
PI-33	6

FIGURE 3.8   Chemical structures of side chain azo-co-polyimides.

TABLE 3.4   Content of *Cis*-Isomer in Side Chain Co-polyimides with Azo Moieties ($P = 100$ W; $\lambda = 350$ nm; $t = 1$ min).

Polymer code	Content of *cis*-isomer	Relaxation time (in visible light)
PI-30	52%	38 min
PI-31	53%	40 min
PI-32	57%	>140 min
PI-33	50%	55 min

The maximum of absorption of side chain PIs was at ca. 360 nm. Application of UV light excited $\pi–\pi^*$ transitions of azochromophore. Flexible attachment of azobenzene resulted in high content of *cis*-isomer in the range 50–57% (Table 3.4), which was generated within 20 s–1 min of irradiation. Increasing the number of methylene groups resulted in an increase in the *cis*-isomer population in the sample. PI-32 with five methylene groups in the aliphatic chain exhibited the highest content of *cis*-isomer. The *cis–trans* isomerization was in the range 38–140 min in natural visible light. The highest stability of *cis–trans* conversion showed for PI, PI-32. Relaxation to the *trans*-form takes place in two steps for co-polyimides with long spacer; fast one, where 15–20% of *cis*-isomer recovered to *trans*-form, and a second with linear kinetics.

Side chain PIs exhibited larger content of *cis*-isomer than main chain and T-type azopolyimides. Flexible spencer between azobonds and main chain of PI increased the mobility of azobenzenes resulted in high *cis*-content generated within a short time. On the other hand, incorporation of azoben-zenes in the side chain decreased the stability of the *cis–trans* process in comparison to the main chain and T-type azopolyimides.

3.4.4 "GUEST–HOST" AZOSYSTEMS

The isomerization of guest–host azopolyimides was described in two works.[44,45] All azosystems were based on commercially available poly(ether imide) denoted as Ultem. Different azobenzenes (Az-1 and Az-2) or azopyridine (AzPy) were used to preparation of azopolyimides. Azobenzene, Az-1, has a phenolic group, while Az-2 and AzPy contain alkoxyl linkage (Fig. 3.9). Methyl substituent is attached to the structure of compounds, Az-1 and Az-2. Azosystems contain 2 or 50 wt% of azochromophore.

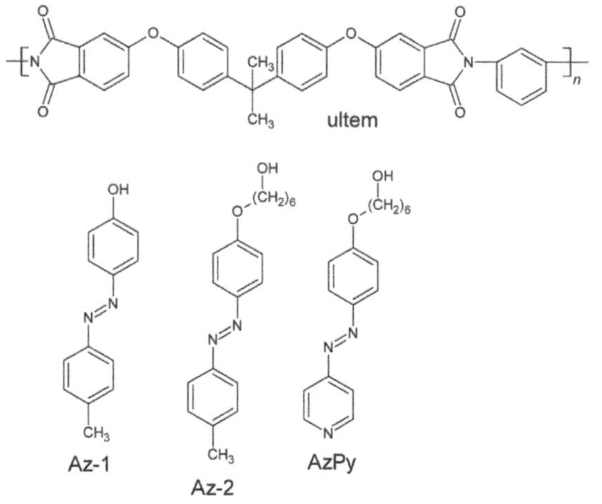

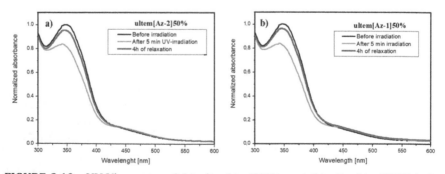

FIGURE 3.9   Chemical structures of guest–host azosystems based on polyimide (ultem).

FIGURE 3.10   UV-Vis spectra of (a) ultem[Az-1]50%, and (b) ultem[Az-2]50% before irradiation (black spectra), after 405 nm light exposition (light gray spectra), and after 4 h of relaxation in the dark (dark gray spectra).

TABLE 3.5 Content of *Cis*-Isomer and Relaxation Time in the Dark for Guest–Host Azosystems.

Polymer code	Content of *cis*-isomer	Relaxation time (in the dark)	Irradiation conditions
Ultem[Az-1]50%	18%	>4 h	$I = 150$ mW/cm^2; $\lambda = 405$ nm; $t = 5$ min
Ultem[Az-2]50%	18%	>4 h	$I = 150$ mW/cm^2; $\lambda = 405$ nm; $t = 5$ min
Ultem[AzPy]2%	10%	–	$\lambda = 405$ nm; $t = 1$ min

All guest–host azosystems characterized similar UV–Vis spectra with λ_{max} located at ca. 360 nm (Fig. 3.10). The used 405 nm excitation light generated both $\pi–\pi^*$ and $n–\pi^*$ transitions. Ultem[Az-1]50% and Ultem[Az-2]50% exhibited the same content of *cis*-isomer ca. 18%. After 4 h of relaxation, the full recovery was not observed for Ultem[Az-1]50% and Ultem[Az-2]50%. The *cis*-isomer content was ca. 3%. Ultem[AzPy]2% characterized lower *cis*-isomer content than their azobenzene analogues (Table 3.5), which can be connected to the lower chromophore content and/or lower irradiation time. After 1 min of light exposure, the photostationary state may not be reached. It should be noticed that the photoisomerization reaction for Ultem[AzPy]2% was investigated for polymers based on ultem matrices with different weight-average molar masses, that is, 4300, 28,000, and 63,000 g/mol.[45] The *cis–trans* relaxation in the dark was observed only for azosystem based on the higher molar mass polymer matrix. The possible reason for hindering of the *cis–trans* isomerization for azosystems based on the matrix with lower molar masses may be connected to the fragility parameter of PI. In general, glass-forming liquids can be classified into two groups: (1) strong glasses with the fragility parameter $m \approx 16$[46] and (2) fragile glasses exhibit high values of m ($m \geq 200$).[47] Polymers with significant differences in the stiffness of individual parts exhibit higher fragility parameters, because of difficulty in the packing of the polymer chains, resulting from significant differences in the rotation energy barrier and higher sensitivity of structural relaxation to the changes in temperature.[48] Azosystem based on the highest molar mass matrix exhibited the highest fragility parameter; it means that its packing ability is the lowest. In this case, higher free-volume favors the *cis–trans* isomerization in comparison to azosystems based on low molar masses matrices.

Guest–host azosystems exhibited lower *cis*-isomer content and lower *cis–trans* stability than their functionalized analogues. It seems that the content of azochromophore may have the highest impact on the *cis* population than the kind of azo compound. Despite low *cis*-isomer and low stability of *cis–trans* back reaction, guest–host azosystems may be interesting materials for applications that required fast *trans–cis–trans* isomerization.

3.5 APPLICATIONS

Azopolymers may be applied in different fields of photonics, optoelectronics, medical devices, and gas separation applications.[12,49,50,55] The potential application of azpolyimides will be described in details for photosensitive layers for the construction of LC nematic cells, azomembranes for gas separation, and for the preparation of devices based on the photomechanical effect.

Photochromic PIs can be applied as the photoalignment liquid crystal layers. These types of materials can be an interesting alternative for the rubbing technique, which uses PIs as layers for the alignment of liquid crystal. Rubbing of polymers turned out to be a very convenient technique and it is now widely used in both small scientific labs and large LCD factories.[52] The rubbing technique has advantages like it is less expensive and easy control method. Despite these advantages, this technique has some serious drawbacks that became appeared to be crucial for the production of LCDs and miniature LC telecommunication devices. These drawbacks follow from the contact type of rubbing technology. Additional difficulties connected with the precise control of the rubbing characteristics over huge substrate areas. Using this method is also problematic in the case of projection displays because rubbing traces become visible when the former image is magnified. Other problems may arise when rubbing is used in miniature telecommunication devices where there is a need to align LC in the thin gaps of the light waveguides. Finally, this method in principle cannot be used to align LCs in closed volumes. Limitations of the rubbing technique caused the search for alternative methods of LC alignment. The interesting method is the application of photosensitive materials in which photoalignment of liquid crystal is based on irradiation of the photosensitive polymer layer by the laser beam.[51–53] Photo-ordering can be used in devices where it is impossible to use mechanical techniques, that is, in the production of glass microtubes, photonic glass fibers, and planar optical fibers.[54] Oriented high-purity LC semiconductor layers can be used in photovoltaic cells, organic light-emitting diodes (OLEDs), or organic thin-film transistors (OTFTs).[51]

a)

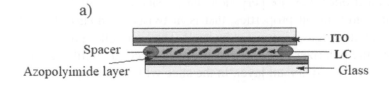

b)

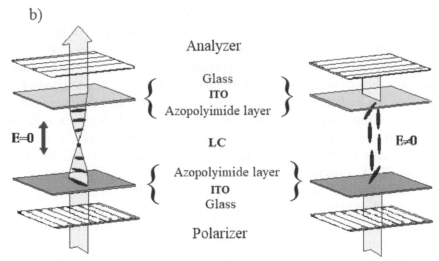

FIGURE 3.11   (a) Schematic of LC nematic cell; (b) operation diagram of a liquid crystal cell, based on the principle of the twisted nematic effect.

The liquid crystal cells are constructed based on the twisted nematic effect. A typical procedure preparation of a liquid crystal cell involves several stages,[55,56] that is, centrifuging the photosensitive polymer solution onto a properly prepared layer (glass, silicon, or polymeric), covered with a conductive layer of indium zinc oxide (ITO); drying the layer at an appropriate temperature; and irradiating a beam of linearly polarized ultraviolet light. Polarized light indicates *trans–cis–trans* isomerization of the azomolecules and their perpendicular reorientation to the vector of the light. The layers prepared in this way are put together and limited by spacers so that a gap of several micrometers is created between them. This gap is filled with the liquid crystal mixture. The schematic picture of LC nematic cell is shown in Figure 3.11a. Polymer orienting layers force a homogeneous arrangement of liquid crystal molecules, that is, those where their long axes are parallel to the orienting layers (Fig. 3.11b). Twisting the orienting layers by an angle of 90 degree causes a change in the position of the long axis of the molecules so that at the electrodes they are perpendicular to each other. Such a liquid crystal cell has birefringent properties, that is, it twists the polarization of the light beam by the angle it is twisted by itself. The light falling on the cell passes through the polarizer, the glass layer of the electrode until it comes across the ordering liquid crystal layer. In the state without applied external electric field, the plane of polarization of the light beam is also twisted by an angle of 90 degree. The beam passes through the analyzer without any obstacles. When the electric field is applied, the LC molecules orientate themselves according to the field force lines. Then, the plane of polarization of the light beam is perpendicular to the axis of the analyzer, thus it is not passed through it (Fig. 3.11b). Azopolyimides were successfully applied as a layer to the orientation of the LCs in three papers.[56,57,58] It was showed that azopolymers dedicated for the preparation of LC cells and their applications in optoelectronics devices should characterize relatively high (ca. 0.02) and stable photoinduced birefringence. In a few previous works, it is shown that LC cells can be used for the preparation of 1D or 2D LC diffraction gratings,[58] LC Fresnel lens,[59] or construction of Vortex.[59] 1D and 2D LC diffraction gratings were tested for changes in the light diffraction under the influence of an applied electric field (0–1.5 V/μm). Prepared diffraction gratings diffracted the light, while an applied electric field disturbed the effect. The maximum diffraction efficiency of diffracted in the first diffraction order was 22 and 12%, respectively for 1D and 2D structures ($E = 0.2$ V/μm). It was shown that the diffraction efficiency can be controlled by different values of the applied electric field.[58]

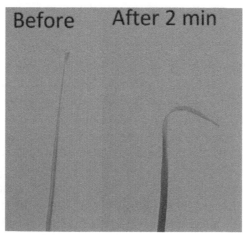

FIGURE 3.12 The cantilever bending upon 405 nm laser beam ($I = 150$ mW/cm²) for Ultem(Az-1)50%.

Another interesting possibility of the application shows azopolyimides able to form the free-standing cantilevers. Irradiation of azopolyimides with light polarized parallel to the long axis of the cantilever, the chromophore ordering in the perpendicular directions results in a contraction of the sample along the vertical direction, visualized as bending toward the laser source (Fig. 3.12). Application of light as a force controlling the movement of the cantilevers gave the opportunity to prepare new devices, that is, micropipettes,[60] light-driven plastic motors,[61] and polymer oscillators.[62] This phenomenon is described in general for semicrystalline or liquid crystalline materials, for example, polymeric liquid crystal polymers (LCPs), liquid crystals (PLCs), liquid crystal elastomers (LCEs), cross-linked LCPs (CLCPs), and LCP networks (LCNs).[37] Within the last few years, a mechanical response upon blue light irradiation was also observed for amorphous materials.[34,35,63–66] The photomechanical effect in amorphous polymers was investigated generally for homo- and co-polyimides,[35,63–65] polyimide gels,[34] poly(amic acid)s,[34] polyamides,[37] and star polyimides.[66] It was shown that azopolyimides are able to bend of cantilevers in the range 5–110 degree depending on the chemical structure and content of azochromophore.[36,37,63,64] The mechanism of cantilever bending in glassy polymers is complicated and not fully understood. The proposed mechanism in our previous work[44] showed that the cantilever movement may be the result of (1) *cis*-isomer formation, (2) *trans*-chromophore reorientation, and (3) polymer backbone migration. The effect of bending is stable from a few hours for functionalized main chain PIs[37,63–65,67] to at least 6 months for guest–host azosystems.[44]

Photosensitive azopolyimides may be also used in different areas of membrane technologies. Smart membranes are in high demand in many advanced fields of biotechnology, including drug delivery, biosensors, microfluidics, light-controlled molecular machines, molecular shuttles, and data storage. Azopolymer membranes are systems that sense the changes in their environment as a stimulus and make the desired response. In all cases, either a controlled porosity or a texture and chemical composition are coupled with adaptive properties, such as light-, electric-, pH-, ionic strength-, thermo-, magnetic-response, and molecular-recognition.[68] Light can be considered as a clean stimulus that allows remote control without physical contact or a mechanical apparatus. It is attractive because it enables one to change the geometry and dipole moment of photo-switching molecules causing macroscopic variations of molecularly organized structures by small perturbations.

FIGURE 3.13   Azomembrane for gas separation prepared from PI-5.

Although typical polymers such as hydrogenated acrylic butadiene rubber (HNMR), block copolymers made up of rigid polyamide blocks and soft polyether blocks (PEBAX®), polymethacrylate, and polyamides, they are still the main materials in membrane technology. Despite a lot of advantages of polymers such as good membrane-forming ability, flexibility, and low cost of preparation, their performance is limited by a trade-off between permeability and selectivity, that is, higher permeability is gained at the expense of lower selectivity and vice versa. The glassy polymers (i.e., PIs) have proved to be particularly attractive as membrane materials due to their excellent gas separation and physical properties, such as high thermal stability, chemical tolerance, and mechanical strength. Thus, the variation in structure–transport property relationships could be studied relative to the type of functional groups incorporated, to their content as well as to the position in the macromolecular chain. The introduction of the light-stimuli azobenzene moieties to the polymer backbone can open the new possibilities of application of glassy polymers. For the first time, the concept of azomembranes containing derivatives of azobenzene for gas separation was described by Weh et al.[69] The team prepared polymethacrylate membranes with azobenzene moieties, as thin films on porous ceramic supports. They measured the flux rates and *trans/cis* permselectivities for hydrogen, methane, *n*-butane, sulfur hexafluoride, and methanol. The photoisomerization of azobenzene groups caused changes in their molecular size and geometry as well as in their dipole moment. It was expected, therefore, that the transport behavior of gases through such a membrane could be influenced. For a polymethacrylate membrane with chemically bounded azobenzene, characteristic changes of the permeation properties were found by photochemical *trans/cis* switching of the azopolymer. It should be noticed that there have been only three works, in which the azomembrane for gas separation was prepared from the polymers containing covalently bonded azobenzene moieties.[7,69,70] Where only in our previous work, azopolyimide (PI-5) was used as azomembrane.[7] The picture of azomembrane prepared from PI-5 was shown in Figure 3.13. Most of the publications in the literature on azomembranes concern membranes filled with modified porous metal–organic frameworks (MOF) or membranes made of LCPs.[71–73] The MOF structure was modified by the attachment of the azobenzene groups, and then the azoMOF was introduced to the polymer matrix. A few studies have been reported previously about the benefit of having azo-functionality for CO_2/N_2 separation in inorganic porous membranes. Azobenzene has been known to have a good affinity to CO_2 because of its Lewis acid–base interaction[74] and dipole–quadrupole interaction.[75] Furthermore, from the nitrogen transport point of view, Patel et al. recently coined the term of nitrogen phobic environment existing in the series of the azo porous organic framework that they synthesized.[76] Their careful analysis of the available azo-based porous framework has shown that bonding between N_2 and the azo-group is less favorable than it is for CO_2 and that the presence of azo-group has resulted in unprecedentedly high CO_2/N_2 selectivity. Such a correlation has also been proven to be valid for various azo-based porous organic polymers.[77–79] In addition, light-responsive azo-groups are able to change their isomer-form. *Trans* and *cis* isomers exhibit different physicochemical properties that can influence the permeability of gases. It was described that *cis*-isomer shows worse permeability properties than *trans*-form. Reversible isomerization of azobenzene molecules gives the possibility of programming molecular transport for ions, liquids, and gases and opens the way to build up miniaturized nanodevices. Azopolyimide membranes may be applied in new devices that work based on controlled dosing/removal/separation of different gases. Azochromophores applied for the preparation of the azomembranes should characterize the long life of *cis*-isomer and slow *cis–trans* isomerization in the dark.

3.6 CONCLUSIONS

Azopolyimides are promising materials for the application in different light-sensitive applications, where the high thermal stability, good mechanical and electrical properties, and low susceptibility to a laser light damage are needed. For all practical applications, the knowledge about *trans–cis–trans* isomerization in a solid state is necessary. The review shows that the chemical structure of azopolyimides influences the *cis*-isomer content, which was in the range of 5–57% depending on the chemical structure of polymers. In general, the photostationary state is generated faster for side-chain, T-type PIs, and guest–host azosystems than polymers with azo moieties incorporated to the main chain. The influence of the chemical structure the stability of *cis–trans* isomerization was observed. Main chain PIs characterized the slowest *cis–trans* back reaction in the dark. Time of *cis*-isomer recovery was higher than 7 days. High stability of isomerization can be the result of low free volume of azochromophore. In main chain PIs, two phenyl rings are hinder by incorporation to the polymer backbone. The less stability of *cis–trans* isomerization was observed for T-type PIs 7.5–39.5 h. Attachment of one of the phenyl ring to the polymer backbone increased the free-volume, which results in the stability of isomerization that is significantly lower than for functionalized main chain PIs. The lowest stability of isomerization was observed for side chain and guest–host azosystems. The full recovery was observed in the range of 38 min to more than 4 h. It is seen that incorporation azo moieties to the side chain or depression in polymer matrix increase the *cis–trans* isomerization. It seems that main chain PIs may be promised materials for application in devices where high stability of *cis–trans* isomerization is necessary, while the side chain, T-type, and guest–host azosystems are better choices for technologies where the fast changes of azobenzene isomers are desired.

KEYWORDS

- **azopolyimides**
- **classification**
- ***cis–trans* isomerization**
- **photoalignment layers**
- **birefringence**
- **azomembranes**

REFERENCES

1. Yesodha, K.; Pillai, C. K.; Tsutsumi, N. Stable Polymeric Materials for Nonlinear Optics: A Review Based on Azobenzene Systems. *Prog. Polym. Sci.* **2004**, *29*, 45–74.
2. Natansohn, A.; Rochon, P. Photoinduced Motions in Azo-Containing Polymers. *Chem. Rev.* **2002**, *102*, 4139–4173.

3. Priimagi, A.; Lindfors, K.; Kaivola, M.; Rochon, P. Efficient Surface-Relief Gratings in Hydrogen-Bonded Polymer-Azobenzene Complexes. *ACS Appl. Mater. Interf.* **2009**, *1*, 1183–1189.

4. He, T. C.; Wang, C. S.; Pan, X.; Zhang, C. Z.; Lu, G. Y. The Nonlinear Optical Property and Photoinduced Anisotropy of a Novel Azobenzene-containing Fluorinated Polyimide. *Appl. Phys. B* **2009**, *94*, 653–659.

5. Xue, X.; Zhu, J.; Zhang, Z.; Zhou, N.; Tu, Y.; Zhu, X. Soluble Main-Chain Azobenzene Polymers via Thermal 1,3-Dipolar Cycloaddition: Preparation and Photoresponsive Behavior. *Macromolecules* **2010**, *43*, 2704–2712.

6. Kim, T.-D.; Lee, K.-S.; Lee, G. U.; Kim, O.-K. Synthesis and Characterization of a Novel Polyimide-based Second-order Nonlinear Optical Material. *Polymer*, **2000**, *41*, 5237–5245.

7. Bujak, K..; Nocoń, K.; Jankowski, A.; Wolińska-Grabczyk, A.; Schab-Balcerzak, E.; Janeczek, H.; Konieczkowska, J. Azopolymers with Imide Structures as Light-Switchable Membranes in Controlled Gas Separation. *Europ. Polym. J.* **2019**, *118*, 186–194.

8. Schab-Balcerzak, E.; Siwy, M.; Jarzabek, B.; Kozanecka-Szmigiel, A.; Switowski, K.; Pura, B. Post and Prepolimeruzation Strategies to Develop Novel Photochromic Poly(esterimide)s. *J. Appl. Polym. Sci.* **2011**, *120*, 631–643.

9. Liou, G.-S.; Yen, H.-J. *Polyimides*. In *Polymer Science: A Comprehensive Reference*; Matyjaszewski, K., Möller M., Eds.; Elsevier Science: Amsterdam, 2012, vol. 5; pp 497–535.

10. Pączkowski, J. *Fotochemia polimerów. Teoria i zastosowanie., wyd. Uniwersytetu Mikołaja Kopernika*; Toruń, 2003, ISBN 83-231-1615-6.

11. Shibaev, V.; Bobrovsky, A.; Boiko, N.; Photoactive Liquid Crystalline Polymer Systems with Light-Controllable Structure and Optical Properties. *Prog. Polym. Sci.* **2003**, *28*, 729–836.

12. Yager, K. G.; Barrett, C. J. Novel Photo-switching Using Azobenzene Functional Materials. *J. Photochem. Photob. A: Chem.* **2006**, *182*, 250–261.

13. Yu, H.; Kobayashi, T. Photoresponsive Block Copolymers Containing Azobenzenes and Other Chromophores. *Molecules* **2010**, *15*, 570–603.

14. Wang, D.; Wang, X. Amphiphilic Azo Polymers: Molecular Engineering, Self-assembly and Photoresponsive Properties. *Prog. Polym. Sci.* **2013**, *38*, 271–301.

15. Georgi, U.; Reichenbach, P.; Oertel, U.; Eng, L. M.; Voit, B. Synthesis of Azobenzene-containing Polymers and Investigation of their Substituent-dependent Isomerisation Behaviour. *React. Funct. Polym.* **2012**, *72*, 242–251.

16. Oliveira Jr, O. N.; dos Santos Jr, D. S.; Balogh, D. T.; Zucolotto, V.; Mendonc, C. R. Optical Storage and Surface-relief Gratings in Azobenzene-containing Nanostructured Films. *Adv. Colloid Interf. Sci.* **2005**, *116*, 179–192.

17. Priimagi, A.; Shevchenko, A. Azopolymer-based Micro- and Nanopatterning for Photonic Applications. *J. Polym. Sci. Part B: Pol. Phys.* **2014**, *52*, 163–183.

18. Wanic, A.; Siwy, M.; Schab-Balcerzak, E. Studies on Azobenzene Chromophores Content and Its Localization in Functionalized Poly(esterimide)s. *Pol. J. Appl. Chem.* **2009**, *53*, 277–294.

19. Fang, L.; Zhang, H.; Li, Z.; Zhang, Y.; Zhang, Y.; Zhang, H. Synthesis of Reactive Azobenzene Main-Chain Liquid Crystalline Polymers via Michael Addition

Polymerization and Photomechanical Effects of their Supramolecular Hydrogen-Bonded Fibers. *Macromolecules* **2013**, *46*, 7650–7660.

20. He, M.; Zhou, Y.; Dai, J.; Liu, R.; Cui, Y.; Zhang, T. Synthesis and Nonlinear Optical Properties of Soluble Fluorinated Polyimides Containing Hetarylazo Chromophores with Large Hyperprolarizability. *Polymer* **2009**, *50*, 3924–3931.

21. Schab-Balcerzak, M. Siwy, Kawalec, M.; Sobolewska, A.; Chamera, A.; Miniewicz, A. Synthesis, Characterization, and Study of Photoinduced Optical Anisotropy in Polyimides Containing Side Azobenzene Units. *J. Phys. Chem. A* **2009**, *113*, 8765–8780.

22. Yan, X.; Wang, F.; Zheng, B.; Huang, F. Stimuli-responsive Supramolecular Polymeric Materials. *Chem. Soc. Rev.* **2012**, *41*, 6042–6065.

23. Vappvuori, J.; Mahimwalla Z.; Chromik R. R.; Kaivola, M.; Priimagi, A.; Barrett, C. J. Nanoindentation Study of Light-induced Softening of Supramolecular and Covalently Functionalized Azo Polymers. *J. Mater. Chem. C*, **2013**, *1*, 2806–2810.

24. Hu, D.; Chen, K.; Zou, G.; Zhang, Q. Preparation and Characterization of Hydrogen-bonded P4VP-bisazobenzene Complexes. *J. Polym. Ras.* **2012**, *19*, 9983.

25. Koskela, E. J.; Vappvuori, J.; Hautala, J.; Priimagi, A.; Faul, C. H. J.; Kaivola, M.; Ras, R. H. A. Surface-Relief Gratings and Stable Birefringence Inscribed Using Light of Broad Spectral Range in Supramolecular Polymer-Bisazobenzene Complexes. *J. Phys. Chem. C* **2012**, *116*, 2363–2370.

26. Gao, J.; He, Y.; Liu, F. Azobenzene-Containing Supramolecular Side-Chain Polymer Films for Laser-Induced Surface Relief Gratings. *Chem. Mater.* **2007**, *19*, 3877–3881.

27. Jin, C.; Zhao, Y.; Wang, H.; Lin, K.; Yin, Q. Synthesis and Self-assembly of Side-chain Azocomplex for Preparation of Solid and Hollow Nanosphere. *Colloid Polym. Sci.* **2012**, *290*, 741–749.

28. Wu, S.; Duan, S.; Lei, Z.; Su, W.; Zhang, Z.; Wang, K.; Zhang, Q. Supramolecular Bisazopolymers Exhibiting Enhanced Photoinduced Birefringence and Enhanced Stability of Birefringence for Four-dimensional Optical Recording. *J. Mater. Chem.* **2010**, *20*, 5202–5209.

29. Hu, D.; Hu, Y.; Huang, W.; Zhang, Q. Two-photon Induced Data Storage in Hydrogen Bonded Supramolecular Azopolymers. *Opt. Commun.* **2012**, *285*, 4941–4945.

30. Koskela, J. E.; Vapaavuori, J.; Ras, R. H. A.; Priimagi, A. Light-Driven Surface Patterning of Supramolecular Polymers with Extremely Low Concentration of Photoactive Molecules. *Macromolecules* **2014**, *3*, 1196–1200.

31. Priimagi, A.; Kaivola, M.; Virkki, M.; Rodriguez, F. J.; Kauranen, M. Suspension of Chromophore Aggregation in Amorphous Polymeric Materials: Towards More Efficient Photoresponsive Behavior. *J. Nonlin. Opt. Phys. Mater.* **2010**, *19*, 57–73.

32. Saiz, L. M.; Oyanguren, P. A.; Galante, M. J.; Zucchi, I. A. Light Responsive Thin Films of Micelles of PS-b-PVP Complexed with Diazophenol Chromophore. *Nanotechnology* **2014**, *25*, 065601.

33. Lee, K. M.; Wang, D. H.; Koerner, H.; Vaia, R. A.; Tan, L.-S.; White, T. J. Enhancement of Photogenerated Mechanical Force in Azobenzene-Functionalized Polyimides. *Angew. Chem. Int. Ed.* **2012**, *51*, 4117–4121.

34. Hosono N.; Yoshikawa, M.; Furukawa, H.; Totani, K.; Yamada, K.; Watanabe, T.; Horie, K. Photoinduced Deformation of Rigid Azobenzene-Containing Polymer Networks. *Macromolecules* **2013**, *46*, 1017–1026.

35. Wie, J. J.; Wang, D. H.; Lee, K. M.; Tan, L.-S.; White, T. J. Molecular Engineering of Azobenzene-Functionalized Polyimides to Enhance Both Photomechanical Work and Motion. *Chem. Mater.* **2014,** *26,* 5223–5230.

36. Wang, D. H.; Wie, J. J.; Lee, K. M.; White, T. J.; Tan, L.-S. Impact of Backbone Rigidity on the Photomechanical Response of Glassy, Azobenzene-Functionalized Polyimides. *Macromolecules* **2014,** *47,* 659–667.

37. Wie, J. J.; Wang, D. H.; Lee, K. M.; White, T. J.; Tan, L.-S. The Contribution of Hydrogen Bonding to the Photomechanical Response of Azobenzene-functionalized Polyamides. *J. Mater. Chem. C,* **2018,** *6,* 5964–5974.

38. Sava, I.; Sacarescu, L.; Stoica, I.; Apostol, I.; Damian, V.; Hurduc, N. Photochromic Properties of Polyimide and Polysiloxane Azopolymers. *Polym. Int.* **2009,** *58,* 163–170.

39. Sava, I.; Hurduc, N.; Sacarescu, L.; Apostol, I.; Damian, V. Study of the Nanostructuration Capacity of Some Azopolymers with Rigid or Flexible Chains. *High Perform. Polym.* **2013,** *25,* 13–24.

40. Sava, E.; Simionescu, B.; Hurduc, N.; Sava, I. Considerations on the Surface Relief Grating Formation Mechanism in Case of Azo-polymers, Using Pulse Laser Irradiation Method *Opt. Mater.* **2016,** *53,* 174–180.

41. Sava, I.; Damaceanua, M.-D.; Nitschke, P.; Jarząbek, B. The First Evidence of Redox Activity of Polyimide Systems Modified with Azo Groups with Photo-induced Response. *React. Funct. Polym.* **2018,** *129,* 64–75.

42. Bujak, K.; Orlikowska, H.; Małecki, J. G.; Schab-Balcerzak, E.; Bartkiewicz, S.; Bogucki, J.; Sobolewska, A.; Konieczkowska, J. Fast Dark Cis-trans Isomerization of Azopyridine Derivatives in Comparison to their Azobenzene Analogues: Experimental and Computational Study. *Dyes Pigm.* **2019,** *160,* 654–662.

43. Sava, I.; Lisa, G.; Sava, E.; Hurduc, N. Synthesis and Characterization of Some Azo-copolyimides. *Rev. Roum. Chim.* **2016,** *61,* 419–426,

44. Konieczkowska, J.; Bujak, K.; Nocoń, K.; Schab-Balcerzak, E. The Large and Stable Photomechanical Effect in the Glassy Guest-Host Azopolymers. *Dyes Pigm.* **2019,** *171,* 107659.

45. Bujak, K.; Orlikowska, H.; Sobolewska, A.; Schab-Balcerzak, E.; Janeczek, H.; Bartkiewicz, S.; Konieczkowska, J. Azobenzene *vs* Azopyridine and Matrix Molar Masses Effect on Photoinduced Phenomena. *Europ. Polym. J.* **2019,** *115,* 173–184.

46. Ngai, K. L.; Rendell, R. W.; Pye, L. D.; LaCourse, W.C.; Stevens, H. J. *The Physics of Non-Crystalline Solids*; Taylor and Francis: London, 1992.

47. Vilgis, T. A. Strong and Fragile Glasses: A Powerful Classification and Its Consequences. *Phys. Rev. B* **1993,** *47,* 2882–2885.

48. Agapov, A. L.; Novikov, V. N.; Sokolov, A. P. Fragility and Other Properties of Glassforming Liquids: Two Decades of Puzzling Correlations. In *Fragility of Glass Forming Liquids*; Greer, L. A., Kelton, K., Sastry, S., Eds.; Hindustan Book Agency, 2013.

49. Vapaavuori, J.; Geraldine Bazuin, C.; Priimagi, A. Supramolecular Design Principles for Efficient Photoresponsive Polymer–azobenzene Complexes. *J. Mater. Chem. C,* **2018,** *6,* 2168–2188.

50. Barrett, C. J.; Mamiya, J.-i.; Yagerc, K. G.; Ikeda, T. Photo-mechanical Effects in Azobenzene-containing Soft Materials. *Soft Matter,* **2007,** *3,* 1249–1261.

51. Yaroshchuk, O.; Reznikov, Y. Photoalignment of Liquid Crystals: Basics and Current Trends. *J. Mater. Chem.* **2012,** *22,* 286–300.

52. Stohr, J.; Samant, M. G. Liquid Crystal Alignment by Rubbed Polymer Surfaces: A Microscopic Bond Orientation Model. *J. Electr. Spectr. Rel. Phenom.* **1999,** *98–99,* 189–207.
53. Yua, Y.; Ikeda, T. Alignment Modulation of Azobenzene-containing Liquid Crystal Systems by Photochemical Reactions. *J. Photochem. Photob. C Photochem. Rev.* **2004,** *5,* 247–265.
54. Chychłowski, M.; Ertman, S.; Tefelska, M.; Wolinski, T.; Nowinowski-kruszelnicki, E.; O. Yaroshchuk, O. Mikro kapilary (światłowodowe) ze zorientowaną strukturą anizotropową". XII Konferencja Naukowa—Światłowody i ich zastosowania; Krasnobród, 2009.
55. Węgłowski, R.; Piecek, W.; Kozanecka-Szmigiel, A.; Konieczkowska, J.; Schab-Balcerzak, E. Poly(esterimide) Bearing Azobenzene Units as Photoaligning Layer for Liquid Crystals. *Opt. Mater.* **2015,** *49,* 224–229.
56. Chrzanowski, M. M.; Zielinski, J.; Nowinowski-Kruszelnicki, E.; Olifierczuk, M. Fotoporządkowanie—alternatywna technika dla szybkich przetworników ciekłokrystalicznych. *Biuletyn WAT* **2010,** *59,* 375–394.
57. Akiyama, H.; Kudo, K.; Ichimura, K.; Yokoyama, S.; Kakimoto, M.; Imai, Y. Azimuthal Photoregulation of a Liquid Crystal with an Azobenzene-Modified Polyimide Langmuir-Blodgett Monolayer. *Langmuir* **1995,** *11,* 1033–1037.
58. Węgłowski, R.; Kozanecka-Szmigiel, A.; Piecek, W.; Konieczkowska, J.; Schab-Balcerzak, E. Electro-optically Tunable Diffraction Grating with Photoaligned Liquid, Crystals. *Opt. Commun.* **2017,** *400,* 144–149.
59. Węgłowski, R.; Kozanecka-Szmigiel, A.; Piecek, W.; Konieczkowska, J.; Schab-Balcerzak, E. Photonics Devices Based on Photoaligned LC Cells International School, Photonic Integration: Advanced Materials, New Technologies and Applications, Erince, Italy September 25–October 1, 2016.
60. Lee, J.; Oh, S.; Pyo, J.; Kim, J.-M.; Je, J. H. A Light-driven Supramolecular Nanowire Actuator. *Nanoscale* **2015,** *7,* 6457–6451.
61. Mamiya J-i. Photomechanical Energy Conversion Based on Cross-linked Liquid-crystalline Polymers. *Polym. J.* **2013,** *45,* 239–246.
62. Serak, S.; Tabiryan, N.; Vergara, R.; White, T. J.; Vaiab, R. A.; Bunning, T. J. Liquid Crystalline Polymer Cantilever Oscillators Fueled By Light. *Soft Matter,* **2010,** *6,* 779–783.
63. Wie, J. J.; Wang, D. H.; Tondiglia, V. P.; Tabiryan, N. V.; Vergara-Toloza, R. O.; Tan, L.-S.; White, T. J. Photopiezoelectric Composites of Azobenzene Functionalized Polyimides and Polyvinylidene Fluoride. *Macromol. Rapid Commun.* **2014,** *35,* 2050–2056.
64. Baczkowski, M. L.; Wang, D. H.; Lee, D. H.; Lee, K. M.; Smith, M. L.; White T. J.; Tan, L.-S. Photomechanical Deformation of Azobenzene-Functionalized Polyimides Synthesized with Bulky Substituents. *ACS Macro Lett.* **2017,** *6,* 1432–1437.
65. Ravi Shankar, M. R.; Smith, M. L.; Tondiglia, V. P.; Lee, K. M.; McConney, M. E.; Wang, D. H.; Tan. L.-S.; White, T. J. Contactless, Photoinitiated Snap-through in Azobenzene-Functionalized Polymers. *Proc. Natl. Acad. Sci. USA,* **2013,** *110,* 18792–18797.
66. Lee, K. M.; Wang, D. H.; Koerner, H.; Vaia, R. A.; Tan, L.-S.; White, T. J. Photomechanical Response of Pre-strained Azobenzene-Functionalized Polyimide Materials. *Macromol. Chem. Phys.* **2013,** *214,* 1189–1194.

67. Kozanecka-Szmigiel, A; Schab-Balcerzak, E.; Szmigiel, D.; Konieczkowska, J.; The Unexpected Photomechanical Effect in Glassy "T-type" Azopolyimides. *J. Mater. Chem. C* **2019**, *7*, 4032–4037.

68. Fiore Pasquale, N.; Cupelli, D.; Formoso, P.; De Filpo, G.; Colella, V.; Gugliuzza, A. Light Responsive Polymer Membranes: A Review. *Membranes* (Basel). **2012**, *2*, 134–197.

69. Weh, K.; Noack, M.; Ruhmann, R.; Hoffmann, K.; Toussaint, P.; Caro, J. Modification of the Transport Properties of a Polymethacrylate-Azobenzene Membrane by Photochemical Switching. *Chem. Eng. Tech.* **1998**, *21*, 408–412.

70. Hauensteina, D. E.; Rethwisch, D. G. Photocontrol of Gas Separation Properties *Sep. Sci.Techn.* **1990**, *25* (13–15), 1441–1453.

71. Knebel, A.; Sundermann, L.; Mohmeyer, A.; Strauß, I.; Friebe, S.; Behrens, P.; Caro, J. Azobenzene Guest Molecules as Light-Switchable CO_2 Valves in an Ultrathin UiO-67 Membrane. *Chem. Mater.* **2017**, *29*, 3111–3117.

72. Wang, Z.; Knebel, A.; Grosjean, S.; Wagner, D.; Brase, S.; Woll, C.; Caro, J.; Heinke, L. Tunable Molecular Separation By Nanoporous Membranes. *Nat. Commun.* **2016**, *7*, 13872.

73. Głowacki, E.; Horovitz, K.; Tang, C. W.; Marshall, K. L. Photoswitchable Gas Permeation Membranes Based on Liquid Crystals. *Adv. Funct. Mater.* **2010**, *20*, 2778–2785.

74. Arab, P.; Parrish, E.; İslamoğlu, T.; El-Kaderi, H. M. Synthesis and Evaluation of Porous Azo-linked Polymers for Carbon Dioxide capture and Separation. *J. Mater. Chem. A* **2015**, *3*, 20586–20594.

75. Yang, Z.; Zhang, H.; Yu, B.; Zhao, Y.; Ma, Z.; Ji, G.; Buxing Han, B.; Liu, Z. Azo-Functionalized Microporous Organic Polymers: Synthesis and Applications in CO_2 Capture and Conversion. *Chem. Commun.* **2015**, *51*, 11576–11579.

76. Patel, H. A.; Je, S. H.; Park, J.; Jung, Y.; Coskun, A.; Yavuz, C. T. Directing the Structural Features of N_2-Phobic Nanoporous Covalent Organic Polymers for CO_2 Capture and Separation. *Chem. A- Polym. J.* **2014**, *20*, 772–780.

77. Zhang, H.; Zhang, X.; Qiu, Z.; Liang, X.; Chen, B.; Xu, J.; Jiang, J.-X.; Li, Y.; Li, H.; Wang, F. Microporous Organic Polymers Based on Tetraethynyl Building Blocks with *N*-functionalized Pore Surfaces: Synthesis, Porosity and Carbon Dioxide Sorption. *RSC Adv.* **2016**, *6*, 113826.

78. Zhao, S.; Dong, B.; Ge, R.; Wang, C.; Song, X.; WMa, W.; Wang, Y.; Hao, C.; Guo, X.; Gao, Y.; Channel-wall Functionalization in Covalent Organic Frameworks for the Enhancement of CO_2 Uptake and CO_2/N_2 Selectivity. *RSC Adv.* **2016**, *6*, 3877438781.

79. Buyukcakir, O.; Je, S. H.; Park, J.; Patel, H. A.; Jung, Y.; Yavuz, C. T.; Coskun, A.; Systematic Investigation of the Effect of Polymerization Routes on the Gas-Sorption Properties of Nanoporous Azobenzene Polymers. *Chem. -A Europ. J.* **2015**, *21*, 15320–15328.

CHAPTER 4

From Insulating to Conducting Polyimides

GÖKNUR DÖNMEZ[1], AYÇA ERGÜN[1], MERVE OKUTAN[2], and HÜSEYİN DELİGÖZ[1*]

[1]*Faculty of Engineering, Department of Chemical Engineering, İstanbul University-Cerrahpaşa, Avcılar 34320, Istanbul, Turkey*

[2]*Faculty of Engineering, Department of Chemical Engineering, Hitit University, Çorum 19030, Turkey*

Corresponding author. E-mail: hdeligoz@istanbul.edu.tr

ABSTRACT

Polyimides (PIs) prepared from a dianhydride and a diamine generally stand out in many applications due to their superior properties such as high-temperature stability, low dielectric constant, tensile and high compressive strength, good resistance to solvents and moisture, good adhesion properties for inorganic, metallic, and dielectric materials, and easy casting of thin or thick films. Due to these unique properties, PIs are prevalently used in microelectronics, films, adhesives, and membranes. In addition to the use of PIs as an insulating material, their electrical properties can also be controlled from semiconducting to conducting state using some different approaches including the addition of metallic particles, introduction of some functional groups into the PI backbone, or doping it with the ionic liquids. These research efforts pave the way for widely using PIs not only in the insulating materials but also in the industries in which semiconducting/conducting material properties are required. As a consequence, it is expected that these multipurpose materials will be used more widely in various technological applications of microelectronic, separation, energy, and aviation in the near future.

4.1   HIGH-TEMPERATURE POLYMERS

Polymers are inexpensive, chemically inert, noncorrosive, and easily ductile materials with sufficient mechanical properties.[1] These superior features facilitated the use of polymers in various applications and also encouraged the researchers to improve the current properties of polymers and to synthesize novel polymeric materials.

Among the polymeric materials, a special group classified as *"high-performance polymers"* contains some polymers such as polyetheretherketone (PEEK), polyethersulfone, polyphenylene sulfide, polysulfone, polybenzimidazole (PBI), and polyimide (PI). These polymers are generally superior in terms of chemical/hydrolysis resistance and mechanical and as well as thermal stability.[2] "High-performance" term is generally explained by the good resistance of the material to harsh environments. Depending on the material, this harsh environment may be temperature, mechanical stress, high/low pH, pressure, radiation, voltage, and so on. If a polymer is resistant to high temperatures during its synthesis or when it is exposed, it may be referred to as "high-temperature polymers." The required properties expected to be in high-temperature polymers are generally listed under five headings:

- Durable to 10,000 h working at 177°C,
- Decomposition above 450°C,
- Low mass losses even at high temperatures,
- Higher heat deflection temperature than 177°C, and
- Glass transition temperature above 200°C (due to the presence of aromatic groups and rigid segments).[3]

A high-performance polymer must possess the above-mentioned properties as well as retain its superior properties when exposed to one or more of mechanical, chemical, electrical actions, and so on.[3] The need for high-performance materials in some industries such as aerospace,[4,5] automotive,[6] electronics,[7] fuel cells,[8] and medicine[9] is the driving force for the development of such polymers. To fulfill this requirement, mainly two different approaches can be applied. The first one is the modification of existing polymers[10,11] and the second one is the preparation of novel polymers.[12,13] One of the commonly used thermally stable polymeric materials is PIs. Therefore, this chapter will give some basic insights on the history, chemistry, synthesis, applications of PIs, and PI-based products. Subsequently, it will review their conductivity/insulation properties and application areas in detail.

4.2 POLYIMIDES

4.2.1 *HISTORY AND STATE OF THE ARTS OF POLYIMIDES*

Indeed, the first paper related to the preparation of aromatic PIs was published by Marston Bogert et al. in 1908.[14] Then, PIs with high molecular weight were synthesized by the polycondensation of pyromellitic dianhydride (PMDA) and diamines in 1955.[15] Since PI was not processable by melt polymerization, there was no significant improvement in the synthesis and processing of PIs until the 1960s. However, the production of Kapton by Dupont in the 1970s led to gain the scientific and commercial importance of PIs and they appeared as one of the most desirable high-performance polymers for many industrial applications.

The changes in the number of papers related to PIs were followed by years and they are given in Figure 4.1. As clearly seen from the figure, preliminary studies were published in the late 1950s and the number of publications on PIs slightly increased after 1970s due to the launching of DuPont's commercial product. In fact, an observable change in the publications about PIs started by 1980s. One of the pioneering studies on PIs was published by Dine-Hart and Wright[16] in 1967. This work mainly focused on the synthesis of aromatic PIs. Also, film and molding fabrication techniques and the effect of variables on PI polymerization were studied in this paper. From past to present, it is possible to see publications examining the synthesis of PI and the effect of parameters (e.g., monomer type and solvent type) on the properties of the final product. It is apparent that the importance of this material has not diminished over the years. On the contrary, it attracts more and more attention by the huge growth of technological demands in terms of material science. Nowadays, the researches on PIs are still growing. In the first half of 2019, the number of published studies reached 825. It is obvious that the studies will continue to rise in the next years.

4.2.2 *INTRODUCTION*

PI is an important member of thermally resistant polymers. This type of polymer has corrosion/fatigue/abrasion/impact resistance, low density, and excellent performance at low and high temperature without deformation. They can withstand temperatures up to 600°C.[17] Scientific and technological importance of PIs are mainly attributed to their high glass transition temperature and high thermal stability as well as mechanical toughness. Besides, they also have chemical resistance and outstanding dielectric properties.[18] The main properties of a general type PI are given in Table 4.1. As a result of these outstanding properties, PIs are suitable to be used as plastics, films, laminating resins, insulating coatings, and high-temperature adhesives.

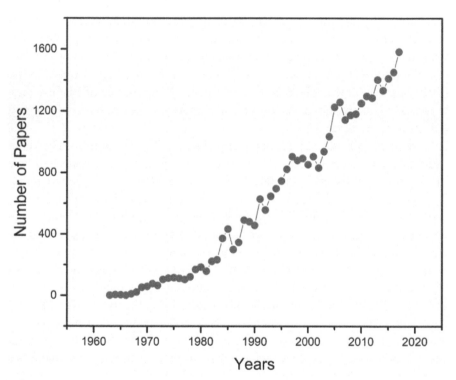

FIGURE 4.1   The change in the number of papers by years (the papers were searched in the Scopus database with "PI" keyword in "abstract-title-keyword section").

TABLE 4.1   Some Properties of PIs.[19]

Dimensional stability	
Coefficient of linear thermal expansion	5.5×10^{-5} 1/°C
Shrinkage	0.2–1.2%
Water absorption 24 h	1.34–1.43%
Physical properties	
Density	1.31–1.43 g/cm^3
Glass transition temperature	250°C–340°C

TABLE 4.1 *(Continued)*

Electrical performances

Dielectric constant	3.1–3.55
Dielectric strength	22–27.6 kV/mm
Dissipation factor	$18–50 \times 10^{-4}$
Volume resistivity	$14–18 \times 10^{15}$ Ω cm

Mechanical properties

Elongation at break	90%
Elongation at yield	4–10%
Flexibility (flexural modulus)	2.48–4.1 GPa
Hardness Rockwell M	110
Stiffness (flexural modulus)	2.48–4.1 GPa
Strength at break (tensile)	72–120 MPa
Strength at yield (tensile)	120 MPa
Toughness (notched Izod impact at room temperature)	60–112 J/m
Young modulus	1.3–4 GPa

Service temperature

Heat deflection temperature (at 1.8 MPa)	240°C–360°C
Max continuous service temperature	260°C–360°C

In contrast to all superior properties and advantages of PIs, they have some limitations such as high manufacturing cost and difficult processing.[19] The rigid backbone and crystallinity (especially in wholly aromatic PIs) limits their solubility in most of the organic solvents, and their high glass transition temperature or melting temperature can cause processing restrictions.[20] The solubility and processability problem of PIs can be solved without sacrificing their thermal stabilities by increasing the amorphous structure in virtue of the addition of the triphenylamine (TPA) and its derivatives to the backbone.[21] In addition, TPA units allow PI to be used as photoactive and electroactive materials due to the easy oxidation of the nitrogen center and the formation of continuous hole transport due to the radical cations.[22]

4.2.3 CHEMISTRY OF POLYIMIDES AND THEIR SYNTHESIS ROUTES

PIs are polymers of imide monomers containing two acyl groups (C=O) bonded to nitrogen and they can exist in both thermosetting and thermoplastic forms. They are classically synthesized by the reaction of a dianhydride and a diamine. It can be said that the subject of PI chemistry is tremendous and the monomers used in the synthesis are abundant and diverse. Depending on the monomer type and synthesis routes, the form of the final product can be different. The two-step method is mostly used to obtain PIs over poly(amic acid) (PAA) intermediate.[23] In the first step of this method, the pre-polymer (PAA) is obtained from a dianhydride and a diamine. Then, the PAA is converted to PI via imidization. The imidization can be carried out by chemical or thermal ways. The thermal imidization process is generally used to obtain cast products of PI such as films, fibers, and coatings. For example, in the preparation of a film, the PAA solution is cast with the aid of an applicator onto a glass plate and subsequently, it is heated to 250°C–400°C for a while. Initially, the volatile solvent is removed depending on the solvent type. Subsequently, the cyclization occurs at high temperatures and the product turns into PI structure. During thermal imidization, heating can be applied gradually to avoid obtaining products with dark colors. That's why the following procedure, heating at 100°C for 1 h, at 200°C for 1 h, and finally at 300°C for 1 h, is mostly applied for thermal imidization.[24] On the other hand, PAA can be treated with water-capturing agents in the presence of catalysts to convert PAA to PI structure by chemical imidization process.[25] It is noted that the chemical imidization of PAA can be achieved by using aliphatic acid anhydrides and tertiary amines together. For instance, Francis et al.[26] used acetic anhydride and β-picoline as dehydrating agent and catalyst, respectively, in the chemical imidization of PAA. Since the selected imidization method affects the morphology of the polymer,[27] it is generally expected some differences in final product properties. For example, PIs obtained by chemical imidization are more soluble and have worse thermal properties compared to the ones prepared via thermal imidization.[15] Classical PI formation reactions from PMDA and 4,4'-oxydianiline (ODA) over PAA intermediate are depicted in Figure 4.2.

In addition to this classical two-step method, there are also other approaches to synthesize the PIs. For this purpose, alternative reagents can be used instead of dianhydrides, diamines, or both. The diisocyanates, dithioanhydrides, bis(maleimide)s (BMI), bisdienes, bidienophiles, di(hydroxyalkyl), and diimide compounds can be listed as alternative chemicals to dianhydrides and diamines.[15,28–30] The synthesis routes carried out by using different chemicals were widely explained in the literature and the properties of obtained products were given comparatively.

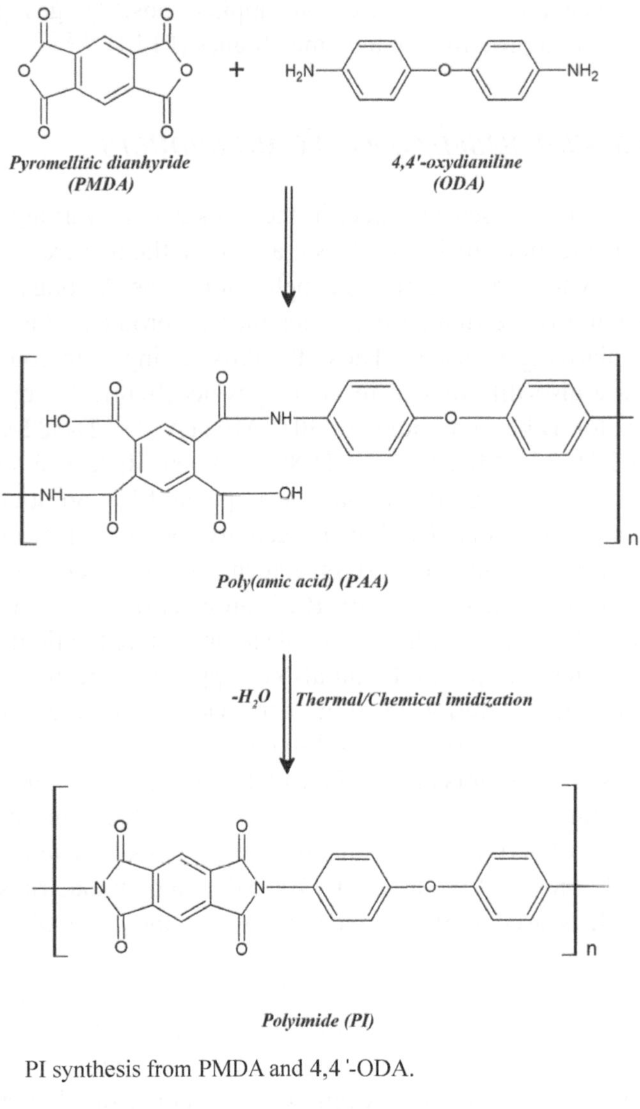

Pyromellitic dianhyride
(PMDA)

4,4'-oxydianiline
(ODA)

Poly(amic acid) (PAA)

-H₂O || Thermal/Chemical imidization

Polyimide (PI)

FIGURE 4.2 PI synthesis from PMDA and 4,4'-ODA.

4.2.4 APPLICATIONS OF POLYIMIDES

It is well-known that PIs are an important member of high-performance polymers due to their excellent physical, chemical, mechanical, dielectric, and thermal stability as well as fire-resistant properties. Thanks to the combination of these properties, PIs found applications in many fields such as microelectronics packaging,[31–35] high-temperature adhesives,[36–38] protective clothing,[39,40] insulators,[41,42] biomedical applications,[43,44] gas separation membranes,[45,46] and proton-exchange membranes (PEM).[47–51]

4.2.5 POLYIMIDE-BASED COMMERCIAL PRODUCTS

There are lots of PI-based products in the market with varying properties depending on the field to be used. So, a few of them were given in the following part with some features and application areas. Among the PI-based products, Kapton is the first prepared commercial product which paves the way for popularizing PI usage. These PI films having numerous commercial names are also different in terms of physicochemical properties. For example, Kapton B has a modulus of 3034 MPa (ASTM D-882-91), tensile strengths of 241–255 MPa (ASTM D-882-91), and dielectric strengths of 1600–2800 V/mil at 23°C. It is a black and opaque film. So, it can be used in applications where opacity, low reflectivity, or aesthetics are needed. These applications include the heaters, antenna, and LED circuitry. Another product of DuPont called as Kapton RS is an electronically conductive PI film and it was developed for heating applications where the lightweight and the uniform heater is required. It can also be applied in the areas of surface deicing, automotive interior heating, aerospace temperature regulation, industrial tube heating, wearables, and so on.[52]

[PI P84®]NT is a thermoplastic product of Evonik and is manufactured in powder or granulates form. It is stated that this product can be processed easily and cost-effectively. It also has a glass transition temperature of 337°C–364°C, a flexural modulus of 3705 MPa, and an electric strength of 558.8 V/mil. It is used in the applications where the other plastics would melt/decompose.[53]

PIs in the foam form are also available in the market and the product of Pyrotek called as Sorbermide is one of them. This product has the capability of sound-proofing and thermal insulation. Due to its numerous advantages such as being fire resistant and extremely lightweight, flexible, and having low outgassing properties, it can be used in navy (e.g., aircraft carriers, cruisers, and minehunters), commercial marine (e.g., yacht and ferries), railway cars, locomotives, aircrafts, duct/piping insulation, and electronic/medical/analytical instruments.[54]

As mentioned above, PI is a high-performance material and preferred in many applications due to its superior properties. In the following headings, the dielectric and conductive properties of PIs (film or bulk) and imidic polymers (maleimide and BMI) will be reviewed from insulating to conducting states in detail as well as the related dielectric and conductivity mechanisms. Furthermore, application areas of these innovative polymers will be discussed.

4.3 POLYIMIDES AS INSULATORS

4.3.1 INSULATION AND DIELECTRIC CONSTANT

The electrical conductivity of a material is determined by the gaps of valance and conductive bands. In order to have high conductivity, it is necessary to partially fill the conductive band with electrons.[55] In a medium where atoms can readily release electrons (e.g., copper, aluminum, and silver), the transfer of electrons occurs more easily. This type of material is called as conductor[56] whose valance and conductance bands collide. If an empty band is placed energetically close on a filled band, the electrons can jump through the small energy gap between the two bands due to their thermal energy $k_B T$, which is called semiconductivity.[55] However, a fully filled band does not contribute to electrical conductivity.[55] If the energy gap between the bands is too large, the material is expressed as insulator.[57] The most important criteria for defining electrical insulation of a material is dielectric constant (ε). Also, the dielectric breakdown voltage of a material indicates the resistance to electrical conduction. The dielectric constant (ε), which is called as relative permittivity (ε_r) is the ratio of the permittivity of a substance to that of the permittivity determined in vacuum.[58] Because of that, ε has no dimension. The dielectric constant is also sometimes defined as the electrostatic energy storage within the solid material. The dielectric constant of most commercial products used as insulation materials ranges from about 2 to 10 and the dielectric value of air is 1. Low values in ε are highly preferred for high-frequency or electronic applications to minimize electric power losses. Whereas higher values in ε are appropriate for capacitance applications.[59] There are other properties determining the electrical insulation performance such as electric breakdown strength, permittivity, and loss tangent. The electrical insulation concerns volume resistivity (a bulk property) and surface resistivity (concerning leakage current across the insulator surface between electrodes having a potential difference). The former is specified in ohm-meters (or megohm-meters) and the latter in ohms per square: the surface resistance between opposite sides of a square surface is independent of the area of the sample. Because of these properties are affected by surface or mass moisture, the measurement of the electrical resistance of a material is often used to assess the dryness condition.[57]

The permittivity of free space (ε_0) is the ratio of the electric flux density in a vacuum to the electric field strength and this is specific to a material under given situations of temperature, frequency, moisture content, etc.[56,57] The dielectric permittivity of a dielectric material is extremely related upon the frequency with a tendency to reduce at higher frequencies. In ferroelectric materials, increasing the temperature will cause a rise in permittivity up to the "Curie point" after which permittivity falls quickly with temperature.[57]

Dielectric strength is the specialty of an insulating material that allows it to resist a given electric field magnitude without a failure/defect. In general, it is expressed in terms of the minimum electric field magnitude (i.e., potential difference per unit thickness) that will reason the dielectrics to failure or breakdown under given circumstances, for example, the shape of the electrodes, the method of applying temperature and voltage, and many other parameters affect the breakdown behavior of the material under electrical voltage. The electric strength of most materials reduces with temperature and hence, it is usual to carry out the relevant tests at elevated temperatures. The loss tangent (tanδ) varies, sometimes significantly, with frequency and temperature. tanδ generally increases with temperature as well as the dielectric permittivity, especially when moisture is present. In short, high temperature is often liable to a considerable increase in dielectric losses.[57]

4.3.1.1   FUNDAMENTALS OF DIELECTRIC THEORY

An ideal dielectric is a material or a medium which has no free electrons, ions, ionizable groups, or some impurities that conduction can take place. If a material is electrically ideal insulator, it can be represented as a pure capacitor. As mentioned above, the dielectric constant is closely associated with the permittivity of the material. Permeability represents the polarization state of a material in response to an applied area and it is defined by the ratio of the permeability of dielectrics to the permeability of a vacuum. That's why greater polarization results in higher dielectric constant.[57]

Let us consider a vacuum capacitor comprising of a pair of parallel electrodes having a cross-section area of A (m^2) and a distance of d (m). When a potential difference (V) is applied between the two electrodes, the electric field intensity (E) at any point between the electrodes is perpendicular to the distance and the relation is given in eq 4.1.[60]

$$E = V / d \qquad (4.1)$$

The capacitance of the vacuum capacitor (C) is expressed as given in eq 4.2,

$$C = \varepsilon_o \times A / d \qquad (4.2)$$

and the charge stored in the capacitor (Q_0) is defined as given in eq 4.3

$$Q_0 = A \times E \times \varepsilon_o \qquad (4.3)$$

in which ε_0 is the permittivity of free space. If a homogeneous dielectric is introduced between the plates keeping the potential constant, the stored charge (Q) is given by eq 4.4.

$$Q = A \times E \times \varepsilon_o \times \varepsilon \qquad (4.4)$$

Dielectrics have two main electrical applications: the first one is the electrical insulation and the second one is energy storage. For insulation, conductors can be coated by electrically insulating materials while some dielectrics can store large amounts of energy by applying the electric field.[57] Dielectrics can have three forms: gas, liquid, and solid. In gases, the atoms/molecules do not interact but may include free electrons which are liable for the conduction. Some gases also have conduction mechanisms that result from positive and negative ions and charged molecules. In liquid dielectrics, there are some interactions between molecules. This renders the conductive process ionic by the addition of additives from the charged particles. Solid dielectrics have conductivity mechanisms (hoping, tunneling, and thermal) managed by free electrons, holes, and ions.[57]

When the atom is situated in an electric field, the charged particles are exposed to an electric force as a result of which the center of the negative charge cloud is displaced with respect to the nucleus. A dipole moment is induced in the atom and the atom is to be electronically polarized. The electron polarizability of an atom is described as the induced dipole moment per unit electric field strength and is a measure of the ease of displacement of charge centers.[60] In the insulation material to which the electric field is applied, a limited charge displacement occurs known as polarization at atomic, molecular, and bulk material levels. There are three different types of polarization[57] and they are given below:

(1) *Electronic polarization*: The electron clouds surround positive ions in atoms and these electrons can react quickly to the motion of an applied electric field because of being too light.

(2) *Molecular polarization*: The ionic bonds in a molecule are deformed by applying an electric field and this concludes with an increase in the dipole moment of the lattice.

(3) *Orientational polarization*: All of the molecules in liquids and gases move in the same direction with the applied electric field. The alignment is generally incomplete under weak static areas. Interfacial polarization occurs at the electrodes and crystallite interfaces in solids, while space charge polarization occurs in the presence of an interface with materials having different electrical properties. Figure 4.3 represents the various cases of orientational polarization.

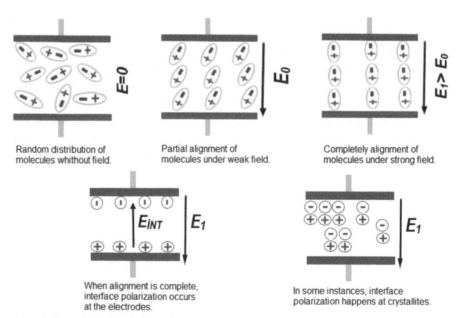

FIGURE 4.3   Orientational polarization cases.

Dielectrics can be grouped according to their structure and the way they react to the action of an electric field.[57]

(1) *Non-polar dielectrics*: These types of dielectrics do not have a permanent dipole moment in their molecules. By virtue of an electric field, dipoles are induced and oriented in the direction of an applied field.

(2) *Polar dielectrics*: If a material has molecules with permanent electric dipole moments when there is no electric field, it is called as a polar dielectric. The dipoles are aligned in the direction of field in the presence of an electric field. Generally, the relative permittivity values of polar dielectrics are low and they only have permanent dipoles at small concentrations.

(3) *Paraelectric dielectrics*: In contrast to nonpolar and polar dielectrics, paraelectrics include a strong dipole in each unit cell with relative permittivity values from 20 to 10,000.

(4) *Ferroelectric dielectrics*: Similar to ferromagnetism, the ferroelectrics have domains (of about 1 μm) where permanent dipoles are aligned in the same direction. When an electric field is applied to ferroelectrics, their domains are aligned in the direction of this electric field. However, they are known to exhibit nonlinear polarization with the applied field. Ferroelectrics have a relative permittivity which increases with temperature until the "Curie temperature." Beyond this temperature, the permittivity decreases and the domains cease to exist.

(5) *Piezoelectricity*: By applying an electric field to a ferroelectric material at paraelectric temperatures and then cooling, the domains in the ferroelectric material are aligned permanently. In such materials, mechanical shear stress may cause charge displacement, known as piezoelectric.[57]

(6) *Electrets:* The dielectrics may include electrical charges at the atomic or molecular level, which may be directed by an external electric field. If the charges are immobilized upon cooling, the dielectric will attitude as an electrical equivalent of a permanent magnet. This is called electret (thermoelectrets in this situation).[57]

4.3.2 DIELECTRIC BEHAVIOR OF POLYIMIDES

PIs are classified as high-performance polymers due to their structure, flexibility/rigidity, usage at high temperatures, excellent mechanical properties, outstanding dielectric behavior, and good chemical resistance.[61–64] PIs are commonly used as electrical insulating materials in microelectronics owing to their dielectric constant ($\varepsilon = 2.0–3.5$).[61–64] In general, PIs were first included in electronic packaging in the early 1980s to prevent package delays due to the interaction of the signal with the dielectric medium.[61] Further, PIs show good electrical insulation with a volume resistivity of $10^{14}–10^{15}$ Ω m and surface resistivity of $10^{15}–10^{16}$ Ω.[61]

Dielectric properties of a polymer are related to dielectric constant and dielectric loss. One of the promising developments in material science is the preparation of novel polymers with low dielectric constant. Hence, the lowering of dielectric constant and dielectric loss of PIs are attracting great attention for 30 years. For this purpose, the following approaches are widely used in the literature and some of the obtained results were comprehensively presented here.

4.3.2.1   FLUORINATION OF POLYIMIDES

Various approaches were applied to reduce the dielectric constant of PIs by physical and chemical alteration of their chemical structure.[65,66] In this respect, one of these methods is the synthesis of fluorinated PIs. Many studies showed that the presence of fluorinated groups reduced the dielectric constant due to an increment in the free volume of fluorines and lower polarizability.[67,68,77–85,69–76] Furthermore, the increase in hydrophobicity by fluorine substitution resulted in the reduction of dielectric constant by eliminating water from the polymer, while the addition of fluorine affected the specific properties of the polymer regardless of the moisture effect.[67]

As it is known from the literature, the most common method for the preparation of fluorinated PIs is the use of flour containing dianhydride and/ or a fluorinated diamine compound. There are too many publications (over 500 in the last 20 years) about the developments of fluorinated PIs using this approach. Generally, it can be said that the use of fluorinated diamine or dianhydride in the preparation of PIs resulted in dielectric constants ranging from 2.6 to 3.2 at various frequencies. But most of them did not report the dielectric loss and dielectric breakdown voltage properties of these polymers. Some of the fresh articles on this issue are summarized below. In a study by Yang et al. a novel fluorinated diamine monomer with the ether–ketone group was synthesized. Dielectric constants of these polymers were found to be in the range of 3.05–3.64 at 1 MHz.[70] In a similar study, a new aromatic diamine including pyridine moieties was synthesized and subsequently dielectric constants of the fluorinated copoly(pyridine ether imide)s were measured as 3.04–3.51 at 1 MHz.[74] In a very similar study by Zhang et al., a series of fluorinated pyridine-containing aromatic poly(ether imide)s were prepared from a newly synthesized diamine monomer. Low dielectric constants of 2.81–2.98 at 1 MHz for these polymers were observed.[75] Using the same approach, a novel fluorinated aromatic tri amine, 1,3,5-tris(4-(2-trifluoromethyl-4-aminophenoxy)phenyl) benzene, was prepared and subsequently, a series of fluorinated hyper-branched PIs with terminal amino or anhydride groups were synthesized by condensation polymerization. It was determined that the dielectric constants of these polymers were in the range of 2.69–2.92.[76] In a different study, fluorinated bis(ether amine)s were prepared from 4,4-bis(3,4-dicarboxyphenoxy)3,3,5,5-tetramethylbiphenyl dianhydride and different fluorinated bis(ether amine)s. These types of fluorinated polymers displayed small dielectric constants of 2.95–3.29 at 1 MHz as well as high thermal and mechanical stability.[78] In a different study, synthesis and properties of soluble aromatic PIs from 4,5-diazafluorene containing dianhydride were reported by Deng et al. The obtained PI films showed low dielectric constant between 2.78 and 3.38.[79] Further, Dong et al. prepared a novel soluble PI containing pyridine and fluorinated units. They found that the dielectric constants of these fluorinated films were in the range of 2.36–2.52 at 1 MHz while they exhibited good thermal and mechanical properties.[82] On the other hand, Constantin et al. studied dielectric and gas transport properties of highly fluorinated PI blends. In that study, the dielectric spectroscopy was performed to investigate the temperature and frequency dependency of dielectric constant, dielectric loss, and electrical resistance. They explained that sub-glass transitions and relaxations mainly arose from the segmental mobility of the polymer chain.[83]

Recently, the use of fluorinated graphene, namely fluorographene (FG), in the preparation of fluorinated PI composites was offered for high-performance materials and devices.[86] In a study by Wang et al., it was reported that the dielectric constant of PI decreased from 3.1 to 2.1 by incorporation of 1 wt.% FG.[87] Furthermore, FG had a positive effect on the mechanical properties of pure PI film. Similarly, Zhang et al. prepared FG/PI composite films and they found that mechanical, dielectric, and thermal properties importantly enhanced in the presence of FG. The FG/PI film with 0.5 wt.% FG loading showed a low dielectric constant of 2.48 and a good electrical insulation of less than 10^{-14} S/m.[86]

4.3.2.2   SILICON CONTAINING POLYIMIDES

Another approach for reducing the dielectric constants of PIs is the introduction of silicon (Si)/siloxane groups into the polymer backbone or addition of silanized agents into polymer matrix.[65,88–90] Si-based materials show good insulating properties owing to their high surface tension and hydrophobicity. That's why some studies were published on the dielectric behavior changes of Si-containing PIs. They are discussed as follow:

In a study, Leu et al. aimed to develop a copolymer to directly synthesize PI-tethered polyhedral oligomeric silsesquioxane (POSS) without adversely affecting the mechanical properties of the PI. It was observed that the dielectric constant of POSS/PI nanocomposites reduced with POSS amount. The maximum reduction in the dielectric constant of POSS/PI nanocomposites was about 29%, by comparison with pure PI (ε = 2.32 vs 3.26).[65] By Deligöz et al., Si-containing poly(urethane-imide) (PUI)s were synthesized by using two different methods. In the first method, isocyanate-terminated PU prepolymer was directly reacted with tetracarboxylic dianhydride. In order to prepare isocyanate-terminated PU prepolymer, diphenylsilanediol compound was reacted with toluene diisocyanate. In conclusion, the dielectric constants of PUIs prepared by the second method were found in the range of 3–3.5 at 100 kHz frequency. In addition, the dielectric breakdown strengths of the Si-containing PUI were determined in the range of 25–150 kV/mm.[88] In another study, Jiang et al. reported the preparation of porous PI silica hybrid film by sol–gel process and subsequently hydrofluoric acid treatment. In this study, they obtained a dielectric constant of 1.84 when the silica concentration in the precursor was 20 wt.%.[89] In a different study by Othman et al., they synthesized a series of poly(siloxane–imide) block copolymers to study the influence on their dielectric behavior and they found that the dielectric constant reduced from 2.90 to 2.43 at 1 kHz with the polysiloxane content in the block copolymer composition.[90]

4.3.2.3   POROUS POLYIMIDES

Porous PIs with a low dielectric constant can be prepared using a templating method. These templates can be removed by chemical oxidation, pyrolysis, or hydrolysis. Then, the remaining pores act as a good barrier for electrical conduction. In this approach, one of the important criteria determining dielectric constant level is the pore size, shape, density, and pore distribution inside the polymer. Therefore, the preparation of porous PIs is extensively studied for achieving lower dielectric constants and dielectric loss in the last years.[91–98] Here, some examples are presented.

In a study, Lee et al. prepared nanoporous PI films with the use of hybrid poly(ethylene oxide) (PEO) polyhedral oligosilsesquioxane (PEO–POSS) nanoparticles as templates. In this study, they managed to lower the dielectric constant of the thin film from 3.25 to 2.25 at 1 MHz.[99] In the study by Krishnan et al., nanoporous PI films were prepared from PUI after the removal of thermally labile PU segments by heat treatment. It was determined that the dielectric constant of porous PI decreased with PU content.[93] In a different study by Wang et al., self-assembled structures of PS-b-P4VP/ PAA blends were developed to obtain low dielectric nanoporous PI films. The dielectric constants of the obtained nanoporous PI films were found to be ranging from 2.82 to 2.41, which was considerably lower than that of the dense PI film (3.60).[94] Lv et al. used PEO, poly(methyl methacrylate), and polystyrene (PS) as templates of pores to obtain nanoporous PI films. The dielectric constant of the nanoporous PI films formed by using PEO was found to be more stable.[95] The aerogel obtained as a porous solid was remarkable for many applications such as thermal insulation, optical, acoustic insulation, and catalysis due to its low density, high porosity, and small pore size.[96] Furthermore, Meador et al. prepared fluorinated aerogel PI and they investigated the effect of 2,2-bis(3,4-dicarboxyphenyl) hexafluoropropane dianhydride (6FDA) content in the PI backbone on the dielectric properties. Very promisingly, the lowest relative dielectric constant was found as 1.084 for the fluorinated aerogels with 0.078 g/cm^3 density.[97] In a study by Wu et al., the porous PI structure obtained from biphenyl-3,3′,4,4′-tetracarboxylic dianhydride (BPDA) and 4,4′-ODA was modified using a versatile diamine, namely 2,2′-bis-(trifluoromethyl)-4,4′-diaminobiphenyl (TFMB), to enhance mechanical and dielectric properties. Dielectric constants and loss tangents of the obtained aerogels varied in the range of 1.27–1.35 at 10 MHz and 0.001–0.004 at 1 GHz, respectively. They also found that more TFMB fractions resulted in a slight decrease in dielectric constant and loss tangent.[98] Wang et al. used silica microspheres as a template to prepare fluorinated PI with controllable porosity. They emphasized that the strong hydrogen bonding between microspheres and PAA resulted in better dispersion of inorganic filler. The obtained results indicated that high-level porosity in fluorinated PI films caused an ultra-low dielectric constant of 1.84 when the porosity increased to 35%.[100]

4.3.2.4   CROSS-LINKED POLYIMIDES

Designing a cross-linked structure is known to be an effective solution for reducing dielectric constant and improving other essential properties such as thermal and dimensional stability, chemical resistance, and excellent mechanical property. Because of the cross-linked structure, the segmental mobility of the PI chains are effectively limited and the polymer segments are prevented from being oriented with the external electric field direction, thereby the dielectric constant of the PIs can be reduced.[101] In a study by Deligöz et al., they developed a novel cross-linked PI film (CPI) with low dielectric constants and a conventional PI by thermal imidization of cross-linked and conventional PAA. The dielectric constant and dielectric loss of the conventional and novel CPI films were found to be frequency and temperature-dependent. They pointed out that the dielectric constant of the new CPI film has lower values than that of the dielectric constant of the conventional PI film.[62] In another study of this group, CPI films were prepared by the reaction of 4,4-diphenylmethane diisocyanate and conventional PAA based on benzophenone tetracarboxylic dianhidryde (BTDA) and diaminodiphenylsulphone. According to their results, the lowest dielectric constant was found to be 2.2 at 1 MHz.[102] Yao et al. synthesized a series of high molecular weight fluorinated cross-linkable co-polyimides (Co-PIs) from 1,3-bis(2-trifluoromethyl-4-aminophenoxy)-5-(2,3,4,5-tetrafluoro-phenoxy)benzene, 1,4-bis(4-amino-2-trifluoromethylphenoxy)benzene, 1,4-bis(4-amino-2-trifluoromethyl-phenoxy)-2-(3'-trifluoro-methylphenyl) benzene, and 4,4'-(hexafluoroisopropylidene)diphthalic anhydride (6FDA). The dielectric constants of the cross-linked films were found to be ranging from 2.37 to 2.70 at 1 MHz.[103] In the study by Song et al., they prepared new soluble cross-linkable fluorinated Co-PIs from a diamine containing double-fluorinated cross-linkable hydrophobic tetrafluorostyrol side-groups and 3,3'-bis(2,3,5,6-tetrafluoro-4-vinylphenoxy)-4,4'-biphenyldiamine. The lowest dielectric constant for the prepared cross-linked films was found to be 2.48 at 1 MHz and the corresponding films showed a considerably reduced water uptake of 0.31%.[101] In another study of this group, cross-linkable fluorinated Co-PI films based on the 2,2-bis(3-amino-4-(2,3,5,6-tetrafluoro-4-vinylphenoxy)-phenyl)hexafluoropropane with cross-linkable tetrafluorostyrol groups were prepared using a high temperature polycondensation. They determined the dielectric constants of the films in the range of 2.32–2.82 at 1 MHz.[104]

4.3.2.5 INORGANIC FILLER CONTAINING POLYIMIDES

The approach for lowering the dielectric constant is the preparation of PI-based composites with the addition of an insulating inorganic additive. Biçen et al. investigated the change in the physical and electrical properties of the modified/unmodified Co-PI film with zeolite 4A (MZA/UMZA). They determined that the dielectric constant values of the PI films were found to be ranging from 4.5 to 8.6 for the 1–5 wt.% UMZA doped PIs at 295K and 100 kHz. Additionally, the obtained PI films with the addition of MZA in 1–5 wt.% showed dielectric constants in the range of 4.2–6.5 at the same temperature and frequency.[63]

Table 4.2 represents the dielectric constant and dielectric breakdown voltage values of some insulating PIs depending on the PI components and used imidization technique.

TABLE 4.2 Dielectric Constant and Dielectric Breakdown Voltage Values of Some Insulating PIs.

PI components	Imidization technique	Dielectric constant	Dielectric breakdown voltage	Reference
Commercial product				
Kapton film (ODA/PMDA)	–	3.67 at 1 MHz	339–268 kV/mm	[104,105]
Fluorinated PIs				
FG/BPDA/ODA	Thermal	2.8–2.1 at 1 MHz	–	[87]
FG/BTDA/ODA	Thermal	2.48–5.5 at 0.1 GHz	–	[86]
Porous PIs				
6FDA/ODA+BPDA/ODA+TAB	Chemical	1.084 at 11–12 GHz	–	[97]
TFMB/ODA/BPDA+TAB	Chemical	1.35–1.27 at 10 MHz	–	[98]
(PMDA/ODA/PEO), (PMDA/ODA /PS), (PMDA/ODA /PMMA)	Thermal	2.47, 2.65, 2.75 at 10^7 Hz	–	[95]
PS-b-P4VP/ODA/BPDA	Thermal	2.82–2.41	–	[94]

TABLE 4.2 *(Continued)*

PI components	Imidization technique	Dielectric constant	Dielectric breakdown voltage	Reference
Cross-linked PIs				
PMDA/Diamine/MDI, BPDA+Diamine+MDI	Thermal	4.1–2.2 at 1 MHz	117.8–80.4 kV/mm	[102]
6FTFPB/m-3F-6FAPB/6FDA*	Chemical	2.70–2.37 at 1 MHz	–	[103]
TFVBPA/6FAPB/6FDA	Thermal	3.11–2.48 at 1 MHz	–	[101]
Cross-linked PI based 6FATFVP /6FAPB/6FDA	Thermal	5.9–4.2 at 100 kHz	311–269 kV/mm	[104]
Si containing PIs				
PMDA/ODA/SiO$_2$	Chemical	2.85–1.84 at 1 MHz	–	[89]
16% mol POSS in PMDA/ODA	Thermal	2.32–3.26	–	[65]
(PEO-POSS)/PMDA/ODA	Thermal	3.25–2.25 at 1 MHz	–	[99]
Si-PU/PMDA/BTDA	Thermal	3–3.5 at 100 kHz	25–150 kV/mm	[88]
Poly(siloxane–imide) from (PDMS)/BPDA/BAPP	Thermal	2.90–2.43 at 1 kHz	–	[90]
Inorganic filler containing PIs				
UMZA/ODA/BTDA; 1–5 wt.%	Thermal	8.6–4.5, 5.9–4.2 at 100 kHz	–	[63]
MZA/ ODA/BTDA; 1–5 wt.%				

BPDA, biphenyl-3,3',4,4'-tetracarboxylic dianhydride; ODA, 4,4'-oxidianiline; TAB, 1,3,5-triaminophenoxybenzene; BTDA, 3,30,4,40-benzophenonetetracarboxylic dianhydride; PMDA, pyromellitic dianhydride; POSS, polyhedral oligosilsesquioxane; PEO, poly(ethylene oxide); PSVP, poly[styreneco-(4-vinylpyridine)] (for PS-b-P4VP); PMMA, poly(methyl methacrylate); PS, polystyrene; 6FDA, 2,2-bis(3,4-dicarboxyphenyl) hexafluoropropane dianhydride; TFMB, 2,2'-bis-(trifluoromethyl)-4,4'-diaminobiphenyl; MDI, 4,4-diphenylmethane diisocyanate; 6FTFPB, 1,3-bis(2-trifluoromethyl-4-aminophenoxy)-5-(2,3,4,5-tetrafluorophenoxy)benzene;6FAPB,1,4-bis(4-amino-2-trifluoromethylphenoxy)benzene; m-3F-6FAPB, 1,4-bis(4-amino-2-trifluoromethyl-phenoxy)-2-(3'-trifluoromethylphenyl)benzene; 6FDA*, 4,4'-(hexafluoroisopropylidene)diphthalic anhydride; TFVBPA, 3,3'-bis(2,3,5,6-tetrafluoro-4-vinylphenoxy)-4,4'-biphenyldiamine; 6FATFVP, 2,2-bis(3-amino-4-(2,3,5,6-tetrafluoro-4-vinylphenoxy)-phenyl)hexafluoropropane; UMZA, unmodified zeolite 4A; MZA, modified zeolite 4A.

4.3.3 DIELECTRIC BEHAVIOR OF MALEIMIDE/BIS(MALEIMIDE) POLYMERS

Apart from the insulating PIs, BMI which is an imide precursor was used for the preparation of thermally and mechanically stable thermosetting material with low dielectric constant. In this context, some polymers with imidic structure,[106,107] BMI,[108,109] and fluorinated BMI[110,111] were published in the literature. Furthermore, preparation and electrical characterization of maleimide and BMI containing thermoset resins such as triazine, polysiloxane, and unsaturated polyester were reported.[112-120] One of the earlier study reported on the electrical properties of some polymers having imidic structure was presented by Cosutchi et al.[106] In this study, two types of imidic polymers, copolymaleimides bearing azobenzene groups and epiclon-based PIs were prepared and their electrical behaviors were measured in the range of 1–100 kHz. They found that these imidic polymers were very promising for microelectronic applications due to their good insulating properties.[106] In another study of Cosutchi, epiclon-based PIs, fluorinated PIs, and copolymaleimides containing azobenzene groups were synthesized and dielectric properties as well as the optical properties of these polymers were studied. In that study, dielectric constants of the polymers were determined from the refractive indices measurements.[107] Thermosetting resin including dicyclopentadiene or dipentene was synthesized. They found that the BMI with dicyclopentadiene or dipentene showed low dielectric constant (around 3 at 1 MHz) and small dissipation factor (0.1 at 1 MHz). Owing to those outstanding properties, the authors offered these polymers to be used in printed circuit board applications.[108] In another study reported by Lorenzini et al., imidic polymers of bis-furanes and 1,3-BMI propane by Diels-Alders reactions were described. Dielectric constants of these materials were found to be in the range of 4.31–6.75 at room temperature while their dissipation losses were 1–2%.[109]

Similar approach used in lowering dielectric constants of PIs was applied for BMI-based products too. Fluorinated maleimide and BMI were prepared according to these studies.[110,111] A bio-based allylphenol functionalized fluoro-containing maleimide was synthesized and the results indicated that dielectric constant of below 2.9 and dissipation factor of less than 0.007 at the frequency range of 40 Hz–25 MHz.[110] In another study, novel fluorinated BMI resins were prepared by the copolimerization of 2,2-bis[4-(4-maleimidephenoxy)phenil]hexafluoropropane and diallyl hexafluoro bisphenol A to improve their dielectric properties. They found that the incorporation of a hexafluoroisopropyl group decreased the dielectric constant of BMI resin due to the small dipole and the low polarizability of the C-F bonds. They reported that dielectric constant and dielectric loss were 2.88 and 0.009, respectively, at 10 GHz and 25°C.[111]

In another approach, BMI-based or BMI-containing thermoset resins were reported to improve thermal, mechanical, and dielectric properties. Chen et al. presented the preparation of phosphorus-containing maleimide and the combination of these materials with cyanate ester (CE) resin. They indicated that modified CE had dielectric constants varying from 2.96 to 3.02 at 1 GHz, which were smaller than that of pristine CE.[112] In another study, fluorinated bismaleimide-triazine (BT) resins were prepared by Guo et al. The ε of the fluorinated BT was found to be almost 2.9 at 1 MHz while the dielectric loss for this sample was around 0.002–0.004 at the same frequency. Similarly, they explained that small dipole, low polarizability, and high bond energy of C-F due to the existence of hexafluoroisopropyl group were the main reasons for the enhanced dielectric properties.[113] A novel high-performance modified BT resin was developed and its dielectric properties were studied in a study by Guan et al.[114] and also the effect of silanized agent on the dielectric constant and dielectric loss of BT was investigated over a wide frequency range (10 Hz–1 MHz). It can be seen from their results that dielectric constants of cured BT resin were in the range of 3.75–4 depending on applied frequency while dielectric losses were around 0.002–0.005. On the other hand, it was pointed out that the addition of hyper-branched polysiloxane (HBPSi) led to worse dielectric properties. Similarly, Liu et al. described the importance of the interface between fiber and polymer matrix in the presence of HBPSi for modified BT resin. The results indicated that the dielectric loss of BT composite samples exhibited a strong dependence on frequency. Whereas, the dielectric loss of glass fiber/modified BT had better stability on frequency than that of glass fiber/BT composite. Overall, it can be said that the dielectric loss of BT resin was around 0.002 and lower than that of modified BT (tanδ = 0.003–0.0037).[115] In addition to BT resins, POSS including hybrid resins were developed. In 2011, Hu et al. reported a novel kind of high-performance resins of POSS/BT hybrid resins with lowered dielectric loss. They found that dielectric loss of POSS/BT resin was 25% of that of BT resin at the frequency lower than 100 kHz, whereas all hybrid resins had almost the same dielectric loss beyond this frequency compared to BT resin.[116] In a similar study, dielectric properties of BT resins containing octa(maleimidophenyl)silsesquioxanes (OMPS) were investigated. In that study, they found that OMPS accelerated the curing reaction of BT resin and dielectric constant of the materials clearly dropped until 2 wt.% of OMPS loading. Beyond this point, dielectric constant of the material slightly increased. This increase in dielectric constant with higher amount of OMPS content was attributed to the aggregation of OMPS.[117] In another study, a hybrid resin based on hyper-branched polysiloxane containing maleimide (HPMA) and CE was prepared. In that study, the dielectric loss of 15% HPMA including CE hybrid resin was only about 27% of that of neat CE resin.[118] In a different study, BMI resins were modified by some polytriazoles bearing allyl groups. After curing, the sample had dielectric constant ranging from 3.09 to 3.39 at 1 kHZ. Moreover, the obtained results showed that dielectric constant of these samples was almost the same with frequency changes; whereas the dielectric losses showed some fluctuations related to dielectric relaxations.[119] On the other hand, unsaturated polyester resin with low dielectric constant was developed by Chen et al. The dielectric constant of N-phenylmaleimides doped unsaturated polyester resin was around 3.6 if 5 wt.% of -CF$_3$ including N-phenylmaleimides was applied.[120]

There is very little study on the preparation of maleimide containing CPI networks. Park et al. investigated the use of maleimide for end-capping of PI. The final product, PUI, had dielectric constants in the range of 2.39–2.45. It can be concluded that the dielectric constant of PUI was not significantly improved with the use of maleimide as an end-capper.[121]

4.3.4 *DIELECTRIC RELAXATION PROCESS*

Dielectric relaxation processes are generally attributed to the segmental motion of polymer systems. As the molecular dynamics and segmental mobility of the polymers are not our main scope in this chapter, hereby, it will not be discussed comprehensively. By the way, some basic introduction about the relaxation process and some studies reported in the literature will be presented briefly. Aromatic PIs and Co-PIs exhibit several relaxation processes (α, β, and γ) in their dynamical and dielectric behaviors.[62,122] Gamma (γ) relaxation in polymers can be attributed to the frozen water which may be present in the polymer segments. Beta (β) relaxation is called as sub-glass secondary relaxation. It is associated with phenyl ring motions and is also influenced by moisture absorption content, aging history, and morphology. Finally, an alpha (α) relaxation process attributed to glass transition can be normally observed at high temperatures. In literature, mostly the relaxation peaks are determined from dielectric loss graphs. Some of the studies reporting molecular dynamics and dielectric relaxations of imidic polymers and PIs are given as follows. Please note that the dielectric relaxations of polymeric solutions are out of our scope in this chapter.

4.3.4.1   DIELECTRIC RELAXATION PROCESS OF POLYIMIDES

Constantin et al. reported the physicochemical properties of highly fluorinated PI blends. Also, the dielectric characterization of these blends was performed in terms of dielectric constant, dielectric loss, and electrical resistance. They indicated that dielectric loss diagrams showed sub-glass transitions in the form of gamma and beta relaxations meanly due to the segmental mobility of the polymer chain components. Furthermore, it was determined that the blend films had a dielectric constant in the range of 2.49–2.89 and dielectric loss less than 0.02 at room temperature and in the frequency range of 10 Hz–1 MHz.[83] In a different study, broadband dielectric spectroscopy of BPDA/4,4'-ODA PIs was performed in a wide temperature (−150°C to 370°C) and frequency ranges (0.1–1 MHz). From the dielectric loss plots, they observed two sub-glass relaxation processes called as gamma and beta relaxations. The gamma relaxation appeared at a low temperature (around −60°C) and it was corresponded to the existence of water in the PI sample. Moreover, the beta relaxation was observed at higher temperatures around 180°C. Beta relaxation was mainly attributed to the non-cooperative relaxation of the PI molecules.[123] In a different study by Deligöz et al., a novel CPI was prepared and the dielectric properties of this polymer were investigated in terms of dielectric constant, dielectric loss, and relaxation process. They pointed out that beta relaxations clearly appeared at 24°C and 40°C for 50 and 100 kHz, respectively. Whereas conventional PI (BTDA/ODA) had no beta relaxation in the applied frequency ranges. Concerning their findings, beta relaxation was corresponded to the rotation of rigid segments of phenylene or imide groups. In good agreement with the literature, they observed that beta relaxation shifted to higher temperatures in accordance with Arrhenius Law.[62]

4.3.4.2   DIELECTRIC RELAXATION PROCESS OF MALEIMIDE POLYMERS

Molecular dynamics in fluorinated side chain maleimide copolymers were studied by broadband dielectric spectroscopy by Tsuwi. In this study, a series of alternating maleimide copolymers with fluorinated side chains were prepared. These copolymers showed a fast β-relaxation characterized by Arrhenius-type temperature dependence with activation energy. Due to the microphase separated structure of the polymer systems, independent relaxations in the different domains were observed. Furthermore, it was indicated that the fluoroalkyl side chains showed a bimodal relaxation due to the parallel and perpendicular components of the dipole moment of C-F segment in the case of the presence of short side chains. Also, the dynamic glass transition was shifted to lower temperatures by replacing the alkyl side chains with fluoroalkyl ones.[124] In a different study of this group, relaxation processes of poly(ethane-alt-*N*-alkylmaleimide)s were studied in the frequency range of 0.1 Hz–10 MHz and at temperatures between 120 and 500K. These alternating maleimide copolymers had alkyl side chains varying in length from methyl to octadecyl. In this study, four relaxation processes were observed including the β-relaxation, the α'-relaxation, the α-relaxation (dynamic glass transition), and the α(s)-relaxation.[125]

4.3.5 APPLICATIONS OF INSULATING POLYIMIDES

PIs are one of the important high-performance polymeric materials and they are used in various applications such as aerospace, microelectronics, and membrane technologies due to their excellent thermal stability, chemical resistance, and mechanical, dielectric, and adhesion properties.[126] PIs have good insulating properties and therefore, they are widely used in a variety of electrical and electronic applications. PIs used as interlayer dielectric films are expected to withstand the metal sintering temperature at 400°C, the final step of integrated circuit processing, without causing deformation of electrical, chemical, or mechanical properties.[61]

Microelectronic packaging requires low dielectric constant over a wide frequency range. Moreover, low and constant dissipation factor in different frequencies is also necessary for microelectronic applications to reduce the signal propagation delay.[127] Microelectronic packaging is an application area, which requires high thermal conductivity of packaging material with superior electrical properties since it will produce more heat when the microelectronic circuit works. Therefore, polymers with low thermal conductivity are used for microelectronic packaging; they can cause malfunctions as they cannot effectively dissipate the heat when overheating occurs. In such cases, various additives can be added to the medium to increase thermal conductivity.[128]

Hedrick et al. developed foam structured PIs to obtain thin-film dielectric layers with very low dielectric constants for use in microelectronic devices. They formed the foam by thermolysis of the thermally labile block. In this study, nanofoams prepared as matrix materials from a series of polymers and also a series of thermally unstable polymers were investigated and the dielectric constant of matrix PI homopolymer of PMDA/3FDA was found to be 2.9. In addition, a foam structure of PMDA/3FDA obtained from a 24% propylene oxide-based copolymer showed a significant decrease in dielectric constant to 2.3.[129] In a study by Jiang et al., layered silicates/PI nanocomposites were synthesized for evaluating these nanocomposites as advanced dielectric materials for microelectronic applications. In this study, fluorinated PI based on 4,4'-(hexafluoroisopropylidene)-diphthalic anhydride (6FDA) and 4,4'-ODA was prepared. According to their results, the nanostructured silicates/fluorinated PI nanocomposite displayed excellent water-absorption retardation behavior which therefore determined in a lower leakage-current density and lower dielectric constants under higher moisture conditions than those of the pristine fluorinated PI.[130] In the study of Kuntman et al., PI was synthesized from 4,4'-bis(3aminophenoxy)diphenyl sulfone, prepared by nucleophilic aromatic substitution of 4,4'-dichlorodiphenyl sulfone with m-aminophenol and PMDA. With this obtained PI, a metal-PI-silicon structure was designed to demonstrate the dielectric properties of the material. Experimental data showed that the minimum capacitance of C_{Mmin} = 52.7 pF at 10.7 V.[131]

Organic field-effect transistors (OFETs) and circuits are evaluated extensively due to their large potentials in flexible organic displays, sensors, radio frequency identification devices tags, and electronic papers.[132] Ji et al. used PI materials as insulating layers in order to be compatible with the photolithography process to OFETs to produce large-scale flexible circuits. According to their findings, the devices exhibited high performance with field-effect mobility of about 0.55 cm^2/V s and on/off ratio of 10^5.[132] In another study, the nonvolatile flash memory characteristics of N,N'-bis(2-phenylethyl)-perylene-3,4:9,10-tetracarboxylic diimide based transistors containing different pendent conjugation length were also reported.[133] Concerning the organic thin-film transistors, photo-sensitive PIs are used as interlayer dielectrics.[134,135] Jang et al. reported a photo-sensitive, low-temperature processable, soluble PI (PSPI) gate insulator with excellent resistance to the photo-patterning process. They were synthesized PSPI with a simple one-step condensation polymerization of 5-(2,5-dioxytetrahydrofuryl)-3-methyl-3-cyclohexene-1,2-dicarboxylic anhydride and 3,5-diaminobenzyl cinnamate with isoquinoline as a base catalyst. They reported that the prepared PSPI gate insulator exhibited a dielectric constant of 3.3 at 10 kHz, and leakage current density of <1.7 × 10^{-10} A/cm^2.[42]

4.4 POLYIMIDES AS SEMICONDUCTORS/CONDUCTORS

4.4.1 CONDUCTIVITY AND MECHANISM OF CONDUCTION

Depending on the demand of functional and high-performance polymers, PIs can be prepared in a controllable manner of conductivity from insulating to conductive form. In this subchapter, semiconducting and conducting properties of the PIs are reviewed comprehensively. In the literature, the electrical resistance of the thermoplastic PI was reported as 10^{16} Ω cm.[136] In this respect, a few decades ago, new types of PI-based materials were developed with the addition of some electronic conductive agents,[137–139] conductive polymers,[140–142] ionic group containing functional polymers,[143,144] or ionic liquids (IL).[145,146] Indeed, electrical conductivity can be classified as electronic and ionic conductivity depending on the type of used conductive agent. However, the electrical conductivity term is mostly used regardless of the conductivity type. The conductivity of a polymer is mainly affected by its chemical structure and the physicochemical properties of the additives. While additives with high conductivity lead to a conductive region, the insulating polymers block the conductivity as a barrier between the particles.[147–147] By the way, the conductivity of a polymer can be shown in different units. For example, more scientifically conductivities are shown in S/cm or S/m. In addition, the surface resistivities in Ω/cm or Ω/cm^2 can be used for defining the conductivity of a material.

Let's explain the conductivity phenomenon for a polymer. The conductivity of an insulating material like polymers depends on the charge (q), quantity/concentration (N), and mobility (μ) of the charge carriers as shown in eq 4.5. A charge carrier may be a hole, an electron, an ion, or a polar group that allows the transport and movement of the electrical charge. These charge carriers tend to randomly move in the absence of an electrical field. In order to increase the conductivity, it is necessary to raise the amount of charge carriers in the insulating component and ensure their mobility.

$$\sigma = \sum Nq\mu \tag{4.5}$$

where σ is the conductivity, N is the concentration or quantity, q is the charge number and the μ is the mobility of the charge carriers.

In the polymers, ion is injected into the film surface by performing discharge from a needle to examine the charge transport. In this process, called as surface potential decay experiments, electrons or holes tend to act toward the back electrode due to their own electrical fields. If PI was electrically charged for any reason, it would release its charge with spontaneous discharges and it would be inevitable the electrical field sensitive devices and circuit malfunctions. Controlling this situation will improve the performance of the PI to be used in microelectronic packaging, etc. In many studies, transport of the surface charges was investigated and still surface potential decay experiments are frequently used to examine the charge transport in polymers. The mobility of the carriers in this case is given by the following eq 4.6.[150]

$$\mu_{eff}(t) = \mu_0 \eta(t) \qquad (4.6)$$

where μ_{eff} is the effective mobility, μ_0 is the proper mobility and $\eta(t)$ is the carrier ratio at a certain time (t).

The mobility of the carriers, in this case, is controlled by the action of traps formed between the conduction and valence band and other charge carriers. Aragoneses et al. suggested an explanation for the surface potential decay observed in the PI structure. According to their evaluations, the charge transport along the sample might be caused by two reasons. The first one is nondispersive mobility, a mechanism specifying the injection of charge into the bulk after corona injection and the second one is dispersive mobility identifying the abnormal transport observed in some photosensitive materials.[150] Therefore, the effective mobility formula was rearranged like eq 4.7 where the N and D are used for describing the nondispersive and dispersive mobility, respectively.

$$\mu_{eff}(t) = \mu_N(t) + \mu_D(t) = \mu_0 \left(\eta_N(t) + \eta_D(t) \right) \qquad (4.7)$$

where μ_{eff} is the effective mobility, μ_0 is the proper mobility, μ_N is the nondispersive process mobility and μ_D is the dispersive process mobility at a certain time (t).

This charge transfer mechanism is important for polymeric systems and it must be known that the conduction mechanism of polymeric or hybrid systems is not simple and insignificant. The electronic conductivity of the polymers containing σ-bond is mostly due to the disjointed ions in the structure. Therefore, the desired conductivity can be achieved by charge doping into the structure. On the other hand, it is possible to directly provide an electrical conductivity with the polymers containing π-bond.[147] The estimation of the conductivity mechanism requires the understanding of the conductive agent-polymer interaction considering the additive ratio, film thickness, temperature, and also the examination of I–V characteristics.[151] In addition, the mechanisms and theories used for the identification of electrical conductivity describe the dependence of molecular structure, temperature, electrical field, morphology, etc. over the concentration and mobility of conductive agent.[147]

Although the basic electronic conductivity theories of the polymer differ from metal and semiconductors, the band theory is the first step in understanding the conductivity of conjugated polymers. According to this theory, the electrical conductivity of the material improves as the distance between the conduction and valence bands decreases.[147,152] However, the use of this theory is restricted in situations such as low bandwidth, increased trap density, and the deterioration of the crystalline structure. By the way, it is necessary to understand the electron–electron and electron–space interactions because of the delocalization of electrons inside the polymer structure. In this case, the hopping conduction will assist in the identification of the charge transport. According to hopping theory, the charge can be generated by thermal stimulation of electrons along the band and this charge can be transported between local regions. Since this type of conductivity requires electrons to jump from one to another along the energy barrier, it must have sufficient thermal energy to overcome the barrier. If the tunneling is concerned, the site separation (space from one site to the next) must be small. Similarly, transfer can be occurred by providing sufficient energy to remove the barrier in front of the tunnel. Consequently, the conductivity can be achieved by passing an electron both over the energy barrier and through the tunnel depending on the barrier shape, separated areas, and thermal energy.[147]

In addition to these basic approaches, the conductivity behavior observed in these systems arises from mechanisms such as space charge limited current (SCLC) theory,[153–155] Schottky and Poole-Frenkel mechanisms.[151,154,156,157]

If there is a direct proportion between the current density and electrical field (J and E^n), the conductivity mechanism is assumed to fit the SCLC. n is a parameter indicating the trap distribution. If n is equal to 2, samples will correlate to the trap-free SCLC mechanism. When the value of n is changed between 2 and 6, it shows that the trapped SCLC mechanism is valid. On the other hand, the numerical value of n is greater than or equal to 3 when the current density is related to the sample thickness ($J \propto d^{-n}$). In this case, if the slope does not match (slope ≤ 3), SCLC mechanism cannot be mentioned.[154]

In many studies, it was reported that PI exhibited ohmic mechanism in the low field while it displayed nonohmic mechanism with nonlinear Schottky behavior in the high voltage range.[141,151,158] As it is known from the literature, $I = f(V)$ characteristic is a straight line from the origin and it is linear if ohm law is valid for conductivity mechanism. Exceeding this value (e.g., 100 kV/cm for Kapton) causes a deviation from linearity and ultimately leads to superlinearity. In the steady-state of the superlinear region, the conductivity can be explained by Schottky, Poole-Frenkel, ionic hopping or thermally assisted tunneling mechanisms.[159]

The next mechanisms to be examined are the Schottky (eq 4.8) and Poole-Frenkel (eq 4.9). It pointed out the conformity of either the conduction mechanism to the Schottky or Poole-Frenkel mechanism when $\ln I-V^{1/2}$ graphs is changed linearly.[154] These two mechanisms are not likely to be valid at the same time. The Schottky effect is controlled by the properties of the electrode–polymer interface whereas the Poole-Frenkel effect is adjusted by the bulk polymer properties.[160] However, it is clear that the conduction is electronic at room temperature in both cases. In order to determine which mechanism is valid, the beta coefficient is calculated first, and then the dielectric constant is estimated from the eq 4.10.[154,156]

$$J = A_R T^2 \exp\left(-\frac{\phi_S}{kT}\right)\exp\left(\frac{\beta_S E^{1/2}}{kT}\right) \tag{4.8}$$

$$J = A_R T^2 \exp\left(-\frac{\phi_{PF}}{kT}\right)\exp\left(\frac{\beta_{PF} E^{1/2}}{kT}\right) \tag{4.9}$$

$$\beta_S = \left(\frac{q^3}{4\pi\varepsilon_0\varepsilon_r d}\right)^{1/2} = \frac{\beta_{PF}}{2} \tag{4.10}$$

where J is the current density, A_R is the Richardson constant, T is the absolute temperature, ϕ is the barrier height of electrode–polymer interface, k is the Boltzmann constant, E is the electrical field, β_S is the Schottky coefficient, β_{PF} is the Poole-Frenkel coefficient, d is the polymeric film thickness, q is the electronic charge, ε_r is the dielectric constant, and ε_0 is the permittivity in vacuum.

Normally, these data are not sufficient to determine the mechanism definitively, but the mechanism can be estimated by comparing with the literature.[161] For this purpose, ionic jump distance, activation energy, and dielectric constant can be estimated by using $I = f(V)$ characteristics. Among these parameters, ionic jump distance can give information about whether the conductivity mechanism is ionic or not.[159]

As mentioned above, electrical conductivity is the movement of the charge via electrons and holes in response to an electric field. If the conduction takes place by the movement of the ionic charge, this type of conductivity is defined as ionic conductivity. Similar to electronic conductivity parameters given in eq. 4.1, ionic conductivity can be explained by a charge carrier, concentration, and mobility of ionic species.[162] In addition, the ionic conductivity is similar to the concept of equivalent conductivity used in the electrochemistry as shown in eq 4.11

$$\lambda_i = \frac{\sigma_i}{|z_i|c_i} = Fb_i \tag{4.11}$$

where λ_i is the equivalent conductivity, σ_i is the ionic conductivity, z_i is the charge number, c_i is the molar concentration, F is the Faraday constant, and b_i is the mobility of ionic charges. Since the total ionic conductivity is obtained by the contributions of all ions in the sample, eqs 4.5 and 4.11 can be rearranged as follow,[162]

$$\sigma = \sum Nq\mu = \sum |z_i|c_i\lambda_i \tag{4.12}$$

For ionic conductivity, one or more ion types should be diffused through the material, the diffusion coefficient is one of the required parameters to define the ionic conduction. The diffusion coefficient is given by the eq 4.13,

$$\sigma = \frac{nq^2D}{k_BT} \tag{4.13}$$

where D is the diffusion coefficient, a is the jump distance, υ_0 is the frequency, k_B is the Boltzmann constant, E_m is the energy, $(1 - c)Z$ is the number of unoccupied sites and γ is the correlation factor.[163]

In addition, the Nernst–Einstein equation is related to the diffusion coefficient of transported ions. So the ionic conductivity can be expressed as in eq 4.14.[164]

$$\sigma = \frac{nq^2 D}{k_B T} \tag{4.14}$$

where n is the ion number and q is the ion charge. Consequently, the ionic conductivity becomes as the following expression (eq 4.15),

$$\sigma_{ion} = \gamma \left[\frac{N(Z_i q)^2}{k_B T} \right] c(1-c) Z a^2 \upsilon_0 \exp\left(\frac{\Delta S}{k_B} \right) \exp\left(\frac{-E_m}{k_B T} \right) = \left(\frac{\sigma_0}{T} \right) \exp\left(\frac{-E}{k_B T} \right) \tag{4.15}$$

In accordance with the above-mentioned explanations, it is known that the ionic conductivity in solids is always associated with an ion in the structure jumping from the minimum energy position to another location of similar energy level.[165] Therefore, one of the factors necessary for defining ion conduction is related to providing the energy (activation energy) required by the ion to jump from one region to another and is expressed by Arrhenius equation (eq 4.16).[164]

$$\sigma = \left(\frac{A}{T} \right) \exp\left(\frac{-E_a}{k_B T} \right) \tag{4.16}$$

where E_a is the activation energy and A is the pre-exponential factor.

Besides, ionic conductivity of a material is decisive in applications such as batteries, fuel cells, electrochromic devices (ECDs), sensors, and ion conductive electrolytes. PEM and functional thin films became key components for these applications. In addition to rapid ion conduction, solid ion-conducting membranes and films have some advantages over liquid electrolytes in terms of leakage problems, stability, and the ability to downsize electrochemical devices.[163,166] The following equation (eq 4.17) is used to describe the ion conduction of a solid electrolyte used in such systems.[167–170]

$$\sigma = \frac{d}{R A} \tag{4.17}$$

where d is the thickness of solid electrolyte, R is the bulk resistance in impedance spectra and A is the effective area.

It is possible to determine the conduction mechanism (electronic or ionic) with the complex impedance measurement. For example, the half circles show that the charge carriers are ions rather than electrons. Ionic jump conductivity is caused by ionic impurities, such as polar and partially dissociated molecules. As it is well known, an electron can move through long distances in a lattice, but the ions are particles that are able to jump to only adjacent site and they are not capable of freely moving along a longer distance.[159] The ionic hopping conductivity model expresses the current density when the ions in the PI are thermally activated and incorporated into the conductivity process. The suitability for ionic hopping is examined by comparing the current densities considering the applied electric field (eq 4.18). Since the U in the eq 4.18 is independent of the electronic field, the jump distance can be determined from the slope of $\ln J–E$ graph. After comparison, it can be considered that the ionic jumping movement is facilitated for the sample which has a smaller hopping distance. Another proof of the suitability of the ionic hopping mechanism can be obtained from calculation of the activation energy (eq 4.19). The linearity of this graph $(\ln(J/\sinh(eE\alpha/2kT))$ vs $1/T$) confirms that the conductivity mechanism does not change in the corresponding temperature range. The ionic conductivity mechanism would be more favorable for the sample with a lower E_a value.[154] The equation was organized as in equation under the low-field or high-field assumptions (eq 4.20).[171]

$$J = n\alpha \upsilon \exp\left(-\frac{U}{kT}\right)\exp\left(\frac{eE\alpha}{2kT}\right) \tag{4.18}$$

$$J = 2n\alpha \upsilon \exp\left(-\frac{U}{kT}\right)\sin h\left(\frac{eE\alpha}{2kT}\right) \tag{4.19}$$

$$J = en\upsilon\alpha\left(\frac{eE\alpha}{2kT}\right)\exp\left(-\frac{U}{kT}\right)\ (low\ field; e\alpha F \ll 2kT) \tag{4.20}$$

where n is the carrier density, α is the hopping distance, υ is the frequency, U is the activation energy, k is the Boltzmann constant, e is the electron charge and T is absolute temperature, E is the applied electrical field.[171,172]

Nevin and Summe tried to clarify the conductivity mechanism of commercially available Pyralin PI-2555 at low fields (0–5 V). They stated that the conductivity behavior of this film (thickness 1100 Å, electrode Al) could be explained by Schottky and Poole-Frenkel below 60°C. They also explained that there is an ionic conductivity resulting from the dissociation of weakly bonded water molecules between 60°C and 100°C at which the relationship between I and $V^{1/2}$ corresponds to the rate theory[173] (eq 4.21).[156]

$$I = I_0 \sinh\frac{qV\lambda}{2kTd} \tag{4.21}$$

where I is the current, I_0 is the zero voltage intercept current, q is the electronic charge, V is the voltage, λ is the ionic jump distance, k is the Boltzmann constant, T is the absolute temperature, and d is the polymer thickness.

In contrast to Nevin and Summe's study, Sharma and Pillai examined the electronic conductivity of poly (4,4'-diphenylene) pyromellitimide film (Kapton, Dupont) under the steady-state conditions of 100°C–200°C and 50–450 kV/cm electric field and they tried to explain the conductivity mechanism of Kapton under the influence of the high electrical field. Kapton exhibited ohmic behavior under the electrical field of less than 100 kV/cm.[159] Whereas high electrical field conductivity is caused by electron migration with tunneling.[174] Sharma and Pillai suggested that the Kapton conductivity mechanism is ionic under these conditions by eliminating the Poole-Frenkel and thermally assisted tunneling mechanisms. Because they found that the reported ionic distance values were partly dependent on temperature and there was no linear change with temperature.[159]

4.4.2 POLYIMIDE-BASED CONDUCTIVE COMPOSITES

The simplest way is to incorporate the electron-conducting agents into the polymer matrix to prepare PI-based conductive composites. For this purpose, metallic,[175,176] polymeric[142,143] or carbon-based[138,177] materials are frequently used. However, in the preparation of conductive composites, homogeneous distribution of conductive additives in the polymer matrix is an important problem.[148,149] In addition, it should not be neglected that the final physico-chemical and mechanical properties of the composite are strongly affected by the structure of the polymer matrix and conductive agent, processing technique[178] and their compatibility as well as the size of the conductive particles. It is possible to use additives of very different character to make the PI conductive. These additives can be classified in three groups as conductive polymers, metals and/or inorganic additives, and carbon-based fillers.

4.4.2.1 CONDUCTIVE POLYMER CONTAINING POLYIMIDE COMPOSITES

There are some studies in the literature on the PI-based conductive composites including conducting polymer such as polyaniline (PAni),[142,179] polypyrrole,[142] poly(3,4-ethylenedioxythiophene) (PEDOT),[142,180,181] and polythiophene (PT).[182] These polymers were expected to impart a suitable viscosity of the material as well as the conductivity and it was stated that particularly PAni meets this demand.[183] Among these studies, Ding et al. determined the conductivity value of PAni/PI fabric as 1.10 Ω/sq and the electrical conductivity value of Ni–W–P/PAni coated PI-based fabric as 0.08 Ω/sq. In addition, PAni not only increased the fabric conductivity but also it acted as a reducing agent which allowed the fabric to be coated with Ni–W–P.[184] Besides, it was reported that new PI-based material with pH sensitivity was obtained with PAni insertion.[185] Another conducting polymer, PEDOT can be an opposite electrode in an optoelectronic device[141] or it can increase the conductivity of the sulfonated PI (sPI) up to 10 times.[15,186]

4.4.2.2 METALLIC FILLER INCLUDING POLYIMIDE COMPOSITES

Regarding the metallic conductive agents, some of the metal and/or inorganic additives used in the literature for PI are: titanate nanowire,[187] iron oxide,[188,189] metallo-organic compound,[190] silver/silver nanowire (AgNW),[139,176,191–193] TiO_2,[194] copper,[175] ZnO,[195] and so on. In addition, some of the aforementioned additives were used together to benefit from the synergistic effect on the conductivity of PI (Ag-CNTs,[196] reduced graphene oxide-PAni (RGO-PAni),[197] AgNW-PAni,[198] Ag-CNT,[191,196] carbon-PT,[199] Ni-graphene,[200] silane grafted multi-wall CNT (MWCNT)[138]). Among the recommended additives for applications of PI requiring a high electrical conductivity, metallic nanowires are standing out and AgNW is widely used.[139,198,201] According to Nquyen et al., the percolation threshold can be effectively reduced with the use of this filler (0.48 vol.%) to reach acceptable conductivity values.[139] But it should be noted that the use of a high amount of AgNW leads to decay in the mechanical properties of the PI matrix even if it increases the electrical conductivity.[198,201] Therefore, studies are being carried out to reduce the filler ratio in the composite systems.

4.4.2.3 C-BASED FILLER INCLUDING POLYIMIDE COMPOSITES

Carbon-based materials include carbon black (CB), graphite, carbon nanotube (CNT), graphene (RGO), etc. These materials have wide applications in different technologies because of their various properties like electronic and thermal conductivity. Some carbon-derived materials used as additives in PI are as follows: graphene,[140,202–207] single-wall or multi-wall CNT,[138,202,208–210] CB,[143,149,211–213] expanded graphite.[203] Graphene is frequently preferred among other carbon-based materials due to its high electrical conductivity.

When aromatic PI-based composites containing graphene were used as a cathode material, it was reported that π–π interaction occurred between PI and graphene.[205] In other words, this interaction in the main backbone meant that electrical and ionic conduction came true faster. Another important point is the well-dispersion of graphene into the polymer matrix. Hence, mostly chemically modified graphene is used for the fine distribution of graphene. If the thermal reduction of graphene can be applied in the production of the graphene/PI nanocomposites, it is possible to have a higher electrical conductivity due to the defective sites of the modified graphene.[206]

Recently, IL was introduced into PIs to enhance the electrical conductivity, effectively. Deligöz et al. reported the incorporation of three imidazolium-based ILs (RIm^+BF4^-) with different alkyl chain lengths (R = methyl, ethyl, and butyl) into PAA and subsequent conversion to PI by thermal imidization. They determined that acid-doped PI/IL films exhibited the conductivity of 10^{-4}–10^{-5} S/cm at 180°C, whereas the conductivities of acid-free PI/IL films were at around 10^{-9}–10^{-10} S/cm.[145] In a similar study of this group, an acid doped highly conductive film based on sPI and IL was presented. They claimed that the conductivity of sPI/IL film increased with temperature and reached to 5.59×10^{-2} S/cm at 180°C and 100 kHz.[146]

Table 4.3 shows a comparison for some studies in the literature considering composite types, preparation method, electrical characteristics (maximum conductivity or resistivity), and percolation threshold.

TABLE 4.3   Electrical Properties of Conductive Filler Including PI Composites.

Composite	Preparation method	Specific conductivity/ resistivity	Loading	Percolation threshold (%)	Reference
Conductive polymer containing PI composites					
PEDOT/SPI PAni/SPI PPy/SPI	In-situ	5.94×10^{-1} S/cm 2.51 S/cm 5.63×10^{-1} S/cm	–	–	[142]
PAni/PI	In-situ	3.8×10^{-2} S/cm	–	–	[179]
Metallic filler including PI composites					
AgNW/PI	Solvent mixing	10^2 S/m	–	0.48	[139]
Co_3O_4/PI	Surface modification and ion-exchange technique	1.26×10^7 Ω/cm	21%-atomic conc. of Co(2p)	–	[214]

TABLE 4.3 *(Continued)*

Composite	Preparation method	Specific conductivity/ resistivity	Loading	Percolation threshold (%)	Reference
AgNW/PI	Spray coating	100 S/cm	0.0974 vol.%	–	[191]
Pt/PI	Electrochemical deposition	20 S/cm	3 wt.%	2.5–5	[199]
C-based filler including PI composites					
(RGO/ MWCNT)/PI	In-situ	4.4×10^{-4} S/m	0.2 wt.%	–	[137]
CB/PEEK/ TPI	Melt blending	5.33×10^{-1} S/m	12.5 wt.%	0.05	[143]
CNFs/PI	In-situ	2.03×10^{-6} S/cm	0.5 wt.%	0.0049	[158]
CNT/PI	Spray coating	6.3 S/cm	–	–	[191]
SWCNT/PI	In-situ	1×10^{-6} S/cm	1 wt.%	0.05	[215]
MWCNT/PI	Electrophoretic deposition	4.2×10^{2} Ω	1.2 wt.%	-	[216]
RGO/PI	In-situ	1.4×10^{-2} S/m	2 wt.%	0.15	[204]
RGO–PI– HT–Fs	Dip-coating, chemical and thermal reduction	0.4 S/m	–	–	[217]
CMG/PI	Solution blending	10^{-3} S/m	1 wt.%	0.2	[206]
Carbon/PI	Electrochemical deposition	1.2 S/cm	12 wt	2.5–5	[199]
MWCNT/PI	In-situ	55.6 S/cm	40 wt.%	0.5	[209]
MWCNT/PI	In-situ	10^{-1} S/cm	3.7 vol.%	0.15	[210]

PEDOT, poly(3,4-ethylenedioxythiophene); sPI, sulfonated polyimide, PAni, polyaniline; PPy, polypyrole; AgNW, silver nanowire; PEEK, polyetheretherketone; TPI, thermoset polyimide, RGO-PI–HT–Fs, reduced graphene oxide-polyimide heat-treated foam; CMG, chemically modified grapheme; SWNT Single-wall carbon nanotubes; MWCNT, Multi-wall carbon nanotube

On the other hand, fully aromatic PIs may be useful for electronic applications without conductive additives and they are known as materials comprising intra/interchain charge transfers and electronic polarization that are mainly dominated by strong interactions of symmetrical and polar groups. The formation of charge transfers and electronic polarization is supported by the imide structure that has electron acceptor feature and the amine structure acting as an electron donor.[15] However, the use of these types of PIs, either by adding conductive agents or agent-free, is limited by the problem of solubility and processability due to the high crystallinity and stiffness in the polymer backbone. In addition, the problem of heating under the high potential of PI-based conductive composites, which are preferred for use in electronic applications, requires the use of thermoset PI instead of thermoplastic PI and this leads to processing challenge. This problem can be eliminated by chemical modifications without sacrificing the thermal and mechanical properties of fully aromatic PIs. The addition of aryl ether,[218] ketone,[143] ester,[219] methyl,[220] and especially sulfonic acid groups[142,146,225–228,180,181,186,199,221–224] to the structure can increase the ionic conductivity in the selected application area beside solubility.[229]

4.4.3   PERCOLATION THEORY AND CONDUCTION MECHANISMS OF CONDUCTIVE FILLER INCLUDING POLYIMIDES

Matrix and filler interactions of conductive polymer composites are presented by various mechanisms. The most extensive theory is the percolation theory.[158,230,231] This theory can simply be expressed as the sudden change in the properties of the composite such as conductivity and physical properties after a certain critical value of inorganic/organic additives.[230] The critical value at which this sudden change occurs is defined as the percolation threshold. Above this value, the endless conductive cluster is formed and the polymer becomes conductive while the conductivity of the composite is mainly dominated by the polymer matrix below this point. While there is no contact between the filler particles at low loading, continuous paths occur between the filler particles with an increasing amount of conductive filler inside the composite. As a result, when a potential is applied to the composite, the current passes through a less resistant path containing conductive additives.[136,230] In Figure 4.4, a schematic representation of percolation behavior for conductive filler containing polymeric systems is given.

The critical concentration value (electrical percolation threshold) of the conductive filler must be determined to detect the point at which the composite materials become conductive. The percolation threshold is not only affected by the filler concentration but also depends on the aspect ratio of filler and its distribution in the composite.[143] It is known from the literature that the percolation threshold is lower if conductive fillers with high aspect ratio or surface area are used in the preparation of composites.[137,232] Considering the shape of the conductive fillers, the spherical shape of conductive agent does not significantly affect the percolation threshold, whereas the especially random orientation of fiber-shaped fillers reduce the percolation threshold due to the aforementioned reasons.[198,215,233,234] Besides, the filler distribution may be improved by the addition of another polymeric component having a lower melt viscosity into the polymer matrix. That's why the percolation threshold can be reduced by the aid of some polymers such as PEEK.[143,216] On the other hand, if metal salts are used as conductive fillers, the thermal curing process is preferred after mixing filler and polymer to in-situ metal oxide formation, better dispersion, and impurity removal.[214] The relationship between electrical conductivity and percolation threshold can be given by the following equation (eq 4.22)[143,158,204]:

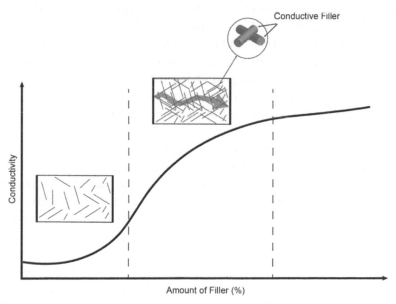

FIGURE 4.4 A schematic representation of percolation behavior for conductive filler containing polymeric systems.

$$\sigma \alpha \sigma_0 (m - m_c)^t \tag{4.22}$$

where σ is the electrical conductivity, σ_0 is the bulk electrical conductivity of filler, m is the filler fraction, m_c is the filler fraction at the percolation threshold, and t is the critical exponent.

In addition to basic approaches, some researchers tried to explain the conductivity of composites with mechanisms. Zhang et al. classified the mechanisms between the polymer matrix and inorganic filler as tunneling effect (low loading), conductive network mechanism (high loading of filler), and doping (loading of filler at saturation value), while Pourrahimi et al. sorted them as "Lewis model, Tanaka model, multi-region structure model, and induced dipole moment model."[149,235] Nguyen et al. reported that the conductivity mechanism for the homogeneous AgNW-PI system fitted to the tunneling mechanism below −50°C, while hopping mechanism was valid above the 50°C according to the Mott theory.[139] The conductivity of a composite material (vapor-grown carbon fiber (CF)/PI) shows an increased regime with temperature. This behavior is examined with hopping and tunneling mechanisms.[236] Electron hopping can be achieved either mobilizing of electrons that are stimulated at high temperature to the conduction band, or jumping of electrons to outlying sites that have a similar energy level with conduction band at low temperature which is called variable range hopping theory. The electrons tend to tunneling between the nearest conductive regions. Therefore, large voltage fluctuations are observed at the junction points. According to Sheng, studying the effect of voltage fluctuation and temperature on electrical conductivity; classical tunneling theory is valid at low temperatures, while tunnel conductivity caused by voltage fluctuation is effectual at high temperatures.[237]

In a different study, Fang and colleagues[198] described the temperature dependence for the AgNW/PAni-PI film in accordance with the following eq 4.23. In this study, PI bandwidth increased by using PAni. Also, they reported that the electrical conductivity values for PI, AgNW/PI, and AgNW/PAni-PI depending on the loading of filler ratio were found to be 5.5×10^{-13} S/cm, 175 S/cm, and 1528 S/cm, respectively. In this study, it was reported that the interaction between the components led to the change in conductivity. Also, it was indicated that the tunneling effect should be examined due to the short distance between AgNWs. According to eq 4.24 (Richardson–Schottky model), the tunneling resistance can be reduced by decreasing the distance (d) between the AgNW and the height of the barrier (ϕ) between the electrode and PI-based film. The ϕ value can be found using the ln (I/T^2)–$1/T$ plot. Since increasing the amount of AgNWs inside the composite will lead to reducing the distance between nanowires, changes in the composite matrix should be directed to regulate barrier height.

$$I = RT^2 \exp\left(-\frac{\phi}{kT}\right) \tag{4.23}$$

where I, R, T, k, and ϕ are the current density, the Richardson constant, temperature, the Boltzmann constant, the barrier height between the electrode and film, respectively.

$$R_{tunnel} = \frac{h^2 d}{Ae^2 \sqrt{2m\phi}} \exp\left(\frac{4\pi d}{h}\sqrt{2m\phi}\right) \tag{4.24}$$

in which R_{tunnel} is tunneling resistance, h is the Planck constant, d is distance, A is the cross sectional area of tunnel, e is the electron charge, m is the electron mass, and ϕ is the barrier height between the electrode and polymer film.

Concerning the ionic species addition into the PI structure, it is possible to use IL for a significant enhancement in ionic conductivity without adversely affecting the thermal stability of the PI membranes. However, they can act as a plasticizer for polymeric systems.[145,146,217,238] If PI membranes are specifically prepared for use in fuel cells, proton conduction (or ionic conduction) is the most important criterion for cell performance. Mostly accepted mechanisms are hopping and vehicle mechanisms for ion conduction.[238] Ma et al. reported that dielectric spectroscopy was an important technique in determining the charge transport mechanism in ion-containing polymers.[239] In the case of the vehicle mechanism, proton transport is carried out with the aid of H_3O^+ as a vehicle consisting of $-SO_3H$ and water for a sPI.[240] Generally, the type of conduction in the polymeric membrane where protons jump in a structure containing hydrogen bonds is called the Grotthuss mechanism.[241] However, when the mechanism of ionic conductivity is examined around the glass transition temperature (T_g), the ionic conductivity largely depends on structural relaxation.[239] In this situation, researchers generally use the Arrhenius ($\sigma = f(1/T)$) ($T > T_g$) and Vogel–Tamman–Fulcher ($T < T_g$) equations in order to estimate the ionic conductivity change of polyelectrolyte membranes with temperature dependence.[242,243]

As it can be seen from the aforementioned examples, research on polymer composites containing conductive additives is increasing day by day in order to fulfill the increasing demand for use in advanced technology applications. Consequently, studies on PI-based conductive composites were carried out intensively and significant improvements were achieved in this area. According to the studies, the effects of many parameters such as preparation procedure, filler type, filler concentration, filler morphology, and the interactions between components in the preparation of PI-based composites were studied on conductivity. Among these parameters, the selection of the production process is important to prevent deterioration of the composite components. It is known that multi-phase polymer matrices to be prepared with the use of low viscosity additives are important because of the compatibility of filler-matrix and formation of a homogeneous dispersion medium will improve conductivity. The aspect ratio, specific surface area, morphology, and orientation of the preferred conductive fillers in the composite are also important in improving the composite conductivity.

4.4.4   MALEIMIDE/BIS(MALEIMIDE)-BASED CONDUCTIVE COMPOSITES

The conductive properties of maleimide polymers and maleimide group containing functional materials are reported in the literature.[224–249] Especially, these types of imidic polymers were applied in ionic conducting media as solid electrolytes. Toppare et al. reported a new monomer including the maleimide group, namely *N*-(2-(thiophen-3-yl)methylcarbonyloxiethyl) maleimide, and it was polymerized to prepare maleimide polymers.[247] The electrical conductivity of this polymer was measured as 2.8×10^{-5} S/cm by using the four-probe technique. In a different study, poly(ethylene glycol) monomethyl ether (PEGME) grafted poly(ethylene-alt-maleimide) was synthesized by Kang et al. The conductivity of this polymer electrolyte consisting of 66 wt.% PEGME was 3×10^{-4} S/cm at 30°C.[248] Similarly, propylene carbonate (PC) plasticized polyelectrolyte based on poly[lithium-*N*(4-sulfophenyl)maleimide-co-methoxy oligo(oxyethylene)metacryalate] was prepared and the ionic conductivity of this polyelectrolyte was found to be 1.8×10^{-5} S/cm with 54.5 wt.% PC loading.[249] Concerning the application potential of maleimide polymers and maleimide group containing functional materials, it was reported that they can be used as polymer electrolytes for electroactive systems such as polymer actuators,[244] optoelectronics,[245] and electrochromics.[247–249]

Regarding the conductive composites of BMI resins, different methods were used to improve the electrical conductivity of these materials. Among them, (1) the incorporation of carbon-based materials,[250–254] (2) addition of carbon-based filler and reinforcement with epoxy/triazine[255–257] are attracting more interest. In the first approach, mostly CNTs are introduced into BMI resins to improve their conductivity as well as mechanical properties. In one of these studies, it was claimed that an ultrastrong BMI/CNT composite film with superaligned and tightly compacted structure was prepared. The electrical conductivity of this composite film was determined as 2.45×10^4 S/m if CNT loading was up to 65.5 wt.%.[250] In a similar study, BMI-based CF reinforced composite material showed a conductivity of 27.9 S/m, while its mechanical strength was not deteriorated.[251] Further increment in conductivity of CNT/BMI composites, hyper-branched polyethyleneimine (PEI) grafted CNTs were used to reinforce BMI composites. The electrical conductivity of PEI grafted CNT including BMI composites was improved by 10^{14} times compared to neat BMI.[252] In a different study reported by Wang et al., the effect of CNT length on electrical properties of CNT/BMI composites was studied. They found that the longer CNTs (length of 1.3 mm) resulted in higher electrical conductivities.[253]

In the second common approach, the addition of carbon-based filler and reinforcement with epoxy/triazine was applied. In a study, CF reinforced epoxy, BMI, and PEEK were used. The electrical conductivities of CF/epoxy (59 vol.% fiber), CF/BMI (67 vol.% fiber), and CF/PEEK (59 vol.% fiber) were found to be 8100, 7610, and 3120 S/m, respectively.[255] Mandhakini et al. reported the toughening of brittle epoxy matrix with C8 ether-linked BMI. It was found that the electrical conductivity was increased using 5 wt.% of CB.[256] In a different study, electrically conductive adhesive based on BT resin filled with microcoiled CFs as the conducting filler was developed. The electrical conductivity of this material improved to 3.16×10^{-2} S/cm.[257]

Apart from the above-mentioned studies, BMI hyper-branched polymeric systems were designed to be used as an electrolyte in sustainable energy systems[258] and electrical vehicles.[259] Chu et al. announced the cross-linking of sulfonated PEEKs with hyperbranched BMI oligomer. The prepared membranes displayed an acceptable proton conductivity around 10^{-2} S/cm at room temperature. Furthermore, the single direct methanol fuel cell performance was found to be 30 mW/cm^2 at 60°C for electrolytes with 20 wt.% sulfonated PEEKs.[258] In another study, a hyperbranched network gel copolymer electrolyte was prepared from BMI oligomer in the presence of lithium hexafluorophosphate (LiPF$_6$). According to their results, the highest ionic conductivity was determined as 7.72×10^{-3} S/cm at 23°C if 5 wt.% of 1.1 M LiPF$_6$ was added to the polymer matrix.[259]

4.4.5   APPLICATION AREAS OF SEMI/CONDUCTIVE POLYIMIDES

Generally speaking, semi/conductive PIs have many technological applications such as electromechanically responsive device,[201,213,217,260,261] electrochemical sensing,[144,223,262] gas sensing[208,262,263] aerosol jet printing,[264] ion-track membrane,[265] sodium ion batteries,[266] CF reinforced plastics,[183] lithium-ion batteries,[168,182,205,212] fuel cell,[145,146,170,224–227,238,267,268] ECD,[177,269–271] electromagnetic interference shielding,[158,184,188,272,273] memory device,[157,270,274,275] solar cell,[193] metal adsorption,[179] and so on. Some applications of conductive PIs are reviewed in detail in this section.

Molla-Abbasi and Shabanian prepared an MWCNT and polyetherimide-based conductive polymer composite containing aromatic bulk pendant groups such as xanthene. Then, they designed a gas sensor so that this composite could be used to identify cancer biomarkers in the lungs. It was determined that this composite had some selectivity against some vapors, such as water, methanol, ethanol, acetone, and especially toluene.[144]

In the past decade, high-performance PI films with TPA and/or derivatives were studied for electrochromic applications by different researchers. Su and colleagues designed a diphenylamine-pyrene based PI ECD and determined the coloration/decolorization time as 1.7/0.9 s.[276] When this result was compared to inorganic-based devices, it was determined that PI-based ECD had faster response time than that of a WO_3/NiO-based ECD (3.2/1.1 s)[277] which exhibited a very fast switching time.

Lithium-ion batteries (LIB) are widely used materials, especially with the development of portable electronic devices. The high solubility of small molecules, which are used as active substances in LIB limits their use in commercial devices. Therefore, various polymers such as PI can be used in the construction of electrolytes to prevent loss of active substance. Zhang et al. prepared the PI/CB composite for use as a cathode in LIB. It was reported that the proper electrochemical properties (capacity of 182/mAh/g at 0.1 A/g) can be achieved by controlled suppression of the crystallization of PI.[212] Lyu et al.[205] reported that the LIB exhibited a reversible capacity of 232.6/mAh/g due to π–π interactions and accordingly fast electron and ion transport capability of the PI when the aromatic PI/graphene composite containing 10 wt.% graphene is used as a cathode.

In addition, PI can be used in the production of conductive aerogels with high carbon content. Zhang et al. prepared carbon aerogels using CPI-graphene nanoribbon which had a specific surface area of 2413 m²/g. Also, they produced a high energy density electrode (54.8/Wh/kg) with fast ion and electron transfer using this carbon aerogel for supercapacitor applications.[211]

The polymers exhibit relatively low thermal emissivity ($0.9 > \varepsilon > 0.7$). Nonmetallic semiconductor materials are the most suitable materials for the applications requiring low thermal conductivity. These materials have high optical transparency and IR reflectivity as well as their electronic conductivity. Babrekar et al. reported that the energy density of ZnO/PI composite film prepared using laser pulse deposition was controlled by means of synthesis method and low thermal emissive PI films were prepared. They reported that the surface resistance of ZnO/PI composite film significantly reduced compared to a nonconductor PI film.[195]

In another study, Bobinger et al. and Romero et al. used the photothermal ablation process to form an electrically conductive layer of graphene on the surface of the commercial Kapton® film.[140,278] Both groups obtained a comparable material with graphene-derived materials in terms of not only thermal but also electrical properties with the laser ablation process. Further, Bobinger et al. reported that a flexible and cost-effective heater was integrated based on this commercial PI film.[140] The surface resistance (<250 Ω/sq for laser-induced graphene on the Kapton® film) of the film could be changed by laser power and electrode size. In addition, resistance change of the modified-PI film led to the current induced heating of the film and consequently, the contact resistance was decreased with temperature. The contact resistance was independent of the laser power and it can be reduced by up to 2 Ω (Ag electrode) depending on the electrode type.[278]

Another important application of PIs is in the preparation of PEM for fuel cells which are defined as one of the sustainable energy sources. PEMs are a very important component for fuel cells, but the commercial PEM, namely Nafion, has some disadvantages such as high cost, high methanol permeability, and working at relatively low temperatures up to 90°C. Therefore, different polymeric materials are being developed to find a promising alternative to Nafion as a proton-conducting membrane. There are lots of studies on the use of polymers such as sulfonated poly(aryl ether),[218] polyphosphazene,[279] PIs,[145,146,221,222] poly(phenylene)s,[280] PS,[268] PBI[281] as PEM. Among these polymers, PI exhibits lowered hydrolytic stability at high temperature.[166] Zhang and Xu reported to synthesize a sPIs consisting of sulfonated unit and naphthalenediimide. With the addition of the naphthalenediimide structure to the polymer backbone, the membrane with improved hydrolytic stability, high proton conductivity (σ), and limited swelling can be achieved.[222] Deligöz and colleagues investigated the effects of aliphatic linkages on some performance parameters such as proton conductivity, ion exchange capacity, and solubility of sulfonated homo/co-PI membranes.[224,225,227] They determined that the ionic conductivity of the sPI membrane was in the range of 7.8–11 mS/cm if the 2,4-diaminobenzenesulfonic acid monomer ratio was increased above 30 mole% in the comonomer composition.[225] In addition, they reported that the ionic conductivity of sulfonated membranes prepared from monomers with -SO_3H groups was higher than that obtained from post-sulfonation of PI.[224] In another studies, Deligöz and colleagues studied the thermomechanical and ionic conductivity properties of complex membranes composed of the PI-COOH/IL (1-methyl imidazolium tetrafluoroborate (MeIm-BF_4) and 1-ethyl 3-methyl imidazolium tetrafluoroborate) and the sPI-SO_3H/IL (MeIm-BF_4). In these studies, they reported an improvement in the long-term conductivity stability of PEMs owing to the interaction between the cationic groups of the ILs and the anionic groups of the sPI. For example, among the PEMs they prepared, the films containing sPI and MeIm-BF_4 had an ionic conductivity of 5.59×10^{-2} S/cm at 180°C (conductivity of acid-free PI/IL was reported as 10^{-9}–10^{-10} S/cm).[145,146]

It is known that nanoparticles with magnetic properties are frequently used in several applications. A new hybrid material was prepared by Zeng et al. to be used in areas such as solar cells and biosensors. For this purpose, they used a nanocomposite prepared from graphene and CNT as a filler and PI film. While CNT is weaker than graphene in terms of the create a route for electron transport, graphene tends to cluster easily within PI.[197,202,207] They reported that the film obtained had a higher electrical conductance (1×10^{-5} S) with less filler amount compared to the CNT-loaded (1×10^{-10} S) film. In addition, they modified the surface of this film with silver to increase the surface conductance and it reached to 2.19×10^{-4} S.[202] Zhan et al. published two different studies, they reported that the electrical and superparamagnetic properties of the PI/Fe_2O_3 and PI/iron (III) 2,4-pentanedionate (Fe(acac)$_3$) composite products were improved by covering the surface with silver in-situ single-stage self-metallization technique and these films can be evaluated as potential material for electromagnetic interference shielding.[188,189] Yonessi and coworkers prepared PI films containing nickel-bound graphene nanoparticles by altering the effect of magnetic field. They reported that the electrical conductivity of the film with 1.3% Ni-graphene loading was 2.5 times higher (0.0367 S/cm) than that of the film containing the same percentage of nanomaterials prepared without applying the magnetic field.[200]

Studies with silver are particularly notable for the preparation of semi/conducting PI composites. For example, Wang et al. used Ag and PI to prepare the circuit board with selective electroless deposition method. They reported that the surface resistance of the PI-based film was found to be 2.82 × 10⁻⁶ Ω cm which is 1.6 times of bulk copper. Consequently, they claimed that this polymer can be used as printed circuit boards, screen, organic light emitting diodes (OLED), solar cell, etc.[282] Surface conductivity or resistances of the circuit elements prepared with the surface silverized principle in the literature are shown in Table 4.4.

Similarly, it is possible to make circuit drawings on the PI sheet by using conductive inks consisted of Ag salts. For example, Tai et al.[287] prepared a silver (I) solution from AgNO₃ (a conductive ink), coated this ink on the PI substrate and they sintered the film at 200°C. They reported that the resistivity of this PI-based film was 6.6 μΩ cm within the first 60 min and that the conductivity mechanism could be explained by the percolation theory. Using the same technique, Mou et al. found the resistivity value of silver-coated PI which is sintered at 110°C, as 12.1 μΩ cm.[288]

AgNW/titania/PAA sandwich structure prepared by Spechler et al. for use in the construction of OLED was transformed into a transparent PI film by thermal imidization at 360°C. It was reported that this film exhibited a much better performance than indium tin oxide (ITO)-based OLED device.[289] Lee et al. pointed out that EQE increased by 74 and 68%, respectively, when PI was used as monochromatic (green) and white OLED substrate.[290] Lin et al. showed that the electrical conductivities of PI composites prepared with CNT and AgNW were 6.3 and 100 S/cm, respectively. In addition, the mechanical behavior of the films was 1200 times higher than those of ITO/PI and ITO/AgNW/PI according to the bending test.[191] As a different application area, while Lee et al. used the transparent conductive PI film as a strain sensor, Kim et al. investigated the silver-based conductive PI film as a capacitive sensor.[201,261]

4.5 CONCLUDING REMARKS

As stated above, wholly aromatic PIs are one of the important members of the thermally stable polymer class. Besides they have many superior properties such as high tensile and compressive strength, low dielectric constant and dielectric loss, good resistance to solvents and moisture, good adhesion properties for inorganic, metallic and dielectric materials, and easy casting of thin or thick films. Due to these unique properties, PIs are prevalently used in microelectronics, films, adhesives, and membranes.

TABLE 4.4 Surface Resistivity/Conductivity of the PI Films Prepared by Surface Silverized Process.

PI components	Conductive filler	Silver source	Conductivity/ resistivity	Application areas	Reference
ODA-ODPA	CNTs	Silver nitrate	1.723×10^{-10} S	Sensors	[202]
	Gr-CNTs (2:1)		1.101×10^{-5} S		
	Ag/Gr-CNTs (2:1)		2.19×10^{-4} S		
ODA-PMDA	Fe$_2$O$_3$	Silver nitrate	0.7 Ω/sq	EMI-shielding	[189]
ODA-PMDA	—	Silver nitrate	0.4 Ω/sq	Microelectronic and space	[283]
ODA-PMDA	—	Silver nitrate	0.6 Ω/sq	Electronic and space	[284]
ODA-ODPA	—	(hexafluoroacetylacetonato)silver(I) (trifluoroacetylacetonato)/ silver(I) tetrafluoroborate	0.05 Ω/sq	—	[285]
ODA-PMDA	—	(1,1,1-trifluoro-2,4-pentadionato) silver(I)	<7 Ω/sq	—	[286]
ODA-PMDA	—	Silver nitrate	About 10^2 Ω/cm	—	[176]

ODA, 4,4'-oxydianiline; ODPA, 4,4'-oxydiphthalic anhydride; PMDA, pyromellitic dianhyride

In addition to the use of PIs as an insulating material, their electrical properties can also be controlled from semiconducting to conducting state using some different approaches. These research efforts pave the way for widely using of PIs not only in the insulating materials but also in the industries where semiconducting/conducting material properties are required. As a consequence, it is expected that these multipurpose materials will be used more widely in various technological applications of microelectronic, separation, energy, and aviation in the near future.

KEYWORDS

- **polyimide**
- **maleimide**
- **bis(maleimide)**
- **polyimide composites**
- **dielectric constant**
- **insulator**
- **conductivity**
- **relaxation**

REFERENCES

1. Saçak, M. *Polimer Kimyası*, 7th ed.; Gazi Kitabevi Tic. Ltd. Şti.: Ankara, 2015.
2. McKeen, L. W. *Fatigue and Tribological Properties of Plastics and Elastomers*; 2000.
3. Hergenrother, P. M. The Use, Design, Synthesis, and Properties of High Performance/ High Temperature Polymers: An Overview. *High Perform. Polym.* **2003,** *15* (1), 3–45. https://doi.org/10.1177/095400830301500101.
4. Nair, S.; Pitchan, M. K.; Bhowmik, S.; Epaarachchi, J. Development of High Temperature Electrical Conductive Polymeric Nanocomposite Films for Aerospace Applications. *Mater. Res. Express* **2019,** *6* (2). https://doi.org/10.1088/2053-1591/aaefa8.
5. Zegaoui, A.; Derradji, M.; Dayo, A. Q.; Medjahed, A.; Zhang, H.; Cai, W.; Liu, W.; Ma, R.; Wang, J. High-Performance Polymer Composites with Enhanced Mechanical and Thermal Properties from Cyanate Ester/Benzoxazine Resin and Short Kevlar/ Glass Hybrid Fibers. *High Perform. Polym.* **2019,** *31* (6), 719–732. https://doi. org/10.1177/0954008318793181.
6. Volpe, V.; Lanzillo, S.; Affinita, G.; Villacci, B.; Macchiarolo, I.; Pantani, R. Lightweight High-Performance Polymer Composite for Automotive Applications. *Polymers (Basel).* **2019,** *11* (2), 326. https://doi.org/10.3390/polym11020326.

7. Feng, C. P.; Bai, L.; Bao, R.-Y.; Wang, S.-W.; Liu, Z.; Yang, M.-B.; Chen, J.; Yang, W. Superior Thermal Interface Materials for Thermal Management. *Compos. Commun.* **2019,** *12*, 80–85. https://doi.org/10.1016/j.coco.2019.01.003.

8. Ding, Y.; Hou, H.; Zhao, Y.; Zhu, Z.; Fong, H. Electrospun Polyimide Nanofibers and Their Applications. *Prog. Polym. Sci.* **2016,** *61*, 67–103. https://doi.org/10.1016/j.progpolymsci.2016.06.006.

9. Zoidis, P.; Papathanasiou, I.; Polyzois, G. The Use of a Modified Poly-Ether-Ether-Ketone (PEEK) as an Alternative Framework Material for Removable Dental Prostheses. A Clinical Report. *J. Prosthodont.* **2016,** *25* (7), 580–584. https://doi.org/10.1111/jopr.12325.

10. Iqbal, H. M. S.; Bhowmik, S.; Benedictus, R. Surface Modification of High Performance Polymers by Atmospheric Pressure Plasma and Failure Mechanism of Adhesive Bonded Joints. *Int. J. Adhes. Adhes.* **2010,** *30* (6), 418–424. https://doi.org/10.1016/j.ijadhadh.2010.02.007.

11. Bhowmik, S.; Bonin, H. W.; Bui, V. T.; Weir, R. D. Modification of High-Performance Polymer Composite through High-Energy Radiation and Low-Pressure Plasma for Aerospace and Space Applications. *J. Appl. Polym. Sci.* **2006,** *102* (2), 1959–1967. https://doi.org/10.1002/app.24230.

12. You, D.-J.; Yin, Z.; Ahn, Y.; Cho, S.; Kim, H.; Shin, D.; Yoo, J.; Kim, Y. S. A High-Performance Polymer Composite Electrolyte Embedded with Ionic Liquid for All Solid Lithium Based Batteries Operating at Ambient Temperature. *J. Ind. Eng. Chem.* **2017,** *52*, 1–6. https://doi.org/10.1016/j.jiec.2017.03.028.

13. Chang, G.; Luo, X.; Zhang, L.; Lin, R. Synthesis of Novel High-Performance Polymers via Pd-Catalyzed Amination of Dibromoarenes with Primary Aromatic Diamines. *Macromolecules* **2007,** *40* (24), 8625–8630. https://doi.org/10.1021/ma070679s.

14. Thiruvasagam, P. Synthesis and Characterization of AB-Type Monomers and Polyimides: A Review. *Des. Monomers Polym.* **2013,** *16* (3), 197–221. https://doi.org/10.1080/15685551.2013.771307.

15. Liaw, D. J.; Wang, K. L.; Huang, Y. C.; Lee, K. R.; Lai, J. Y.; Ha, C. S. Advanced Polyimide Materials: Syntheses, Physical Properties and Applications. *Prog. Polym. Sci.* **2012,** *37* (7), 907–974. https://doi.org/10.1016/j.progpolymsci.2012.02.005.

16. Dine-Hart, R. A.; Wright, W. W. Preparation and Fabrication of Aromatic Polyimides. *J. Appl. Polym. Sci.* **1967,** *11* (5), 609–627. https://doi.org/10.1002/app.1967.070110501.

17. Introduction Of Polyimide–News–Solver Polymide. http://www.chinapolyimide.com/news/introduction-of-polyimide-3541.html (accessed Aug 5, 2019).

18. Zha, J.-W.; Sun, F.; Dang, Z.-M. Fabrication and Properties of High Performance Polyimide Nanofibrous Films by Electrospinning. In *2013 IEEE International Conference on Solid Dielectrics (ICSD)*; IEEE, 2013; pp 923–926. https://doi.org/10.1109/ICSD.2013.6619776.

19. SpecialChem. Polyimide (PI) Plastic: Uses, Structure, Properties & Applications https://omnexus.specialchem.com/selection-guide/polyimide-pi-plastic (accessed Aug 5, 2019).

20. Chen, C.-J.; Yen, H.-J.; Hu, Y.-C.; Liou, G.-S. Novel Programmable Functional Polyimides: Preparation, Mechanism of CT Induced Memory, and Ambipolar Electrochromic Behavior. *J. Mater. Chem. C* **2013,** *1* (45), 7623. https://doi.org/10.1039/c3tc31598c.

21. Ma, L.; Niu, H.; Cai, J.; Zhao, P.; Wang, C.; Bai, X.; Lian, Y.; Wang, W. Photoelectrochemical and Electrochromic Properties of Polyimide/Graphene Oxide Composites. *Carbon N. Y.* **2014**, *67*, 488–499. https://doi.org/10.1016/j.carbon.2013.10.021.

22. Cai, J.; Niu, H.; Zhao, P.; Ji, Y.; Ma, L.; Wang, C.; Bai, X.; Wang, W. Multicolored Near-Infrared Electrochromic Polyimides: Synthesis, Electrochemical, and Electrochromic Properties. *Dye. Pigment.* **2013**, *99* (3), 1124–1131. https://doi.org/10.1016/j.dyepig.2013.08.019.

23. Zhuang, Y.; Seong, J. G.; Lee, Y. M. Polyimides Containing Aliphatic/Alicyclic Segments in the Main Chains. *Prog. Polym. Sci.* **2019**, *92*, 35–88. https://doi.org/10.1016/j.progpolymsci.2019.01.004.

24. Deligöz, H.; Yalcinyuva, T.; Özgümüs, S.; Yildirim, S. Electrical Properties of Conventional Polyimide Films: Effects of Chemical Structure and Water Uptake. *J. Appl. Polym. Sci.* **2006**, *100* (1), 810–818. https://doi.org/10.1002/app.23174.

25. Haruhiko Ohya, Vladislav V. Kudryavsev, S. I. S. *Polyimide Membranes: Applications, Fabrications and Properties*; CRC Press: Tokyo, 1997.

26. Dabral, M.; Xia, X.; Gerberich, W. W.; Francis, L. F.; Scriven, L. E. Near-Surface Structure Formation in Chemically Imidized Polyimide Films. *J. Polym. Sci. Part B Polym. Phys.* **2001**, *39* (16), 1824–1838. https://doi.org/10.1002/polb.1157.

27. Lee, H. J.; Won, J.; Park, H. C.; Lee, H.; Kang, Y. S. Effect of Poly(Amic Acid) Imidization on Solution Characteristics and Membrane Morphology. *J. Memb. Sci.* **2000**, *178* (1–2), 35–41. https://doi.org/10.1016/S0376-7388(00)00459-2.

28. Yeganeh, H.; Tamami, B.; Ghazi, I. A Novel Direct Method for Preparation of Aromatic Polyimides via Microwave-Assisted Polycondensation of Aromatic Dianhydrides and Diisocyanates. *Eur. Polym. J.* **2004**, *40* (9), 2059–2064. https://doi.org/10.1016/j.eurpolymj.2004.05.022.

29. Liou, G.-S.; Hsiao, S.-H.; Ishida, M.; Kakimoto, M.; Imai, Y. Synthesis and Properties of New Aromatic Poly(Amine-Imide)s Derived from N,N?-Bis(4-Aminophenyl)-N,N?-Diphenyl-1,4-Phenylenediamine. *J. Polym. Sci. Part A Polym. Chem.* **2002**, *40* (21), 3815–3822. https://doi.org/10.1002/pola.10430.

30. Di Bella, S.; Consiglio, G.; Leonardi, N.; Failla, S.; Finocchiaro, P.; Fragalà, I. Film Polymerization—A New Route to the Synthesis of Insoluble Polyimides Containing Functional Nickel(II) Schiff Base Units in the Main Chain. *Eur. J. Inorg. Chem.* **2004**, *2004* (13), 2701–2705. https://doi.org/10.1002/ejic.200300959.

31. Liao, B.; Wu, X. Y.; Zhang, X.; Liang, H.; Zhang, H. X. The Study of Polyimide Modified by Ni Plasma and Its Adhesion to Cu Films. *Nucl. Instr. Method Phys. Res. Sect. B Beam Interact. Mater. Atoms* **2013**, *307*, 580–585. https://doi.org/10.1016/j.nimb.2012.11.064.

32. Zhao, X.; Geng, Q. F.; Zhou, T. H.; Gao, X. H.; Liu, G. Synthesis and Characterization of Novel Polyimides Derived from Unsymmetrical Diamine: 2-Amino-5-[4-(2′-Aminophenoxy)Phenyl]-Thiazole. *Chinese Chem. Lett.* **2013**, *24* (1), 31–33. https://doi.org/10.1016/j.cclet.2012.11.013.

33. Zhao, X.; Li, Y. F.; Zhang, S. J.; Shao, Y.; Wang, X. L. Synthesis and Characterization of Novel Polyimides Derived from 2-Amino-5-[4-(4′-Aminophenoxy)Phenyl]-Thiazole with Some of Dianhydride Monomers. *Polymer (Guildf).* **2007**, *48* (18), 5241–5249. https://doi.org/10.1016/j.polymer.2007.07.001.

34. Sun, S.; Chen, S.; Luo, X.; Fu, Y.; Ye, L.; Liu, J. Mechanical and Thermal Characterization of a Novel Nanocomposite Thermal Interface Material for Electronic

Packaging. *Microelectron. Reliab.* **2016,** *56,* 129–135. https://doi.org/10.1016/J. MICROREL.2015.10.028.

35. Hsiao, Y.-S.; Whang, W.-T.; Wu, S.-C.; Chuang, K.-R. Chemical Formation of Palladium-Free Surface-Nickelized Polyimide Film for Flexible Electronics. *Thin Solid Films* **2008,** *516* (12), 4258–4266. https://doi.org/10.1016/J.TSF.2007.12.166.

36. Kadiyala, A. K.; Sharma, M.; Bijwe, J. Exploration of Thermoplastic Polyimide as High Temperature Adhesive and Understanding the Interfacial Chemistry Using XPS, ToF-SIMS and Raman Spectroscopy. *Mater. Des.* **2016,** *109,* 622–633. https://doi. org/10.1016/j.matdes.2016.07.108.

37. Akram, M.; Jansen, K. M. B.; Ernst, L. J.; Bhowmik, S. Atmospheric Plasma Modification of Polyimide Sheet for Joining to Titanium with High Temperature Adhesive. *Int. J. Adhes. Adhes.* **2016,** *65,* 63–69. https://doi.org/10.1016/j.ijadhadh.2015.11.005.

38. Naito, K.; Onta, M.; Kogo, Y. The Effect of Adhesive Thickness on Tensile and Shear Strength of Polyimide Adhesive. *Int. J. Adhes. Adhes.* **2012,** *36,* 77–85. https://doi. org/10.1016/j.ijadhadh.2012.03.007.

39. Yang, W.; Liu, F.; Zhang, J.; Zhang, E.; Qiu, X.; Ji, X. Influence of Thermal Treatment on the Structure and Mechanical Properties of One Aromatic BPDA-PDA Polyimide Fiber. *Eur. Polym. J.* **2017,** *96* (September), 429–442. https://doi.org/10.1016/j. eurpolymj.2017.09.015.

40. Gu, J.-D. Microbial Colonization of Polymeric Materials for Space Applications and Mechanisms of Biodeterioration: A Review. *Int. Biodeterior. Biodegr.* **2007,** *59* (3), 170–179. https://doi.org/10.1016/J.IBIOD.2006.08.010.

41. Zhang, F.; Saleh, E.; Vaithilingam, J.; Li, Y.; Tuck, C. J.; Hague, R. J. M.; Wildman, R. D.; He, Y. Reactive Material Jetting of Polyimide Insulators for Complex Circuit Board Design. *Addit. Manuf.* **2019,** *25* (August 2018), 477–484. https://doi.org/10.1016/j. addma.2018.11.017.

42. Jang, K. S.; Suk, H. J.; Kim, W. S.; Ahn, T.; Ka, J. W.; Kim, J.; Yi, M. H. Direct Photo-Patternable, Low-Temperature Processable Polyimide Gate Insulator for Pentacene Thin-Film Transistors. *Org. Electron. Physics, Mater. Appl.* **2012,** *13* (9), 1665–1670. https://doi.org/10.1016/j.orgel.2012.05.024.

43. Feili, D.; Schuettler, M.; Doerge, T.; Kammer, S.; Stieglitz, T. Encapsulation of Organic Field Effect Transistors for Flexible Biomedical Microimplants. *Sensors Actuators A Phys.* **2005,** *120* (1), 101–109. https://doi.org/10.1016/J.SNA.2004.11.021.

44. Moya, A.; Zine, N.; Illa, X.; Prats-Alfonso, E.; Gabriel, G.; Errachid, A.; Villa, R. Flexible Polyimide Platform Based on the Integration of Potentiometric Multi-Sensor for Biomedical Applications. *Procedia Eng.* **2014,** *87,* 276–279. https://doi.org/10.1016/J. PROENG.2014.11.661.

45. Dujardin, W.; Van Goethem, C.; Zhang, Z.; Verbeke, R.; Dickmann, M.; Egger, W.; Nies, E.; Vankelecom, I.; Koeckelberghs, G. Fine-Tuning the Molecular Structure of Binaphthalene Polyimides for Gas Separations. *Eur. Polym. J.* **2019,** *114* (January), 134–143. https://doi.org/10.1016/j.eurpolymj.2019.02.014.

46. Sanaeepur, H.; Ebadi Amooghin, A.; Bandehali, S.; Moghadassi, A.; Matsuura, T.; Van der Bruggen, B. Polyimides in Membrane Gas Separation: Monomer's Molecular Design and Structural Engineering. *Prog. Polym. Sci.* **2019,** *91,* 80–125. https://doi. org/10.1016/j.progpolymsci.2019.02.001.

47. Zhang, B.; Ni, J.; Xiang, X.; Wang, L.; Chen, Y. Synthesis and Properties of Reprocessable Sulfonated Polyimides Cross-Linked via Acid Stimulation for Use as

Proton Exchange Membranes. *J. Power Sources* **2017**, *337*, 110–117. https://doi. org/10.1016/j.jpowsour.2016.10.102.

48. Ganeshkumar, A.; Bera, D.; Mistri, E. A.; Banerjee, S. Triphenyl Amine Containing Sulfonated Aromatic Polyimide Proton Exchange Membranes. *Eur. Polym. J.* **2014**, *60*, 235–246. https://doi.org/10.1016/j.eurpolymj.2014.09.009.

49. You, P. Y.; Kamarudin, S. K.; Masdar, M. S. Improved Performance of Sulfonated Polyimide Composite Membranes with Rice Husk Ash as a Bio-Filler for Application in Direct Methanol Fuel Cells. *Int. J. Hydrog. Energy* **2019**, *44* (3), 1857–1866. https://doi. org/10.1016/j.ijhydene.2018.11.166.

50. Deligöz, H.; Vatansever, S.; Öksüzömer, F.; Koç, S. N.; Özgümüş, S.; Gürkaynak, M. A. Preparation and Characterization of Sulfonated Polyimide Ionomers via Post-Sulfonation Method for Fuel Cell Applications. *Polym. Adv. Technol.* **2008**, *19* (8), 1126–1132. https://doi.org/10.1002/pat.1096.

51. Deligöz, H.; Yılmazoğlu, M.; Yılmaztürk, S.; Şahin, Y.; Ulutaş, K. Synthesis and Characterization of Anhydrous Conducting Polyimide/Ionic Liquid Complex Membranes via a New Route for High-Temperature Fuel Cells. *Polym. Adv. Technol.* **2012**, *23* (8), 1156–1165. https://doi.org/10.1002/pat.2016.

52. Polyimide films | DuPont https://www.dupont.com/electronic-materials/polyimide-films.html (accessed Jul 31, 2019).

53. *Polyimide P84®NT.*

54. Polyimide Foam Flexible Thermal Insulation, Sorbermide|Pyrotek https://www. pyroteknc.com/products/sorber/sorbermide/ (accessed Jun 25, 2019).

55. Huebener, R. P. *Conductors, Semiconductors, Superconductors : An Introduction to Solid State Physics*; Springer International Publishing: Switzerland, 2015.

56. Hughes, E.; Hiley, J.; Brown, K.; Mckenzie Smith, I. *Hughes Electrical and Electronic Technology*; 2008.

57. Laughton, M. A.; Warne, D. J. *Handbook, Electrical Engineer's Reference Book*; 16th ed., 2003.

58. Zhang, Y.; Huang, W. *Soluble and Low-κ Polyimide Materials*; Elsevier Inc., 2018. https://doi.org/10.1016/b978-0-12-812640-0.00008-1.

59. Schoch, K. F. Plastics, Elastomers, and Composites. In *Electronic Materials and Processes Handbook*; Harper, C., Third Ed.; The McGraw-Hill Companies, Inc., 2004; pp 2.13. https://www.accessengineeringlibrary.com/content/book/9780071402149

60. Raju, G. *Dielectrics in Electric Fields*; 2010. https://doi.org/10.1201/9780203912270.

61. Lu, Q.-H.; Zheng, F. Polyimides for Electronic Applications. In *Advanced Polyimide Materials Synthesis, Characterization and Applications*; Yang, S.-Y., Ed.; Elsevier Inc., 2018; pp 195–255. https://doi.org/10.1016/b978-0-12-812640-0.00005-6.

62. Deligöz, H.; Özgümüş, S.; Yalçinyuva, T.; Yildirim, S.; Değer, D.; Ulutaş, K. A Novel Cross-Linked Polyimide Film: Synthesis and Dielectric Properties. *Polymer (Guildf).* **2005**, *46* (11), 3720–3729. https://doi.org/10.1016/j.polymer.2005.02.097.

63. Biçen, M.; Kayaman-Apohan, N.; Karataş, S.; Dumludaı, F.; Güngör, A. The Effect of Surface Modification of Zeolite 4A on the Physical and Electrical Properties of Copolyimide Hybrid Films. *Microporous Mesoporous Mater.* **2015**. https://doi. org/10.1016/j.micromeso.2015.06.034.

64. Deligöz, H.; Yalcinyuva, T.; Özgümüs, S.; Yildirim, S. Electrical Properties of Conventional Polyimide Films: Effects of Chemical Structure and Water Uptake. *J. Appl. Polym. Sci.* **2006**, *100* (1), 810–818. https://doi.org/10.1002/app.23174.

65. Leu, C. M.; Chang, Y. Te; Wei, K. H. Synthesis and Dielectric Properties of Polyimide-Tethered Polyhedral Oligomeric Silsesquioxane (POSS) Nanocomposites via Poss-Diamine. *Macromolecules* **2003**. https://doi.org/10.1021/ma034743r.

66. Zhao, X. Y.; Liu, H. J. Review of Polymer Materials with Low Dielectric Constant. *Polym. Int.* **2010,** 597–606. https://doi.org/10.1002/pi.2809.

67. Hougham, G.; Tesoro, G.; Viehbeck, A.; Chapple-Sokol, J. D. Polarization Effects of Fluorine on the Relative Permittivity in Polyimides. *Macromolecules* **1994,** *27* (21), 5964–5971. https://doi.org/10.1021/ma00099a006.

68. Simpson, J. O.; St.Clair, A. K. Fundamental Insight on Developing Low Dielectric Constant Polyimides. *Thin Solid Films,* **1997**. https://doi.org/10.1016/S0040-6090(97)00481-1.

69. Ronova, I. A.; Bruma, M.; Schmidt, H. W. Conformational Rigidity and Dielectric Properties of Polyimides. *Struct. Chem.* **2012,** *23* (1), 219–226. https://doi.org/10.1007/s11224-011-9865-1.

70. Yang, C. P.; Su, Y. Y. Synthesis and Properties of Organosoluble Polyimides Based on 4,4′-Bis(4-Amino-2-Trifluoromethylphenoxy)-Benzophenone. *J. Polym. Sci. Part A Polym. Chem.* **2004,** *42* (2), 222–236. https://doi.org/10.1002/pola.11012.

71. Kurinchyselvan, S.; Hariharan, A.; Prabunathan, P.; Gomathipriya, P.; Alagar, M. Fluorinated Polyimide Nanocomposites for Low K Dielectric Applications. *J. Polym. Res.* **2019,** *26* (9). https://doi.org/10.1007/s10965-019-1852-z.

72. Pu, L.; Huang, X.; Wang, W.; Dai, Y.; Yang, J.; Zhang, H. Strategy to Achieve Ultralow Dielectric Constant for Polyimide: Introduction of Fluorinated Blocks and Fluorographene Nanosheets by in Situ Polymerization. *J. Mater. Sci. Mater. Electron.* **2019,** *30* (15), 14679–14686. https://doi.org/10.1007/s10854-019-01839-3.

73. Simone, C. D.; Vaccaro, E.; Scola, D. A. The Synthesis and Characterization of Highly Fluorinated Aromatic Polyimides. *J. Fluor. Chem.* **2019,** *224*, 100–112. https://doi.org/10.1016/j.jfluchem.2019.05.001.

74. Zhou, Y. T.; Liu, X. L.; Wei, M. H.; Song, C.; Huang, Z. Z.; Sheng, S. R. New Fluorinated Copoly(Pyridine Ether Imide)s Derived from 4,4′-Oxydianiline, Pyromellitic Dianhydride and 4-(4-Trifluoromethylphenyl)-2,6-Bis[4-(4-Amino-2-Trifluoromethylphenoxy)Phenyl]Pyridine. *Polym. Bull.* **2019,** *76* (8), 4139–4155. https://doi.org/10.1007/s00289-018-2585-6.

75. Zhang, X. L.; Sheng, S. R.; Pan, Y.; Huang, Z. Z.; Liu, X. L. Organosoluble, Low Dielectric Constant and Highly Transparent Fluorinated Pyridine-Containing Poly(Ether Imide)s Derived from New Diamine: 4-(4-Trifluoromethyl)Phenyl-2,6-Bis[4-(4-Amino-2-Trifluoromethylphenoxy)Phenyl]Pyridine. *J. Macromol. Sci. Part A Pure Appl. Chem.* **2019,** *56* (3), 234–244. https://doi.org/10.1080/10601325.2019.1565545.

76. Li, Q.; Zhang, S.; Liao, G.; Yi, C.; Xu, Z. Novel Fluorinated Hyperbranched Polyimides with Excellent Thermal Stability, UV-Shielding Property, Organosolubility, and Low Dielectric Constants. *High Perform. Polym.* **2018,** *30* (7), 872–886. https://doi.org/10.1177/0954008317734034.

77. Huang, S. X.; Liu, X. L.; Wei, M. H.; Huang, Z. Z.; Sheng, S. R. Synthesis and Characterization of Novel Fluorinated Pyridine-Based Polyimides Derived from 4-(4-Trifluoromethylphenyl)-2,6-Bis(4-Aminophenyl)Pyridine and Dianhydrides. *High Perform. Polym.* **2018,** *30* (4), 500–508. https://doi.org/10.1177/0954008317705967.

78. Chen, Y. C.; Hsiao, S. H. Optically Transparent and Organosoluble Poly(Ether Imide)s Based on a Bis(Ether Anhydride) with Bulky 3,3′,5,5′-Tetramethylbiphenyl Moiety and

Various Fluorinated Bis(Ether Amine)S. *High Perform. Polym.* **2018**, *30* (1), 47–57. https://doi.org/10.1177/0954008316677154.

79. Deng, B.; Zhang, S.; Liu, C.; Li, W.; Zhang, X.; Wei, H.; Gong, C. Synthesis and Properties of Soluble Aromatic Polyimides from Novel 4,5-Diazafluorene-Containing Dianhydride. *RSC Adv.* **2018**, *8* (1), 194–205. https://doi.org/10.1039/c7ra12101f.

80. Liu, C.; Pei, X.; Mei, M.; Chou, G.; Huang, X.; Wei, C. Synthesis and Characterization of Organosoluble, Transparent, and Hydrophobic Fluorinated Polyimides Derived from 3,3'-Diisopropyl-4,4'-Diaminodiphenyl-4''-Trifluoromethyltoluene. *High Perform. Polym.* **2016**, *28* (10), 1114–1123. https://doi.org/10.1177/0954008315617230.

81. Yi, L.; Li, C.; Huang, W.; Yan, D. Soluble and Transparent Polyimides with High Tg from a New Diamine Containing Tert-Butyl and Fluorene Units. *J. Polym. Sci. Part A Polym. Chem.* **2016**, *54* (7), 976–984. https://doi.org/10.1002/pola.27933.

82. Dong, W.; Guan, Y.; Shang, D. Novel Soluble Polyimides Containing Pyridine and Fluorinated Units: Preparation, Characterization, and Optical and Dielectric Properties. *RSC Adv.* **2016**, *6* (26), 21662–21671. https://doi.org/10.1039/c6ra00322b.

83. Constantin, C. P.; Damaceanu, M. D.; Varganici, C.; Wolinska-Grabczyk, A.; Bruma, M. Dielectric and Gas Transport Properties of Highly Fluorinated Polyimides Blends. *High Perform. Polym.* **2015**, *27* (5), 526–538. https://doi.org/10.1177/0954008315584181.

84. Bu, Q.; Zhang, S.; Li, H.; Li, Y.; Gong, C.; Yang, F. Preparation and Properties of Thermally Stable Polyimides Derived from Asymmetric Trifluoromethylated Aromatic Diamines and Various Dianhydrides. *Polym. Degrad. Stab.* **2011**, *96* (10), 1911–1918. https://doi.org/10.1016/j.polymdegradstab.2011.07.003.

85. Shundrina, I. K.; Vaganova, T. A.; Kusov, S. Z.; Rodionov, V. I.; Karpova, E. V.; Malykhin, E. V. Synthesis and Properties of Organosoluble Polyimides Based on Novel Perfluorinated Monomer Hexafluoro-2,4-Toluenediamine. *J. Fluor. Chem.* **2011**, *132* (3), 207–215. https://doi.org/10.1016/j.jfluchem.2011.01.008.

86. Zhang, P.; Zhao, J.; Zhang, K.; Bai, R.; Wang, Y.; Hua, C.; Wu, Y.; Liu, X.; Xu, H.; Li, Y. Fluorographene/Polyimide Composite Films: Mechanical, Electrical, Hydrophobic, Thermal and Low Dielectric Properties. *Compos. Part A Appl. Sci. Manuf.* **2016**, *84*, 428–434. https://doi.org/10.1016/j.compositesa.2016.02.019.

87. Wang, X.; Dai, Y.; Wang, W.; Ren, M.; Li, B.; Fan, C.; Liu, X. Fluorographene with High Fluorine/Carbon Ratio: A Nanofiller for Preparing Low-κ Polyimide Hybrid Films. *ACS Appl. Mater. Interf.* **2014**, *6* (18), 16182–16188. https://doi.org/10.1021/am5042516.

88. Deligöz, H.; Yalcinyuva, T.; Özgümüs, S. A Novel Type of Si-Containing Poly(Urethane-Imide)s: Synthesis, Characterization and Electrical Properties. *Eur. Polym. J.* **2005**, *41* (4), 771–781. https://doi.org/10.1016/j.eurpolymj.2004.11.007.

89. Jiang, L.; Liu, J.; Wu, D.; Li, H.; Jin, R. A Methodology for the Preparation of Nanoporous Polyimide Films with Low Dielectric Constants. *Thin Solid Films* **2006**. https://doi.org/10.1016/j.tsf.2005.12.216.

90. Othman, M. B. H.; Ramli, M. R.; Tyng, L. Y.; Ahmad, Z.; Akil, H. M. Dielectric Constant and Refractive Index of Poly (Siloxane-Imide) Block Copolymer. *Mater. Des.* **2011**, *32* (6), 3173–3182. https://doi.org/10.1016/j.matdes.2011.02.048.

91. Ishizaka, T.; Kasai, H. Fabrication of Polyimide Porous Nanostructures for Low-k Materials. In *High Performance Polymers–Polyimides Based–From Chemistry to Applications*; 2012. https://doi.org/10.5772/53458.

92. Lee, Y. J.; Huang, J. M.; Kuo, S. W.; Chang, F. C. Low-Dielectric, Nanoporous Polyimide Films Prepared from PEO-POSS Nanoparticles. *Polymer (Guildf).* **2005**, *46* (23), 10056–10065. https://doi.org/10.1016/j.polymer.2005.08.047.

93. Krishnan, P. S. G.; Cheng, C. Z.; Cheng, Y. S.; Cheng, J. W. C. Preparation of Nanoporous Polyimide Films from Poly(Urethane-Imide) by Thermal Treatment. *Macromol. Mater. Eng.* **2003,** *288* (9), 730–736. https://doi.org/10.1002/mame.200300030.

94. Wang, C.; Wang, T. M.; Wang, Q. H. Low-Dielectric, Nanoporous Polyimide Thin Films Prepared from Block Copolymer Templating. *Express Polym. Lett.* **2013,** *7* (8), 667–672. https://doi.org/10.3144/expresspolymlett.2013.63.

95. Lv, F.; Liu, L.; Zhang, Y.; Li, P. Effect of Polymer Structure on the Morphologies and Dielectric Properties of Nanoporous Polyimide Films. *J. Appl. Polym. Sci.* **2015,** *132* (7), 1–7. https://doi.org/10.1002/app.41480.

96. Guo, H.; Ann, M.; Meador, B.; Mccorkle, L.; Quade, D. J.; Guo, J.; Hamilton, B.; Cakmak, M.; Sprowl, G. Polyimide Aerogels Cross-Linked through Amine Functionalized Polyoligomeric Silsesquioxane. *ACS Appl. Mater. Interf.* **2011,** *3*, 546–552. https://doi.org/10.1021/am101123h.

97. Meador, M. A. B.; McMillon, E.; Sandberg, A.; Barrios, E.; Wilmoth, N. G.; Mueller, C. H.; Miranda, F. A. Dielectric and Other Properties of Polyimide Aerogels Containing Fluorinated Blocks. *ACS Appl. Mater. Interf.* **2014,** *6* (9), 6062–6068. https://doi.org/10.1021/am405106h.

98. Wu, T.; Dong, J.; Gan, F.; Fang, Y.; Zhao, X.; Zhang, Q. Low Dielectric Constant and Moisture-Resistant Polyimide Aerogels Containing Trifluoromethyl Pendent Groups. *Appl. Surf. Sci.* **2018,** *440*, 595–605. https://doi.org/10.1016/j.apsusc.2018.01.132.

99. Lee, Y.-J.; Huang, J.-M.; Kuo, S.-W.; Chang, F.-C. Low-Dielectric, Nanoporous Polyimide Films Prepared from PEO-POSS Nanoparticles. *Polymer* **2005.** https://doi.org/10.1016/j.polymer.2005.08.047.

100. Wang, C.; Wang, T.; Wang, Q. Controllable Porous Fluorinated Polyimide Thin Films for Ultralow Dielectric Constant Interlayer Dielectric Applications. *J. Macromol. Sci. Part A Pure Appl. Chem.* **2017,** *54* (5), 311–315. https://doi.org/10.1080/10601325.2017.1294453.

101. Song, N.; Shi, K.; Yu, H.; Yao, H.; Ma, T.; Zhu, S.; Zhang, Y.; Guan, S. Decreasing the Dielectric Constant and Water Uptake of Co-Polyimide Films by Introducing Hydrophobic Cross-Linked Networks. *Eur. Polym. J.*, **2018.** https://doi.org/10.1016/j.eurpolymj.2018.02.024.

102. Deligöz, H.; Yalcinyuva, T.; Özgümüs, S.; Yildirim, S. Preparation, Characterization and Dielectric Properties of 4,4-Diphenylmethane Diisocyanate MDI) Based Cross-Linked Polyimide Films. *Eur. Polym. J.* **2006,** *42* (6), 1370–1377. https://doi.org/10.1016/j.eurpolymj.2005.12.005.

103. Yao, H.; Zhang, Y.; You, K.; Liu, Y.; Song, Y.; Liu, S.; Guan, S. Synthesis and Properties of Soluble Cross-Linkable Fluorinated Co-Polyimides. *React. Funct. Polym.*, **2014.** https://doi.org/10.1016/j.reactfunctpolym.2014.05.011.

104. Song, N.; Yao, H.; Ma, T.; Wang, T.; Shi, K.; Tian, Y.; Zhang, B.; Zhu, S.; Zhang, Y.; Guan, S. Decreasing the Dielectric Constant and Water Uptake by Introducing Hydrophobic Cross-Linked Networks into Co-Polyimide Films. *Appl. Surf. Sci.*, **2019.** https://doi.org/10.1016/j.apsusc.2019.02.141.

105. Dupont™ Kapton® Summary of Properties.

106. Cosutchi, A. I.; Hulubei, C.; Buda, M.; Botila, T.; Ioan, S. Effects of Chemical Structure on the Electrical Properties of Some Polymers with Imidic Structure. *e-Polymers* **2007,** *7* (1). https://doi.org/10.1515/epoly.2007.7.1.793.

107. Cosutchi, A. I.; Hulubei, C.; Ioan, S. Optical and Dielectric Properties of Some Polymers with Imidic Structure. **2007,** *9* (4), 975–980.

108. Hwang, H. J.; Li, C. H.; Wang, C. S. Synthesis and Properties of Bismaleimide Resin Containing Dicyclopentadiene or Dipentene. VI. *Polym. Int.* **2006,** *55* (11), 1341–1349. https://doi.org/10.1002/pi.2092.

109. Lorenzini, R. G.; Sotzing, G. A. Furan/Imide Diels-Alder Polymers as Dielectric Materials. *J. Appl. Polym. Sci.* **2014,** *131* (24). https://doi.org/10.1002/app.40179.

110. Fang, L.; Zhou, J.; Wang, J.; Sun, J.; Fang, Q. A Bio-Based Allylphenol (Eugenol)-Functionalized Fluorinated Maleimide with Low Dielectric Constant and Low Water Uptake. *Macromol. Chem. Phys.* **2018,** *219* (20). https://doi.org/10.1002/macp.201800252.

111. Guo, Y.; Han, Y.; Liu, F.; Zhou, H.; Chen, F.; Zhao, T. Fluorinated Bismaleimide Resin with Good Processability, High Toughness, and Outstanding Dielectric Properties. *J. Appl. Polym Sci.* **2015.** https://doi.org/10.1002/app.42791.

112. Chen, X.; Wang, J.; Huo, S.; Yang, S.; Zhang, B.; Cai, H. Preparation of Flame-Retardant Cyanate Ester Resin Combined with Phosphorus-Containing Maleimide: Flame-Retardant Property Low Dielectric Constant Loss. *J. Therm. Anal. Calorim.* **2018,** *132* (3), 1617–1628. https://doi.org/10.1007/s10973-018-6979-3.

113. Guo, Y.; Chen, F.; Han, Y.; Li, Z.; Liu, X.; Zhou, H.; Zhao, T. High Performance Fluorinated Bismaleimide-Triazine Resin with Excellent Dielectric Properties. *J. Polym. Res.* **2018,** *25* (2). https://doi.org/10.1007/s10965-017-1407-0.

114. Guan, Q.; Gu, A.; Liang, G.; Zhou, C.; Yuan, L. Preparation and Properties of New High Performance Maleimide-Triazine Resins for Resin Transfer Molding. *Polym. Adv. Technol.* **2011,** *22* (12), 1572–1580. https://doi.org/10.1002/pat.1643.

115. Liu, P.; Guan, Q.; Gu, A.; Liang, G.; Yuan, L.; Chang, J. Interface and Its Effect on the Interlaminate Shear Strength of Novel Glass Fiber/Hyperbranched Polysiloxane Modified Maleimide-Triazine Resin Composites. *Appl. Surf. Sci.* **2011,** *258* (1), 572–579. https://doi.org/10.1016/j.apsusc.2011.08.066.

116. Hu, J. T.; Gu, A.; Liang, G.; Zhuo, D.; Yuan, L. Preparation and Properties of High-Performance Polysilsesquioxanes/ Bismaleimide-Triazine Hybrids. *J. Appl. Polym. Sci.* **2011,** *120* (1), 360–367. https://doi.org/10.1002/app.33144.

117. Cao, H.; Xu, R.; Yu, D. Thermal and Dielectric Properties of Bismaleimide-Triazine Resins Containing Octa(Maleimidophenyl)Silsesquioxane. *J. Appl. Polym. Sci.* **2008,** *109* (5), 3114–3121. https://doi.org/10.1002/app.27822.

118. Hu, J. T.; Gu, A.; Liang, G.; Zhuo, D.; Yuan, L. Preparation and Properties of Maleimide-Functionalized Hyperbranched Polysiloxane and Its Hybrids Based on Cyanate Ester Resin. *J. Appl. Polym. Sci.* **2012,** *126* (1), 205–215. https://doi.org/10.1002/app.36688.

119. Ursache, O.; Gaina, C.; Gaina, V.; Musteata, V. E. High Performance Bismaleimide Resins Modified by Novel Allyl Compounds Based on Polytriazoles. *J. Polym. Res.* **2012,** *19* (10). https://doi.org/10.1007/s10965-012-9968-4.

120. Chen, X.; Li, K.; Zheng, S.; Fang, Q. A New Type of Unsaturated Polyester Resin with Low Dielectric Constant and High Thermostability: Preparation and Properties. *RSC Adv.* **2012,** *2* (16), 6504–6508. https://doi.org/10.1039/c2ra20617j.

121. Park, M. H.; Jang, W.; Yang, S. J.; Shul, Y.; Han, H. Synthesis and Characterization of New Functional Poly(Urethane-Imide) Crosslinked Networks. *J. Appl. Polym. Sci.* **2006,** *100* (1), 113–123. https://doi.org/10.1002/app.22678.

122. Qu, W.; Ko, T.-M.; Vora, R. H.; Chung, T.-S. Effect of Polyimides with Different Ratios of Para–to Meta–Analogous Fluorinated Diamines on Relaxation Process. *Polymer (Guildf).* **2001,** *42* (15), 6393–6401. https://doi.org/10.1016/S0032-3861(01)00111-2.

123. Khazaka, R.; Locatelli, M. L.; Diaham, S.; Bidan, P.; Dupuy, L.; Grosset, G. Broadband Dielectric Spectroscopy of BPDA/ODA Polyimide Films. *J. Phys. D. Appl. Phys.* **2013,** *46* (6). https://doi.org/10.1088/0022-3727/46/6/065501.

124. Tsuwi, J.; Appelhans, D.; Zschoche, S.; Zhuang, R. C.; Friedel, P.; Häußler, L.; Voit, B.; Kremer, F. Molecular Dynamics in Fluorinated Side-Chain Maleimide Copolymers as Studied by Broadband Dielectric Spectroscopy. *Colloid Polym. Sci.* **2005,** *283* (12), 1321–1333. https://doi.org/10.1007/s00396-005-1346-x.

125. Tsuwi, J.; Appelhans, D.; Zschoche, S.; Friedel, P.; Kremer, F. Molecular Dynamics in Poly(Ethene-Alt-N-Alkylmaleimide)s as Studied by Broadband Dielectric Spectroscopy. *Macromolecules* **2004,** *37* (16), 6050–6054. https://doi.org/10.1021/ma049370o.

126. Mittal, K. L. *Polyimides and Other High Temperature Polymers: Synthesis, Characterization and Applications,* vol. 3; CRC Press, Boca Raton, 2005.

127. Liu, K.; Zou, F.; Sun, Y.; Yu, Z.; Liu, X.; Zhou, L.; Xia, Y.; Vogt, B. D.; Zhu, Y. Self-Assembled Mn 3 O 4 /C Nanospheres as High-Performance Anode Materials for Lithium Ion Batteries. *J. Power Sources* **2018,** *395* (May), 92–97. https://doi.org/10.1016/j.jpowsour.2018.05.064.

128. Zhou, Y.; Wang, H.; Wang, L.; Yu, K.; Lin, Z.; He, L.; Bai, Y. Fabrication and Characterization of Aluminum Nitride Polymer Matrix Composites with High Thermal Conductivity and Low Dielectric Constant for Electronic Packaging. *Mater. Sci. Eng. B Solid-State Mater. Adv. Technol.,* **2012.** https://doi.org/10.1016/j.mseb.2012.03.056.

129. Hedrick, J. L.; Carter, K. R.; Cha, H. J.; Hawker, C. J.; DiPietro, R. A.; Labadie, J. W.; Miller, R. D.; Russell, T. P.; Sanchez, M. I.; Volksen, W.; et al. High-Temperature Polyimide Nanofoams for Microelectronic Applications. *React. Funct. Polym.* **1996,** *30* (1–3), 43–53. https://doi.org/10.1016/1381-5148(96)00020-X.

130. Jiang, L. Y.; Leu, C. M.; Wei, K. H. Layered Silicates/Fluorinated Polyimide Nanocomposites for Advanced Dielectric Materials Applications. *Adv. Mater.,* **2002.** https://doi.org/10.1002/1521-4095(20020318)14:6<426::AID-ADMA426>3.0.CO;2-O.

131. Kuntman, A.; Kuntman, H. Study on Dielectric Properties of a New Polyimide Film Suitable for Interlayer Dielectric Material in Microelectronics Applications. *Microelectron. J.,* **2000.** https://doi.org/10.1016/S0026-2692(00)00067-7.

132. Ji, D.; Jiang, L.; Cai, X.; Dong, H.; Meng, Q.; Tian, G.; Wu, D.; Li, J.; Hu, W. Large Scale, Flexible Organic Transistor Arrays and Circuits Based on Polyimide Materials. *Org. Electron. Phys. Mater. Appl.* **2013,** *14* (10), 2528–2533. https://doi.org/10.1016/j.orgel.2013.06.028.

133. Chou, Y. H.; You, N. H.; Kurosawa, T.; Lee, W. Y.; Higashihara, T.; Ueda, M.; Chen, W. C. Thiophene and Selenophene Donor-Acceptor Polyimides as Polymer Electrets for Nonvolatile Transistor Memory Devices. *Macromolecules* **2012,** *45* (17), 6946–6956. https://doi.org/10.1021/ma301326r.

134. Pyo, S.; Lee, M.; Jeon, J.; Lee, J. H.; Yi, M. H.; Kim, J. S. An Organic Thin-Film Transistor with a Photoinitiator-Free Photosensitive Polyimide as Gate Insulator. *Adv. Funct. Mater.,* **2005.** https://doi.org/10.1002/adfm.200400206.

135. Pyo, S.; Son, H.; Choi, K. Y.; Yi, M. H.; Hong, S. K. Low-Temperature Processable Inherently Photosensitive Polyimide as a Gate Insulator for Organic Thin-Film Transistors. *Appl. Phys. Lett.,* **2005.** https://doi.org/10.1063/1.1894587.

136. Alva, G.; Lin, Y.; Fang, G. Thermal and Electrical Characterization of Polymer/Ceramic Composites with Polyvinyl Butyral Matrix. *Mater. Chem. Phys.* **2018**, *205*, 401–415. https://doi.org/10.1016/j.matchemphys.2017.11.046.

137. Wang, X.; Fang, X.; Liu, X.; Pei, Q.; Cui, Z. K.; Deng, S.; Gu, J.; Zhuang, Q. Formation of Unique Three-Dimensional Interpenetrating Network Structure with a Ternary Composite. *J. Mater. Sci. Mater. Electron.* **2018**, *29* (21), 18699–18707. https://doi.org/10.1007/s10854-018-9993-0.

138. Yuen, S.; Ma, C. M.; Chiang, C.-L. Silane Grafted MWCNT/Polyimide Composites-Preparation, Morphological and Electrical Properties. *Compos. Sci. Technol.* **2008**, *68* (14), 2842–2848. https://doi.org/10.1016/j.compscitech.2007.10.011.

139. Nguyen, T. H. L.; Cortes, L. Q.; Lonjon, A.; Dantras, E.; Lacabanne, C. High Conductive Ag Nanowire-Polyimide Composites: Charge Transport Mechanism in Thermoplastic Thermostable Materials. *J. Non. Cryst. Solids* **2014**, *385*, 34–39. https://doi.org/10.1016/j.jnoncrysol.2013.11.008.

140. Bobinger, M. R.; Romero, F. J.; Salinas-castillo, A.; Becherer, M.; Lugli, P.; Morales, D. P.; Rodríguez, N.; Rivadeneyra, A. Flexible and Robust Laser-Induced Graphene Heaters Photothermally Scribed on Bare Polyimide Substrates. *Carbon N. Y.* **2019**, *144*, 116–126. https://doi.org/10.1016/j.carbon.2018.12.010.

141. Constantin, C. P.; Damaceanu, M. D.; Bruma, M.; Begunov, R. S. Ortho-CATENATION and Trifluoromethyl Graphting as Driving Forces in Electro-Optical Properties Modulation of Ethanol Soluble Triphenylamine-Based Polyimides. *Dye. Pigment.* **2019**, *163*, 126–137. https://doi.org/10.1016/j.dyepig.2018.11.046.

142. Somboonsub, B.; Thongyai, S.; Scola, D. A.; Sotzing, G. A.; Praserthdam, P. Sulfonated Polyimide as a Thermally Stable Template for Water Processable Conductive Polymers. *Synth. Met.* **2012**, *162* (11–12), 941–947. https://doi.org/10.1016/j.synthmet.2012.03.023.

143. Gao, C.; Zhang, S.; Lin, Y.; Li, F.; Guan, S.; Jiang, Z. High-Performance Conductive Materials Based on the Selective Location of Carbon Black in Poly(Ether Ether Ketone)/Polyimide Matrix. *Compos. Part B Eng.* **2015**, *79*, 124–131. https://doi.org/10.1016/j.compositesb.2015.03.047.

144. Molla-Abbasi, P.; Shabanian, M. A Bulky Aromatic Functional Polyimide Composite as a Sensitive Layer for the Detection of Organic Compound Biomarkers. *Iran. Polym. J.* **2019**, *28*, 203–211.

145. Deligöz, H.; Yilmazoğlu, M.; Yilmaztürk, S.; Şahin, Y.; Ulutaş, K. Synthesis and Characterization of Anhydrous Conducting Polyimide/Ionic Liquid Complex Membranes via a New Route for High-Temperature Fuel Cells. *Polym. Adv. Technol.* **2012**, *23* (8), 1156–1165. https://doi.org/10.1002/pat.2016.

146. Deligöz, H.; Yilmazolu, M. Development of a New Highly Conductive and Thermomechanically Stable Complex Membrane Based on Sulfonated Polyimide/Ionic Liquid for High Temperature Anhydrous Fuel Cells. *J. Power Sources* **2011**, *196* (7), 3496–3502. https://doi.org/10.1016/j.jpowsour.2010.12.033.

147. Blythe, A.; Bloor, D. *Electrical Properties of Polymers*, 2nd ed.; Cambridge University Press, Cambridge, UK, 2005.

148. Qian, Y.; Lan, Y.; Xu, J.; Ye, F.; Dai, S. Fabrication of Polyimide-Based Nanocomposites Containing Functionalized Graphene Oxide Nanosheets by in-Situ Polymerization and Their Properties. *Appl. Surf. Sci.* **2014**, *314*, 991–999. https://doi.org/10.1016/j.apsusc.2014.06.130.

149. Zhang, Q.; Xu, Y.; Yang, Y.; Li, L.; Song, C.; Su, X. Conductive Mechanism of Carbon Black/Polyimide Composite Films. *J. Polym. Eng.* **2017,** *38* (2), 147–156. https://doi.org/10.1515/polyeng-2016-0273.

150. Aragoneses, A.; Mudarra, M.; Belana, J.; Diego, J. A. Study of Dispersive Mobility in Polyimide by Surface Voltage Decay Measurements. *Polymer (Guildf).* **2008,** *49,* 2440–2443. https://doi.org/10.1109/ICSD.2007.4290797.

151. Gul, A.; Tabassam, L.; Bhatti, A. S. Electrical Characterization of Metal Junction Formed with Pure and Polyaniline Blended Poly(Schiff Base) Polymer. *Surf. Rev. Lett.* **2018,** *26* (10), 1950072. https://doi.org/10.1142/s0218625x19500720.

152. Taherian, R. The Theory of Electrical Conductivity. In *Electrical Conductivity in Polymer-Based Composites*; Kausar, A., Taherian, R., Eds.; William Andrew Publishing, 2019; pp 1–18. https://doi.org/10.1016/B978-0-12-812541-0.00001-X.

153. Xia, X.; Yin, J.; Chen, M.; Liu, X. The Role of Interfaces in PI Matrix Nano Composite Films on Electrical Properties. *Polym. Test.* **2018,** *69* (March), 405–409. https://doi.org/10.1016/j.polymertesting.2018.05.026.

154. Kim, T. Y.; Kim, W. J.; Lee, T. H.; Kim, J. E.; Suh, K. S. Electrical Conduction of Polyimide Films Prepared from Polyamic Acid (PAA) and Pre-Imidized Polyimide (PI) Solution. *Express Polym. Lett.* **2007,** *1* (7), 427–432. https://doi.org/10.3144/expresspolymlett.2007.60.

155. Sessler, G. M.; Hahn, B.; Yoon, D. Y. Electrical Conduction in Polyimide Films. *J. Appl. Phys.* **1986,** *60* (1), 318–326. https://doi.org/10.1063/1.337646.

156. Nevin, J. H.; Summe, G. L. DC Conduction Mechanism in Thin Polyimide Films. *Microelectron Reliab.* **1981,** *21* (5), 699–705.

157. Hsiao, Y.-P.; Yang, W.-L.; Lin, L.-M.; Chin, F.-T.; Lin, Y.-H.; Yang, K.-L.; Wu, C.-C. Improving Retention Properties by Thermal Imidization for Polyimide-Based Nonvolatile Resistive Random Access Memories. *Microelectron. Reliab.* **2015,** *55,* 2188–2197. https://doi.org/10.1016/j.microrel.2015.08.013.

158. Nayak, L.; Chaki, T. K.; Khastgir, D. Electrical Percolation Behavior and Electromagnetic Shielding Effectiveness of Polyimide Nanocomposites Filled with Carbon Nanofibers. *J. Appl. Polym. Sci.* **2014,** *131* (24), 40914 (1–12). https://doi.org/10.1002/app.41233.

159. Sharma, B. L.; Pillai, P. K. C. Electrical Conduction in Kapton Polyimide Film at High Electrical Fields. *Polym. Commun.* **1982,** *23,* 17–20.

160. Chohan, M. H.; Mahmood, H.; Shah, F. Electrical Conduction Phenomena in Polyimide Films. *Mod. Phys. Lett. B* **1994,** *8* (25), 1591–1595.

161. Shaikh, R.; Ul Haq, S.; Raju, G. G. Electrical Conduction Processes in Polyimide-Teflon FEF Films-II. *IEEE Trans. Dielectr. Electr. Insul.* **2008,** *15* (3), 671–677. https://doi.org/10.1109/TDEI.2008.4543103.

162. Owen, J. Ionic Conductivity. In *Comprehensive Polymer Science and Supplements*; Allen, G., Bevington, J. C., Eds.; Elsevier, 1996; pp 669–686.

163. Tuller, H. Ionic Conduction and Application. In *Handbook of Electronic and Photonic Materials*; Kasap, S., Capper, P., Eds.; Springer, 2017; pp 247–266.

164. Kumar, P. P.; Yashonath, S. Ionic Conduction in the Solid State. *J. Chem. Sci.* **2006,** *118* (1), 135–154. https://doi.org/10.1007/BF02708775.

165. Bronowski, J. D.C. Conductivity. In *Structural Chemistry of Glasses*; K.J. Rao, Ed.; Elsevier Science Ltd, 2002; pp 203–261. https://doi.org/10.1016/B978-008043958-7/50024-8.

166. Zhou, L.; Zhu, J.; Lin, M.; Xu, J.; Xie, Z.; Chen, D. Tetra-Alkylsulfonate Functionalized Poly (Aryl Ether) Membranes with Nanosized Hydrophilic Channels for Efficient Proton Conduction. *J. Energy Chem.* **2020**, *40*, 57–64. https://doi.org/10.1016/j.jechem.2019.02.013.

167. Shi, C.; Dai, J.; Shen, X.; Peng, L.; Li, C.; Wang, X.; Zhang, P.; Zhao, J. A High-Temperature Stable Ceramic-Coated Separator Prepared with Polyimide Binder/Al2O3 Particles for Lithium-Ion Batteries. *J. Memb. Sci.* **2016**, *517*, 91–99. https://doi.org/10.1016/j.memsci.2016.06.035.

168. Ye, W.; Zhu, J.; Liao, X.; Jiang, S.; Li, Y.; Fang, H.; Hou, H. Hierarchical Three-Dimensional Micro/Nano-Architecture of Polyaniline Nanowires Wrapped-on Polyimide Nanofibers for High Performance Lithium-Ion Battery Separators. *J. Power Sources* **2015**, *299*, 417–424. https://doi.org/10.1016/j.jpowsour.2015.09.037.

169. Zhang, H.; Zhang, Y.; Yao, Z.; John, A. E.; Li, Y.; Li, W.; Zhu, B. Novel Configuration of Polyimide Matrix-Enhanced Cross-Linked Gel Separator for High Performance Lithium Ion Batteries. *Electrochim. Acta* **2016**, *204*, 176–182. https://doi.org/10.1016/j.electacta.2016.03.189.

170. Deligöz, H.; Yilmaztürk, S.; Karaca, T.; Özdemir, H.; Koç, S. N.; Öksüzömer, F.; Durmuş, A.; Gürkaynak, M. A. Self-Assembled Polyelectrolyte Multilayered Films on Nafion with Lowered Methanol Cross-over for DMFC Applications. *J. Memb. Sci.* **2009**, *326* (2), 643–649. https://doi.org/10.1016/j.memsci.2008.10.055.

171. Ito, Y.; Hikita, M.; Kimura, T.; Mizutani, T. Effect of Electrode Metals on Electrical Conduction in Polyimide Thin Films Prepared by Vapor Deposition Polymerization. *Jpn. J. Appl. Phys.* **1990**, *29* (6), 1128–1131. https://doi.org/10.1109/iseim.1995.496516.

172. Suh, K. S.; Nam, J. H.; Lim, K. J. Electrical Conduction in Polyetherimide. *J. Appl. Phys.* **1996**, *80* (11), 6333–6335.

173. Stearn, A. E.; Eyring, H. Absolute Rates of Solid Reactions: Diffusion. *J. Phys. Chem.* **1940**, *44*, 955–980. https://doi.org/10.1021/j150404a001.

174. Tu, N. R.; Kao, K. C. High-Field Electrical Conduction in Polyimide Films. *J. Appl. Phys.* **1999**, *85* (10), 7267–7275. https://doi.org/10.1063/1.337646.

175. Zhou, Y.; Wu, S.; Liu, F. High-Performance Polyimide Nanocomposites with Polydopamine-Coated Copper Nanoparticles and Nanowires for Electronic Applications. *Mater. Lett.* **2019**, *237*, 19–21. https://doi.org/10.1016/j.matlet.2018.11.067.

176. Mu, S.; Wu, Z.; Qi, S.; Wu, D.; Yang, W. Preparation of Electrically Conductive Polyimide/Silver Composite Fibers via in-Situ Surface Treatment. *Mater. Lett.* **2010**, *64*, 1668–1671. https://doi.org/10.1016/j.matlet.2010.05.005.

177. Ma, L.; Niu, H.; Cai, J.; Zhao, P.; Wang, C.; Bai, X.; Lian, Y.; Wang, W. Photoelectrochemical and Electrochromic Properties of Polyimide/Graphene Oxide Composites. *Carbon N. Y.* **2014**, *67*, 488–499. https://doi.org/10.1016/j.carbon.2013.10.021.

178. Yang, L.; Liu, F.; Xia, H.; Qian, X.; Shen, K.; Zhang, J. Improving the Electrical Conductivity of a Carbon Nanotube/Polypropylene Composite by Vibration during Injection-Moulding. *Carbon N. Y.* **2011**, *49* (10), 3274–3283. https://doi.org/10.1016/j.carbon.2011.03.054.

179. Wang, N.; Chen, Y.; Ren, J.; Huang, X.; Chen, X.; Li, G.; Liu, D. Electrically Conductive Polyaniline/Polyimide Microfiber Membrane Prepared via a Combination of Solution Blowing and Subsequent in Situ Polymerization Growth. *J. Polym. Res.* **2017**, *24* (3). https://doi.org/10.1007/s10965-017-1198-3.

180. Romyen, N.; Thongyai, S.; Praserthdam, P.; Sotzing, G. A. Modification of Novel Conductive PEDOT:Sulfonated Polyimide Nano-Thin Films by Anionic Surfactant and Poly(Vinyl Alcohol) for Electronic Applications. *J. Electron. Mater.* **2013,** *42* (12), 3471–3480. https://doi.org/10.1007/s11664-013-2816-4.

181. Sukchol, K.; Thongyai, S.; Praserthdam, P.; Sotzing, G. A. Experimental Observation on the Mixing Systems and Ways to Significantly Enhance the Conductivity of PEDOT-Sulfonated Poly(Imide) Aqueous Dispersion. *Microelectron. Eng.* **2013,** *111,* 7–13. https://doi.org/10.1016/j.mee.2013.04.026.

182. Lyu, H.; Liu, J.; Mahurin, S.; Dai, S.; Guo, Z.; Sun, X. G. Polythiophene Coated Aromatic Polyimide Enabled Ultrafast and Sustainable Lithium Ion Batteries. *J. Mater. Chem. A* **2017,** *5* (46), 24083–24090. https://doi.org/10.1039/c7ta07893e.

183. Li, Y.; Zhang, H.; Liu, Y.; Wang, H.; Huang, Z.; Peijs, T.; Bilotti, E. Synergistic Effects of Spray-Coated Hybrid Carbon Nanoparticles for Enhanced Electrical and Thermal Surface Conductivity of CFRP Laminates. *Compos. Part A Appl. Sci. Manuf.* **2018,** *105,* 9–18. https://doi.org/10.1016/j.compositesa.2017.10.032.

184. Ding, X.; Wang, W.; Wang, Y.; Xu, R.; Yu, D. High-Performance Flexible Electromagnetic Shielding Polyimide Fabric Prepared by Nickel-Tungsten-Phosphorus Electroless Plating. *J. Alloys Compd.* **2019,** *777,* 1265–1273. https://doi.org/10.1016/j.jallcom.2018.11.120.

185. Lv, P.; Zhao, Y.; Liu, F.; Li, G.; Dai, X.; Ji, X.; Dong, Z.; Qiu, X. Fabrication of Polyaniline/Polyimide Composite Fibers with Electrically Conductive Properties. *Appl. Surf. Sci.* **2016,** *367,* 335–341. https://doi.org/10.1016/j.apsusc.2016.01.181.

186. Somboonsub, B.; Invernale, M. A.; Thongyai, S.; Praserthdam, P.; Scola, D. A.; Sotzing, G. A. Preparation of the Thermally Stable Conducting Polymer PEDOT—Sulfonated Poly(Imide). *Polymer (Guildf).* **2010,** *51* (6), 1231–1236. https://doi.org/10.1016/J.POLYMER.2010.01.048.

187. Zhao, H.; Yang, C.; Li, N.; Yin, J.; Feng, Y.; Liu, Y.; Li, J.; Lia, Y.; Yue, D.; Zhu, C.; et al. Electrical and Mechanical Properties of Polyimide Composite Fi Lms Reinforced by Ultralong Titanate Nanotubes. *Surf. Coat. Technol.* **2019,** *360,* 13–19. https://doi.org/10.1016/j.surfcoat.2019.01.013.

188. Zhan, J.; Wu, D.; Chen, C.; Li, J.; Pan, F.; Chen, J.; Zhan, X. Fabrication of Polyimide Composite Film with Both Magnetic and Surface Conductive Properties. *Desalin. Water Treat.* **2011,** *34,* 344–348. https://doi.org/10.5004/dwt.2011.2899.

189. Zhan, J.; Wu, Z.; Qi, S.; Wu, D.; Yang, W. Fabrication of Surface Silvered Polyimide/Iron Oxide Composite Films with Both Superparamagnetism and Electrical Conductivity. *Thin Solid Films* **2011,** *519* (6), 1960–1965. https://doi.org/10.1016/j.tsf.2010.10.034.

190. Li, T.; Hsu, S. L. Preparation and Properties of Conductive Polyimide Nanocomposites with Assistance of a Metallo-Organic Compound. *J. Mater. Chem.* **2010,** *20,* 1964–1969. https://doi.org/10.1039/b916101e.

191. Lin, C.-Y.; Kuo, D.-H.; Chen, W.-C.; Ma, M.-W.; Liou, G.-S. Electrical Performance of the Embedded-Type Surface Electrodes Containing Carbon and Silver Nanowires as Fillers and One-Step Organosoluble Polyimide as a Matrix. *Org. Electron.* **2012,** *13,* 2469–2473. https://doi.org/10.1016/j.orgel.2012.06.047.

192. Kwon, H.-J.; Cha, J.-R.; Gong, M.-S. Preparation of Silvered Polyimide Film from Silver Carbamate Complex Using CO_2, Amine, and Alcohol. *J. CO2 Util.* **2018,** *27,* 547–554. https://doi.org/10.1016/j.jcou.2018.09.005.

193. Kim, Y.; Ryu, T. I.; Ok, K. H.; Kwak, M. G.; Park, S.; Park, N. G.; Han, C. J.; Kim, B. S.; Ko, M. J.; Son, H. J.; et al. Inverted Layer-By-Layer Fabrication of an Ultraflexible and Transparent Ag Nanowire/Conductive Polymer Composite Electrode for Use in High-Performance Organic Solar Cells. *Adv. Funct. Mater.* **2015**, *25* (29), 4580–4589. https://doi.org/10.1002/adfm.201501046.

194. Feng, Y.; Yin, J.; Chen, M.; Song, M.; Su, B.; Lei, Q. Effect of Nano-TiO2 on the Polarization Process of Polyimide/TiO2 Composites. *Mater. Lett.* **2013**, *96*, 113–116. https://doi.org/10.1016/j.matlet.2013.01.037.

195. Babrekar, H. A.; Jejurikar, S. M.; Jog, J. P.; Adhi, K. P.; Bhoraskar, S. V. Low Thermal Emissive Surface Properties of ZnO/Polyimide Composites Prepared by Pulsed Laser Deposition. *Appl. Surf. Sci.* **2011**, *257* (6), 1824–1828. https://doi.org/10.1016/j.apsusc.2010.08.083.

196. Hu, D.; Zhu, W.; Peng, Y.; Shen, S.; Deng, Y. Flexible Carbon Nanotube-Enriched Silver Electrode Films with High Electrical Conductivity and Reliability Prepared by Facile Screen Printing. *J. Manuf. Syst.* **2017**, *33*, 1113–1119. https://doi.org/10.1016/j.jmst.2017.06.008.

197. Feng, H.; Fang, X.; Liu, X.; Pei, Q.; Cui, Z.-K.; Deng, S.; Gu, J.; Zhuang, Q. Reduced Polyaniline Decorated Reduced Graphene Oxide/Polyimide Nanocomposite Films with Enhanced Dielectric Properties and Thermostability. *Compos. Part A* **2018**, *109*, 578–584. https://doi.org/10.1016/j.compositesa.2018.03.035.

198. Fang, F.; Huang, G. W.; Xiao, H. M.; Li, Y. Q.; Hu, N.; Fu, S. Y. Largely Enhanced Electrical Conductivity of Layer-Structured Silver Nanowire/Polyimide Composite Films by Polyaniline. *Compos. Sci. Technol.* **2018**, *156*, 144–150. https://doi.org/10.1016/j.compscitech.2018.01.001.

199. Kinyanjui, J. M.; Hatchett, D. W.; Castruita, G.; Ranasinghe, A. D.; Weinhardt, L.; Hofmann, T.; Heske, C. Synthesis and Characterization of Conductive Polyimide/Carbon Composites with Pt Surface Deposits. *Synth. Met.* **2011**, *161*, 2368–2377. https://doi.org/10.1016/J.SYNTHMET.2011.08.046.

200. Yoonessi, M.; Gaier, J. R.; Peck, J. A.; Meador, M. A. Controlled Direction of Electrical and Mechanical Properties in Nickel Tethered Graphene Polyimide Nanocomposites Using Magnetic Field. *Carbon N. Y.* **2014**, *84*, 375–382. https://doi.org/10.1016/j.carbon.2014.12.033.

201. Kim, D.; Kim, J.; Jung, S.; Kim, Y.; Kim, J. Electrically and Mechanically Enhanced Ag Nanowires-Colorless Polyimide Composite Electrode for Flexible Capacitive Sensor. *Appl. Surf. Sci.* **2016**, *380*, 223–228. https://doi.org/10.1016/j.apsusc.2016.01.130.

202. Zheng, Z.; Wang, Z.; Feng, Q.; Zhang, F.; Du, Y.; Wang, C. Preparation of Surface-Silvered Graphene-CNTs/Polyimide Hybrid Films: Processing, Morphology and Properties. *Mater. Chem. Phys.* **2013**, *138*, 350–357. https://doi.org/10.1016/j.matchemphys.2012.11.067.

203. Cui, T.; Li, P.; Liu, Y.; Feng, J.; Xu, M.; Wang, M. Preparation of Thermostable Electroconductive Composite Plates from Expanded Graphite and Polyimide. *Mater. Chem. Phys.* **2012**, *134* (2–3), 1160–1166. https://doi.org/10.1016/j.matchemphys.2012.04.009.

204. Xu, L.; Chen, G.; Wang, W.; Li, L.; Fang, X. A Facile Assembly of Polyimide/Graphene Core-Shell Structured Nanocomposites with Both High Electrical and Thermal Conductivities. *Compos. Part A Appl. Sci. Manuf.* **2016**, *84*, 472–481. https://doi.org/10.1016/j.compositesa.2016.02.027.

205. Lyu, H.; Li, P.; Liu, J.; Mahurin, S.; Chen, J.; Hensley, D. K.; Veith, G. M.; Guo, Z.; Dai, S.; Sun, X. G. Aromatic Polyimide/Graphene Composite Organic Cathodes for Fast and Sustainable Lithium-Ion Batteries. *ChemSusChem* **2018,** *11* (4), 763–772. https://doi.org/10.1002/cssc.201702001.

206. Huang, T.; Lu, R.; Su, C.; Wang, H.; Guo, Z.; Liu, P.; Huang, Z.; Chen, H.; Li, T. Chemically Modified Graphene/Polyimide Composite Films Based on Utilization of Covalent Bonding and Oriented Distribution. *ACS Appl. Mater. Interf.* **2012,** *4* (5), 2699–2708. https://doi.org/10.1021/am3003439.

207. Longun, J.; Iroh, J. O. Nano-Graphene/Polyimide Composites with Extremely High Rubbery Plateau Modulus. *Carbon N. Y.* **2012,** *50* (5), 1823–1832. https://doi.org/10.1016/J.CARBON.2011.12.032.

208. Yoo, K.-P.; Lim, L.-T.; Min, N.-K.; Lee, M. J.; Lee, C. J.; Park, C.-W. Novel Resistive-Type Humidity Sensor Based on Multiwall Carbon Nanotube/Polyimide Composite Films. *Sensors Actuators B Chem.* **2010,** *145*, 120–125. https://doi.org/10.1016/j.snb.2009.11.041.

209. Huang, Y.-C.; Lin, J.-H.; Tseng, I.-H.; Lo, A.-Y.; Lo, T.-Y.; Yu, H.-P.; Tsai, M.-H.; Whang, W.-Z.; Hsu, K.-Y. An in Situ Fabrication Process for Highly Electrical Conductive Polyimide/MWCNT Composite Films Using 2,6-Diaminoanthraquinone. *Compos. Sci. Technol.* **2013,** *87*, 174–181. https://doi.org/10.1016/j.compscitech.2013.08.008.

210. Jiang, X.; Bin, Y.; Matsuo, M. Electrical and Mechanical Properties of Polyimide–carbon Nanotubes Composites Fabricated by in Situ Polymerization. *Polymer (Guildf).* **2005,** *46*, 7418–7424. https://doi.org/10.1016/J.POLYMER.2005.05.127.

211. Zhang, Y.; Fan, W.; Lu, H.; Liu, T. Highly Porous Polyimide-Derived Carbon Aerogel as Advanced Three- Dimensional Framework of Electrode Materials for High-Performance Supercapacitors. *Electrochim. Acta* **2018,** *283*, 1763–1772. https://doi.org/10.1016/j.electacta.2018.07.043.

212. Zhang, G.; Xu, Z.; Liu, P.; Su, Y.; Huang, T.; Liu, R. A Facile In-Situ Polymerization Strategy towards Polyimide/Carbon Black Composites as High Performance Lithium Ion Battery Cathodes. *Electrochim. Acta* **2018,** *260*, 598–605. https://doi.org/10.1016/j.electacta.2017.12.075.

213. Murugaraj, P.; Mainwaring, D. E.; Mora-Huertas, N. Electromechanical Response of Semiconducting Carbon-Polyimide Nanocomposite Thin Films. *Compos. Sci. Technol.* **2009,** *69* (14), 2454–2459. https://doi.org/10.1016/j.compscitech.2009.06.018.

214. Mu, S.; Wu, Z.; Wang, Y.; Qi, S.; Yang, X.; Wu, D. Formation and Characterization of Cobalt Oxide Layers on Polyimide Films via Surface Modification and Ion-Exchange Technique. *Thin Solid Films* **2010,** *518* (15), 4175–4182. https://doi.org/10.1016/j.tsf.2009.12.004.

215. Ounaies, Z.; Park, C.; Wise, K. E.; Siochi, E. J.; Harrison, J. S. Electrical Properties of Single Wall Carbon Nanotube Reinforced Polyimide Composites. *Compos. Sci. Technol.* **2003,** *63* (11), 1637–1646. https://doi.org/10.1016/S0266-3538(03)00067-8.

216. Wu, D. C.; Shen, L.; Low, J. E.; Wong, S. Y.; Li, X.; Tjiu, W. C.; Liu, Y.; He, C. Bin. Multi-Walled Carbon Nanotube/Polyimide Composite Film Fabricated through Electrophoretic Deposition. *Polymer (Guildf).* **2010,** *51* (10), 2155–2160. https://doi.org/10.1016/j.polymer.2010.03.020.

217. Yang, J.; Ye, Y.; Li, X.; Lü, X.; Chen, R. Flexible, Conductive, and Highly Pressure-Sensitive Graphene-Polyimide Foam for Pressure Sensor Application. *Compos.*

Sci. Technol. **2018**, *164* (December 2017), 187–194. https://doi.org/10.1016/j. compscitech.2018.05.044.

218. Haragirimana, A.; Ingabire, P. B.; Zhu, Y.; Lu, Y.; Li, N.; Hu, Z.; Chen, S. Four-Polymer Blend Proton Exchange Membranes Derived from Sulfonated Poly(Aryl Ether Sulfone) s with Various Sulfonation Degrees for Application in Fuel Cells. *J. Memb. Sci.* **2019**, *583*, 209–219. https://doi.org/10.1016/J.MEMSCI.2019.04.014.

219. Windrich, F.; Kappert, E. J.; Malanin, M.; Eichhorn, K.-J.; Häußler, L.; Benes, N. E.; Voit, B. In-Situ Imidization Analysis in Microscale Thin Films of an Ester-Type Photosensitive Polyimide for Microelectronic Packaging Applications. *Eur. Polym. J.* **2016**, *84*, 279–291. https://doi.org/10.1016/J.EURPOLYMJ.2016.09.020.

220. Shin, S.-R.; Moon, S.-Y.; Park, C.-Y.; Chang, B.-J.; Kim, J.-H. Solution-Processable Methyl-Substituted Semi-Alicyclic Homo- and Co-Polyimides and Their Gas Permeation Properties. *Polymer (Guildf)*. **2018**, *145*, 95–100. https://doi.org/10.1016/J. POLYMER.2018.04.062.

221. Zhang, B.; Ni, J.; Xiang, X.; Wang, L.; Chen, Y. Synthesis and Properties of Reprocessable Sulfonated Polyimides Cross-Linked via Acid Stimulation for Use as Proton Exchange Membranes. *J. Power Sources* **2017**, *337*, 110–117. https://doi. org/10.1016/J.JPOWSOUR.2016.10.102.

222. Zhang, Z.; Xu, T. Proton-Conductive Polyimides Consisting of Naphthalenediimide and Sulfonated Units Alternately Segmented by Long Aliphatic Spacers. *J. Mater. Chem. A* **2014**, *1*, 11583–11585. https://doi.org/10.1039/c4ta00942h.

223. Wang, Y.; Yang, X.; Hou, C.; Zhao, M.; Li, Z.; Meng, Q.; Liang, C. Fabrication of MnOx/Ni(OH)2 Electro-Deposited Sulfonated Polyimides/Graphene Nano-Sheets Membrane and Used for Electrochemical Sensing of Glucose. *J. Electroanal. Chem.* **2019**, *837* (September 2018), 95–102. https://doi.org/10.1016/j.jelechem.2019.02.016.

224. Deligöz, H.; Vatansever, S.; Öksüzömer, F.; Koç, S. N.; Özgümüş, S.; Gürkaynak, M. A. Preparation and Characterization of Sulfonated Polyimide Ionomers via Post-Sulfonation Method for Fuel Cell Applications. *Polym. Adv. Technol.* **2008**, *19*, 1126–1132. https:// doi.org/10.1002/pat.

225. Deligöz, H.; Vatansever, S.; Öksüzömer, F.; Koç, S. N.; Özgümüş, S.; Gürkaynak, M. A. Synthesis and Characterization of Sulfonated Homo- and Co-Polyimides Based on 2,4 and 2,5-Diaminobenzenesulfonic Acid for Proton Exchange Membranes. *Polym. Adv. Technol.* **2008**, *19*, 1792–1802. https://doi.org/10.1002/pat.

226. Ito, G.; Tanaka, M.; Kawakami, H. Sulfonated Polyimide Nanofiber Framework: Evaluation of Intrinsic Proton Conductivity and Application to Composite Membranes for Fuel Cells. *Solid State Ionics* **2018**, *317* (January), 244–255. https://doi.org/10.1016/j. ssi.2018.01.029.

227. Deligöz, H.; Vatansever, S.; Koç, S. N.; Öksüzömer, F.; Özgümüş, S.; Gürkaynak, M. A. Preparation of Sulfonated Copolyimides Containing Aliphatic Linkages as Proton-Exchange Membranes for Fuel Cell Applications. *J. Appl. Polym. Sci.* **2008**, *110*, 1216–1224. https://doi.org/10.1002/app.

228. Pandey, R. P.; Shahi, V. K. Aliphatic-Aromatic Sulphonated Polyimide and Acid Functionalized Polysilsesquioxane Composite Membranes for Fuel Cell Applications. *J. Mater. Chem. A* **2013**, *1* (45), 14375–14383. https://doi.org/10.1039/c3ta12755a.

229. Tsai, C. L.; Yen, H. J.; Liou, G. S. Highly Transparent Polyimide Hybrids for Optoelectronic Applications. *React. Funct. Polym.* **2016**, *108*, 2–30. https://doi. org/10.1016/j.reactfunctpolym.2016.04.021.

230. Antunes, R. A.; Oliveira, M. C. L. De; Ett, G.; Ett, V. Carbon Materials in Composite Bipolar Plates for Polymer Electrolyte Membrane Fuel Cells: A Review of the Main Challenges to Improve Electrical Performance. *J. Power Sources* **2011,** *196*, 2945–2961. https://doi.org/10.1016/j.jpowsour.2010.12.041.

231. Ram, R.; Rahaman, M.; Aldalbahi, A.; Khastgir, D. Determination of Percolation Threshold and Electrical Conductivity of Polyvinylidene Fluoride (PVDF)/Short Carbon Fiber (SCF) Composites: Effect of SCF Aspect Ratio. *Polym. Int.* **2017,** *66*, 573–582. https://doi.org/10.1002/pi.5294.

232. Li, J.; Ma, P. C.; Chow, W. S.; To, C. K.; Tang, B. Z.; Kim, J. K. Correlations between Percolation Threshold, Dispersion State, and Aspect Ratio of Carbon Nanotubes. *Adv. Funct. Mater.* **2007,** *17* (16), 3207–3215. https://doi.org/10.1002/adfm.200700065.

233. Weber, M.; Kamal, M. R. Estimation of the Volume Resistivity of Electrically Conductive Composites. *Polym. Compos.* **1997,** *18* (6), 711–725. https://doi.org/10.1002/pc.10324.

234. Murakami, T.; Ebisawa, K.; Miyao, K.; Ando, S. Enhanced Thermal Conductivity in Polyimide/Silver Particle Composite Films Based on Spontaneous Formation of Thermal Conductive Paths. *J. Photopolym. Sci. Technol.* **2014,** *27* (2), 187–191. https://doi.org/10.2494/photopolymer.27.187.

235. Pourrahimi, A. M.; Olsson, R. T.; Hedenqvist, M. S. The Role of Interfaces in Polyethylene/Metal-Oxide Nanocomposites for Ultrahigh-Voltage Insulating Materials. *Adv. Mater.* **2018,** *30* (4). https://doi.org/10.1002/adma.201703624.

236. Zhang, P.; Bin, Y.; Zhang, R.; Matsuo, M. Average Gap Distance between Adjacent Conductive Fillers in Polyimide Matrix Calculated Using Impedance Extrapolated to Zero Frequency in Terms of a Thermal Fluctuation-Induced Tunneling Effect. *Nature-Polymer J.* **2017,** *49*, 839–850. https://doi.org/10.1038/pj.2017.71.

237. Sheng, P. Fluctuation-Induced Materialss, Tunneling Conduction in Disordered. *Phys. Rev. B* **1980,** *21* (6), 2180–2195.

238. Dahi, A.; Fatyeyeva, K.; Langevin, D.; Chappey, C.; Rogalsky, S. P.; Tarasyuk, O. P.; Marais, S. Polyimide/Ionic Liquid Composite Membranes for Fuel Cells Operating at High Temperatures. *Electrochim. Acta* **2014,** *130*, 830–840. https://doi.org/10.1016/j.electacta.2014.03.071.

239. Wojnarowska, Z.; Knapik, J.; Díaz, M.; Ortiz, A.; Ortiz, I.; Paluch, M. Conductivity Mechanism in Polymerized Imidazolium-Based Protic Ionic Liquid [HSO3-BVIm] [OTf]: Dielectric Relaxation Studies. *Macromolecules* **2014,** *47* (12), 4056–4065. https://doi.org/10.1021/ma5003479.

240. Suzuki, K.; Iizuka, Y.; Tanaka, M.; Kawakami, H. Phosphoric Acid-Doped Sulfonated Polyimide and Polybenzimidazole Blend Membranes: High Proton Transport at Wide Temperatures under Low Humidity Conditions Due to New Proton Transport Pathways. *J. Mater. Chem.* **2012,** *22*, 23767–23772. https://doi.org/10.1039/c2jm34529c.

241. Yasuda, T.; Watanabe, M. Protic Ionic Liquids: Fuel Cell Applications. *MRS Bull.* **2013,** *38*, 560–566. https://doi.org/10.1557/mrs.2013.153.

242. Nilsson-Hallén, J.; Ahlström, B.; Marczewski, M.; Johansson, P. Ionic Liquids: A Simple Model to Predict Ion Conductivity Based on DFT Derived Physical Parameters. *Front. Chem.* **2019,** *7*. https://doi.org/10.3389/fchem.2019.00126.

243. Itoh, T. Hyperbranched Polymer Electrolytes for High Temperature Fuel Cells. In *Polymer Electrolytes Fundamentals and Applications*; Sequeira, C., Santos, D., Eds.; Woodhead Publishing Limited, 2010; p 524.

244. Shioiri, R.; Kokubo, H.; Horii, T.; Kobayashi, Y.; Hashimoto, K.; Ueno, K.; Watanabe, M. Polymer Electrolytes Based on a Homogeneous Poly(Ethylene Glycol) Network and Their Application to Polymer Actuators. *Electrochim. Acta* **2019**, *298*, 866–873. https://doi.org/10.1016/j.electacta.2018.12.142.

245. Yang, Z.; Guo, Y.; Ai, S. L.; Wang, S. X.; Zhang, J. Z.; Zhang, Y. X.; Zou, Q. C.; Wang, H. X. Rational Design and Facile Preparation of Maleimide-Based Functional Materials for Imaging and Optoelectronic Applications. *Mater. Chem. Front.* **2019**, *3* (4), 571–578. https://doi.org/10.1039/c8qm00559a.

246. Li, X.; Yan, L.; Yue, B. Maleimide: A Potential Building Block for the Design of Proton Exchange Membranes Studied by Ab Initio Molecular Dynamics Simulations. *RSC Adv.* **2015**, *5* (98), 80220–80227. https://doi.org/10.1039/c5ra14272e.

247. Ak, M.; Durmus, A.; Toppare, L. Synthesis and Characterization of Poly(N-(2-(Thiophen-3-Yl) Methylcarbonyloxyethyl)Maleimide) and Its Spectroelectrochemical Properties. *J. Appl. Electrochem.* **2007**, *37* (6), 729–735. https://doi.org/10.1007/s10800-007-9308-2.

248. Kang, Y.; Seo, Y.-H.; Lee, C. *Synthesis and Conductivity of PEGME Branched Poly(Ethylene-Alt-Maleimide*, vol. 21; 2000.

249. Xu, W.; Siow, K. S.; Gao, Z.; Lee, S. Y. Electrochemical Characterization of Plasticized Polyelectrolyte Based on Lithium-N(4-Sulfophenyl) Maleimide. *Electrochim. Acta* **1999**, *44* (13), 2287–2296. https://doi.org/10.1016/S0013-4686(98)00348-X.

250. Liu, Y. N.; Li, M.; Gu, Y.; Zhang, Y.; Li, Q.; Zhang, Z. Ultrastrong Carbon Nanotube/ Bismaleimide Composite Film with Super-Aligned and Tightly Packing Structure. *Compos. Sci. Technol.* **2015**, *117*, 176–182. https://doi.org/10.1016/j.compscitech.2015.06.014.

251. Zhao, Z. J.; Xian, G. J.; Yu, J. G.; Wang, J.; Tong, J. F.; Wei, J. H.; Wang, C. C.; Moreira, P.; Yi, X. S. Development of Electrically Conductive Structural BMI Based CFRPs for Lightning Strike Protection. *Compos. Sci. Technol.* **2018**, *167*, 555–562. https://doi.org/10.1016/j.compscitech.2018.08.026.

252. Qiu, J.; Wu, Q.; Jin, L. Effect of Hyperbranched Polyethyleneimine Grafting Functionalization of Carbon Nanotubes on Mechanical, Thermal Stability and Electrical Properties of Carbon Nanotubes/Bismaleimide Composites. *RSC Adv.* **2016**, *6* (98), 96245–96249. https://doi.org/10.1039/c6ra21545a.

253. Wang, X.; Jiang, Q.; Xu, W.; Cai, W.; Inoue, Y.; Zhu, Y. Effect of Carbon Nanotube Length on Thermal, Electrical and Mechanical Properties of CNT/Bismaleimide Composites. *Carbon N. Y.* **2013**, *53*, 145–152. https://doi.org/10.1016/j.carbon.2012.10.041.

254. Cheng, Q.; Wang, B.; Zhang, C.; Liang, Z. Functionalized Carbon-Nanotube Sheet/ Bismaleimide Nanocomposites: Mechanical and Electrical Performance beyond Carbon-Fiber Composites. *Small* **2010**, *6* (6), 763–767. https://doi.org/10.1002/smll.200901957.

255. Kamiyama, S.; Hirano, Y.; Okada, T.; Ogasawara, T. Lightning Strike Damage Behavior of Carbon Fiber Reinforced Epoxy, Bismaleimide, and Polyetheretherketone Composites. *Compos. Sci. Technol.* **2018**, *161*, 107–114. https://doi.org/10.1016/j.compscitech.2018.04.009.

256. Mandhakini, M.; Alagar, M. Synergistically C8 Ether Linked Bismaleimide Toughened and Electrically Conducting Carbon Black Epoxy Nanocomposites. *Mater. Des.* **2017**, 305–321. https://doi.org/10.1016/j.matdes.2014.07.041.

257. Wu, G.; Kou, K.; Li, N.; Shi, H.; Chao, M. Electrically Conductive Adhesive Based on Bismaleimide-Triazine Resin Filled with Microcoiled Carbon Fibers. *J. Appl. Polym. Sci.* **2013,** *128* (2), 1164–1169. https://doi.org/10.1002/app.38348.

258. Chu, P. P.; Wu, C. S.; Liu, P. C.; Wang, T. H.; Pan, J. P. Proton Exchange Membrane Bearing Entangled Structure: Sulfonated Poly(Ether Ether Ketone)/Bismaleimide Hyperbranch. *Polymer (Guildf).* **2010,** *51* (6), 1386–1394. https://doi.org/10.1016/j.polymer.2010.01.042.

259. Wang, F. M.; Wu, H. C.; Cheng, C. S.; Huang, C. L.; Yang, C. R. High Ionic Transfer of a Hyperbranched-Network Gel Copolymer Electrolyte for Potential Electric Vehicle (EV) Application. *Electrochim. Acta* **2009,** *54* (14), 3788–3793. https://doi.org/10.1016/j.electacta.2009.01.072.

260. Li, Q.; Luo, S.; Wang, Q.-M. Piezoresistive Thin Film Pressure Sensor Based on Carbon Nanotube-Polyimide Nanocomposites. *Sensors Actuators A Phys.* **2019,** *295*, 336–342. https://doi.org/10.1016/J.SNA.2019.06.017.

261. Lee, C.; Jun, S.; Ju, B.; Kim, J. Pressure-Sensitive Strain Sensor Based on a Single Percolated Ag Nanowire Layer Embedded in Colorless Polyimide. *Phys. B Phys. Condens. Mater* **2017,** *514*, 8–12. https://doi.org/10.1016/j.physb.2017.03.019.

262. Watcharaphalakorn, S.; Ruangchuay, L.; Chotpattananont, D.; Sirivat, A.; Schwank, J. Polyaniline/Polyimide Blends as Gas Sensors and Electrical Conductivity Response to CO-N2 Mixtures. *Polym. Int.* **2005,** *54*, 1126–1133. https://doi.org/10.1002/pi.1815.

263. Su, T. M.; Ball, I. J.; Conklin, J. A.; Huang, S.-C.; Larson, R. K.; Nguyen, S. L.; Lew, B. M.; Kaner, R. B. Polyaniline/Polyimide Blends for Pervaporation and Gas Separation Studies. *Synth. Met.* **1997,** *84*, 801–802. https://doi.org/10.1016/S0379-6779(96)04153-7.

264. Wang, K.; Chang, Y.-H.; Zhang, C.; Wang, B. Conductive-on-Demand: Tailorable Polyimide/Carbon Nanotube Nanocomposite Thin Film by Dual-Material Aerosol Jet Printing. *Carbon N. Y.* **2016,** *98*, 397–403. https://doi.org/10.1016/J.CARBON.2015.11.032.

265. Kakitani, K.; Koshikawa, H.; Yamaki, T.; Yamamoto, S.; Sato, Y.; Sugimoto, M.; Sawada, S. Preparation of Conductive Layer on Polyimide Ion-Track Membrane by Ar Ion Implantation. *Surf. Coatings Technol.* **2018,** *355*, 181–185. https://doi.org/10.1016/J.SURFCOAT.2018.05.057.

266. Kaliyappan, K.; Li, G.; Yang, L.; Bai, Z.; Chen, Z. An Ion Conductive Polyimide Encapsulation: New Insight and Significant Performance Enhancement of Sodium Based P2 Layered Cathodes. *Energy Storage Mater.* **2019,** in press. https://doi.org/10.1016/J.ENSM.2019.07.010.

267. Feng, S.; Kondo, S.; Kaseyama, T.; Nakazawa, T.; Kikuchi, T.; Selyanchyn, R.; Fujikawa, S.; Christiani, L.; Sasaki, K.; Nishihara, M. Development of Polymer-Polymer Type Charge-Transfer Blend Membranes for Fuel Cell Application. *J. Memb. Sci.* **2018,** *548*, 223–231. https://doi.org/10.1016/j.memsci.2017.11.025.

268. Moharir, P. V.; Tembhurkar, A. R. Comparative Performance Evaluation of Novel Polystyrene Membrane with Ultrex as Proton Exchange Membranes in Microbial Fuel Cell for Bioelectricity Production from Food Waste. *Bioresour. Technol.* **2018,** *266*, 291–296. https://doi.org/10.1016/J.BIORTECH.2018.06.085.

269. Cai, J.; Niu, H.; Zhao, P.; Ji, Y.; Ma, L.; Wang, C.; Bai, X.; Wang, W. Multicolored Near-Infrared Electrochromic Polyimides: Synthesis, Electrochemical, and Electrochromic

Properties. *Dye. Pigment.* **2013,** *99* (3), 1124–1131. https://doi.org/10.1016/j. dyepig.2013.08.019.

270. Chen, C. J.; Yen, H. J.; Hu, Y. C.; Liou, G. S. Novel Programmable Functional Polyimides: Preparation, Mechanism of CT Induced Memory, and Ambipolar Electrochromic Behavior. *J. Mater. Chem. C* **2013,** *1* (45), 7623–7634. https://doi. org/10.1039/c3tc31598c.

271. Hsiao, S. H.; Chen, Y. Z. Electroactive and Ambipolar Electrochromic Polyimides from Arylene Diimides with Triphenylamine N-Substituents. *Dye. Pigment.* **2017,** *144,* 173–183. https://doi.org/10.1016/j.dyepig.2017.05.031.

272. Ma, J.; Wang, K.; Zhan, M. A Comparative Study of Structure and Electromagnetic Interference Shielding Performance for Silver Nanostructure Hybrid Polyimide Foams. *RSC Adv.* **2015,** *5* (80), 65283–65296. https://doi.org/10.1039/c5ra09507g.

273. Yang, H.; Yu, Z.; Wu, P.; Zou, H.; Liu, P. Electromagnetic Interference Shielding Effectiveness of Microcellular Polyimide/in Situ Thermally Reduced Graphene Oxide/ Carbon Nanotubes Nanocomposites. *Appl. Surf. Sci.* **2018,** *434,* 318–325. https://doi. org/10.1016/j.apsusc.2017.10.191.

274. Yeom, S. W.; You, B.; Cho, K.; Jung, H. Y.; Park, J.; Shin, C.; Ju, B. K.; Kim, J. W. Silver Nanowire/Colorless-Polyimide Composite Electrode: Application in Flexible and Transparent Resistive Switching Memory. *Sci. Rep.* **2017,** *7* (1), 1–9. https://doi. org/10.1038/s41598-017-03746-1.

275. Yen, H. J.; Liou, G. S. Solution-Processable Triarylamine-Based High-Performance Polymers for Resistive Switching Memory Devices. *Polym. J.* **2015,** *48* (2), 117–138. https://doi.org/10.1038/pj.2015.87.

276. Su, K.; Sun, N.; Tian, X.; Guo, S.; Yan, Z.; Wang, D.; Zhou, H.; Zhao, X.; Chen, C. Highly Soluble Polyimide Bearing Bulky Pendant Diphenylamine-Pyrene for Fast-Response Electrochromic and Electrofluorochromic Applications. *Dye. Pigment.* **2019,** *171,* 107668. https://doi.org/10.1016/j.dyepig.2019.107668.

277. Xie, Z.; Liu, Q.; Zhang, Q.; Lu, B.; Zhai, J.; Diao, X. Fast-Switching Quasi-Solid State Electrochromic Full Device Based on Mesoporous WO3 and NiO Thin Films. *Sol. Energy Mater. Sol. Cells* **2019,** *200* (June), 110017. https://doi.org/10.1016/j. solmat.2019.110017.

278. Romero, F. J.; Salinas-castillo, A.; Id, A. R.; Albrecht, A.; Id, A. G.; Id, D. P. M.; Rodriguez, N. In-Depth Study of Laser Diode Ablation of Kapton Polyimide for Flexible Conductive Substrates. *Nanomaterials* **2018,** *8* (517), 1–11. https://doi.org/10.3390/ nano8070517.

279. Ouadah, A.; Luo, T.; Gao, S.; Zhu, C. Controlling the Degree of Sulfonation and Its Impact on Hybrid Cross-Linked Network Based Polyphosphazene Grafted Butylphenoxy as Proton Exchange Membrane. *Int. J. Hydrog. Energy* **2018,** *43* (32), 15466–15480. https://doi.org/10.1016/J.IJHYDENE.2018.06.105.

280. Ahmed, F.; Sutradhar, S. C.; Ryu, T.; Jang, H.; Choi, K.; Yang, H.; Yoon, S.; Rahman, M. M.; Kim, W. Comparative Study of Sulfonated Branched and Linear Poly(Phenylene) s Polymer Electrolyte Membranes for Fuel Cells. *Int. J. Hydrog. Energy* **2018,** *43* (10), 5374–5385. https://doi.org/10.1016/J.IJHYDENE.2017.08.175.

281. Huang, B.; Wang, X.; Fang, H.; Jiang, S.; Hou, H. Mechanically Strong Sulfonated Polybenzimidazole PEMs with Enhanced Proton Conductivity. *Mater. Lett.* **2019,** *234,* 354–356. https://doi.org/10.1016/J.MATLET.2018.09.131.

282. Wang, Y.; Hong, Y.; Chen, Q.; Zhou, G.; He, W.; Gao, Z.; Zhou, X.; Zhang, W.; Su, X.; Sun, R. Direct Surface In-Situ Activation for Electroless Deposition of Robust Conductive Copper Patterns on Polyimide Film. *J. Taiwan Inst. Chem. Eng.* **2019,** *97,* 450–457. https://doi.org/10.1016/j.jtice.2019.02.014.

283. Wu, Z.; Wu, D.; Qi, S.; Zhang, T.; Jin, R. Preparation of Surface Conductive and Highly Reflective Silvered Polyimide Films by Surface Modification and in Situ Self-Metallization Technique. *Thin Solid Films* **2005,** *493,* 179–184. https://doi.org/10.1016/j.tsf.2005.07.286.

284. Yang, S.; Wu, D.; Qi, S.; Cui, G.; Jin, R.; Wu, Z. Fabrication of Highly Reflective and Conductive Double-Surface-Silvered Layers Embedded on Polymeric Films through All-Wet Process at Room Temperature. *J. Phys. Chem. B* **2009,** *113,* 9694–9701.

285. Thompson, D. S.; Davis, L. M.; Thompson, D. W.; Southward, R. E. Single-Stage Synthesis and Characterization of Reflective and Conductive Silver -Polyimide Films Prepared from Silver (I) Complexes with. *Appl. Mater. Interf.* **2009,** *1* (7), 1457–1466. https://doi.org/10.1021/am900133a.

286. Qi, S.; Wang, W.; Wu, D.; Wu, Z.; Jin, R. Preparation of Reflective and Electrically Conductive Surface-Silvered Polyimide Films from Silver (I) Complex and PMDA/ODA via an in Situ Single-Stage Technique. *Eur. Polym. J.* **2006,** *42,* 2023–2030. https://doi.org/10.1016/j.eurpolymj.2006.03.009.

287. Tai, Y.; Yang, Z.; Li, Z. A Promising Approach to Conductive Patterns with High Efficiency for Flexible Electronics. *Appl. Surf. Sci.* **2011,** *257* (16), 7096–7100. https://doi.org/10.1016/j.apsusc.2011.03.056.

288. Mou, Y.; Cheng, H.; Wang, H.; Sun, Q.; Liu, J.; Peng, Y.; Chen, M. Facile Preparation of Stable Reactive Silver Ink for Highly Conductive and Flexible Electrodes. *Appl. Surf. Sci.* **2019,** *475,* 75–82. https://doi.org/10.1016/j.apsusc.2018.12.261.

289. Spechler, J. A.; Koh, T.-W.; Herb, J. T.; Rand, B. P.; Arnold, C. B. A Transparent, Smooth, Thermally Robust, Conductive Polyimide for Flexible Electronics. *Adv. Funct. Mater.* **2015,** *25* (48), 7428–7434. https://doi.org/10.1002/adfm.201503342.

290. Lee, S.; Jee, S. S.; Park, H.; Park, S.-H.; Han, I.; Mizusaki, S. Large Reduction in Electrical Contact Resistance of Flexible Carbon Nanotube/Silicone Rubber Composites by Trifluoroacetic Acid Treatment. *Compos. Sci. Technol.* **2017,** *143,* 98–105. https://doi.org/10.1016/j.compscitech.2017.03.004.

CHAPTER 5

Molecular Modeling of Imidic Polymers with Advanced Physico-Chemical Properties

RALUCA MARINICA ALBU*

*"Petru Poni" Institute of Macromolecular Chemistry,
Laboratory of Physical Chemistry of Polymers,
41A Grigore Ghica Voda Alley, 700487 Iasi, Romania*

E-mail: albu.raluca@icmpp.ro

ABSTRACT

Imidic polymers are materials of great importance in various applicative fields. The macromolecular architecture is an essential parameter that affects the physico-chemical properties of polyimides. The chapter describes the need of molecular modeling in imidic polymer investigation, highlighting the importance of understanding the correlation between chemical structure and the performance of the final product. The main computational techniques applied to this type of polymers are presented. The connection between theoretical information extracted from modeling and experimental data is also discussed, including the applicative potential.

5.1 INTRODUCTION

In the past decades, modeling of various kinds of processes in polymer science has been often used in order to understand how to optimize their performance.[1-8] Modeling and computer simulation are essential tools in elucidation of the structural characteristics of small molecules or macromolecular compounds. Such computational procedures provide valuable introspections that enable prediction of certain properties. In this way, it is easier to explain some experimental aspects regarding macromolecular architecture or thermodynamic properties. Based on the latest developments in computing procedures and software, polymer simulations may guide and supplement the data regarding engineering plastic materials design and provide a solid support for the discovery efforts.[9] In order to secure the fact that impact of simulation results are adequately harnessed and significant results are attained, close attention must be paid to guarantee the validity and reproducibility of the performed simulations.

Computational challenges were approached during the past years with many developments concerning the software and algorithms, accompanied by upgrades in computer hardware.[10] So, it is not astonishing that with these advances, in the present it arrives to a stage where molecular simulations of polymers can be performed in a short time on smaller devices. Considering the software and hardware progress, there is a considerable amount of simulation reports of polymers providing the comprehension of novel and already existing macromolecular systems.[11-14] Thus, through the extracted data the modeling data inspire polymer scientists to create other synthetic paths to create new materials which have tremendous potential. Analyses at the microscopic level have a great impact on production of new polymer materials. For instance, to get a good grasp on membrane properties at atomic level one should perform, before synthesis, a molecular simulation of the desired macromolecular system.[15,16]

Polyimides are considered to be an advanced category of thermostable polymers and as a result literature presents a large variety of structures along with their physico-chemical characterization.[17-21] The spatial molecular arrangement of this type of macromolecules is found to be in strong correlation with their ability to undergo charge transfer complex (CTC) interactions.[22] Also, the conformation aspects are connected to free volume formation, which in turn influences a series of properties, like refractive index, dielectric constant, permeability, and so on.[23]

Having in view the above considerations, in this chapter is presented a strategy to select the adequate models and conditions to predict polymer properties using current available simulation software. In particular, simulation methods applied to polyimide materials are described. The most recent investigations are reviewed by highlighting the bridge between computational data and experiments. Such aspects are essential for polymer simulators dealing with such macromolecular structures, helping them to use computational approaches which deliver valid and reproducible information. The importance of simulation in some practical situation is also presented.

5.2 POLYIMIDE MOLECULAR STRUCTURE AND PROPERTIES EVALUATION BY SIMULATION

Polyimide is a polymer having the structural units (SUs) consisting of imide rings. An imide is a group in a molecule that has a general structure as depicted in Figure 5.1. DuPont initiated the production of Kapton polyimide in 1955 and then other structures emerged on market, such as Ultem, Matrimid, PI-2600 Series, Pyralin (PI-2700 series), Upilex R, SE 1211, and so on.[24] The chemical structures of these commercial polyimides are displayed in Figure 5.1.

Polyimides can take one of the following two forms:

- Linear structure: In this situation, the atoms of the imide group are part of a linear chain.
- Heterocyclic structure: The imide unit is a part of a cyclic structure in the main chain. This form is more stable at elevated temperatures and in the presence of corrosive elements comparatively to linear analog.

Aromatic heterocyclic polyimides found use in several applications due to their chemical, thermal, and mechanical resistance combined with good adhesion. This combination of properties reflects the strong intermolecular forces among the chains and from the stiffness of the aromatic units in the repeat units. These forces refer to aromatic stacking, polar interactions, and most relevant here, CTC.[24] The latter holds the chains tightly together, this being one of the reasons why polyimides melt only at high temperatures.

Several methods are known for obtaining polyimides. Among them it is worth mentioning[22,24,25]:

- The reaction between a dianhydride and a diamine;
- The reaction between a dianhydride and a diisocyanate.

The molecular conformation of the polyimide chains is strongly affected by the monomer configuration and also by the nature of the molecular forces exerting among the chains.[26]

Molecular simulation enables a simple correlation between the imidic polymer chemical structure and the resulting properties.[27] Basically, molecular modeling deals with reconstruction of certain structural and energetic parameters of a given macromolecular structure enabling estimation of some measurable characteristics. Therefore, software grants the simulation of electronic features starting from semiempirical approaches using particular algorithms. In this way, one may evidence the effects induced by the substituents (their nature and relative position) on the electro-optic parameters of a set of macromolecular compounds.[28] In addition, it is possible to examine the interactions between two macromolecules or a macromolecule and solvent molecules by monitoring the changes in bond length, energy of the boundary levels, total energy, and dipole moment.[29]

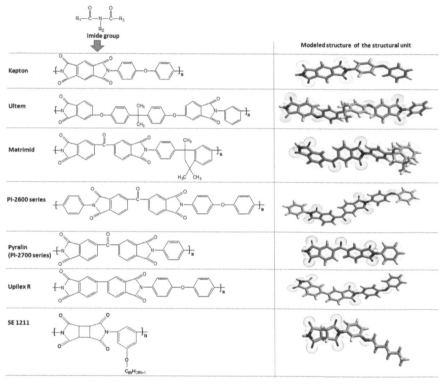

FIGURE 5.1 The chemical structure of imide group and the chemical structures of certain commercial polyimides along with the modeled conformation of their SU.

Based on computing analyses, it is feasible to evaluate some basic properties of a given polyimide structure prior the synthesis stage. For example, a simple modeling experiment of several polyimide SUs enables mainly determination of the following:

- Geometrical properties: Bond length, torsion angle, and bond angle—all describing the shape of the macromolecule.

- Electronic properties: Energy levels, dipole moment, and atomic charges.
- Thermodynamic properties: Heat of formation, enthalpy, or entropy changes.

5.3 COMPUTATIONAL TECHNIQUES APPLIED TO IMIDIC POLYMERS

Molecular modeling is based on theoretical approaches and computational techniques to assess and model the structure of macromolecules and their behavior. [7,8,30,31] The major purpose of this procedure relies on calculation of the basic properties of macromolecules with mathematical approximations, concomitantly with application of a software to a given physico-chemical system. The calculations can be effectuated for the following cases:

- gaseous state or in solution;
- grounded or excited; and
- free macromolecules or clusters.

Molecular simulation techniques are helpful in elucidation and picturing the structure–property relation at molecular level.[9] These approaches are suitable for computing molecular energy and atomic charges in distinct phases. Using a simulation software, one may construct 3D images of molecules constituting the polymer chain that facilitate comprehension of the molecule configuration.[32]

There are numerous theories and computational methods created not only for advancing the design of polymers, but also to allow prediction and optimization of their properties.[7–9] At this point, application of a single computational approach does not cover all scales involved in polymeric systems. It is therefore more useful to bridge the length and time scales via a combination of methods in a multiscale simulation framework. The multiscale method is preferably supposed to predict macroscopic properties of polymeric materials from fundamental molecular processes.[6] For construction of a multiscale simulation, it is necessary to mix models and theories from four characteristics length and time scales.[6,33,34] They are mainly classified as follows[33]:

- The quantum scale ($\sim 10^{-10}$ m, $\sim 10^{-12}$ s): The main components of atom are modeled using quantum mechanics (QM) methods emphasizing formation and rupture of chemical bonds, and the changes in electronic configurations.

- The atomistic scale ($\sim10^{-9}$ m, $\sim10^{-9} \div 10^{-6}$ s): Atoms are considered single sites. The potential energy of the material is evaluated from a number of several bonded or nonbonded interactions that are known as force fields. Molecular dynamics (MD) and Monte Carlo (MC) simulation approaches are utilized to model atomic processes requiring a larger group of atoms in regard to QM.
- *The mesoscopic scale* ($\sim10^{-6}$ m, $\sim10^{-6} - 10^{-3}$ s): The molecule is described by a field or a microscopic particle (bead) and the interactions among beads are monitored. Several methods have been proposed to describe mesoscopic systems, such as Brownian dynamics (BD), dissipative particle dynamics (DPD), lattice Boltzmann (LB), time-dependent Ginzburg-Landau (TDGL) method, and dynamic density functional theory (DDFT).
- *The macroscale* ($\sim10^{-3}$ m, ~1 s): The material is viewed as a continuous medium having a behavior governed by constitutive laws, which sometimes are combined with conservation laws to predict various phenomena. The fundamental hypothesis at this level relies on replacing a heterogeneous system with an equivalent homogeneous one. The most relevant methods used in such situations are finite element method (FEM), finite difference method (FDM), and finite volume method (FVM).

Among the most meaningful computational procedures used to model different scales, one may briefly mention[33]:

- *Quantum mechanism*: Density functional theory (DFT) and ab initio MD (AIMD);
- *Atomistic techniques*: MC and MD;
- *Mesoscale techniques*: BD, DPD, and LB;
- *Macroscale techniques*: FDM, FEM, FVM, smoothed particle hydrodynamics (SPH), and smoothed DPD (SDPD); and
- *Multiscale techniques*: Sequential, concurrent, and adaptive resolution schemes.

The main concerns of each simulation method are described in detail in the work of Gooneie et al.[33] The review also contains information about the manner in which the presented computational methods should be applied to polymer materials.

The most used software in modeling polymer materials are: Chem3D, HyperChem, Gaussian, Materials Studio, Gromacs, ArgusLab, Avogadro, Abalone, Synthia, Culgy, Materials Design Inc., ChemDoodle, Hypercube, ChemAxon, etc.

5.4 BRIDGING MOLECULAR MODELING WITH EXPERIMENTS

The data predicted with computer modeling allow scientists to see in advance what properties would have a novel polyimide structure. Some of these properties can be directly connected with experimental data, while others are able to indirectly explain them. In the following paragraphs, some of the most relevant molecular simulations made on polyimide materials are reviewed.

5.4.1 SPATIAL MOLECULAR CONFORMATION, POLARIZABILITY, AND DIPOLE MOMENT

The spatial molecular conformation of polyimide chain is very important when discussing optical and dielectric performances of these materials. Flexible and randomly disposed macromolecular chains favor the creation of amorphous materials with free volume. These aspects are desirable when designing transparent and low dielectric polyimides.

Sava et al.[35] prepared some azopolyimides that were photo-patterned. The chemical structures of the imidic polymers were modeled with Hyperchem program at the lowest energy. Molecular mechanics approach allowed estimation of certain molecular parameters, like dihedral angles, atomic distances, and bond angles; based on them, the macromolecule geometry is computed and optimized. The azompolyimides have a spatial conformation that is far from being linear rigid rod as widely noticed for aromatic polyimides that lack kinks or pendants in their structure. The adopted shapes of chains impede a close packing and the solvent diffusion among the chains is facilitated resulting in a higher solubility. The chain geometry is less affected by the nature of substituent (para-, ortho-methyl, or chlorine) bonded to the azobenzene segments. The polymers containing -$COCH_3$ or -COCl tend to have an extended macromolecular chain. In other words, the type of kinks and pendant groups affect the spatial conformation of the polyimide and implicitly its response to UV patterning.

Albu et al.[36] investigated a set of new fluorinated aromatic imidic polymers and computed the energies attributed to all spatial molecular conformations.

Based on potential and kinetic energies, it was facile to understand the differences in the flexibility of each simulated polymer structure. By analyzing
them, it could be stated that the main degree of freedom is a result of (1) the
torsion of bonds among imide units and aromatic rings and (2) the presence
of angular bonds (e.g., ether linkages) separating bulky and rigid rings, and
also the flexible and bulky groups (like in hexafluoroisopropylidene and/or
trifluoromethyl ones). Figure 5.2 displays the spatial molecular conformation of seven SUs of a fluorinated aromatic polyimide achieved in conditions
of minimal energy.

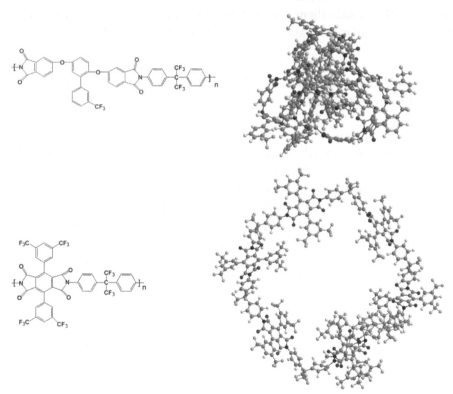

FIGURE 5.2   The spatial molecular conformation of seven structural units of a fluorinated
aromatic polyimide achieved in conditions of minimal energy.

The most flexible polyimides display the highest values of kinetic energy.
As the length and flexibility of the diamine part increases, one may notice a
decrease in refractive index values. The length of the diamine has an impact

on the magnitude of the refractive index since it varies the density of imidic polar rings. However, the backbone flexibility could counteract the lower polarizability created through dilution of the imide units. Besides the backbone flexibility, the variations in the density of fluorinated groups along the main chain significantly influence the refraction properties of the polyimide. The ratio between the molar refraction and molar volume varies from one sample to another depending on the type of diamine or dianhydride moiety. It was also noted that the decrease in dipole moment per unit volume induces diminishment of the dielectric constant.

Barzic et al.[37] used molecular modeling as a tool to assess the relation between polyimide conformation and refraction and dielectric properties for liquid crystal display (LCD) applications. It was attempted to obtain an adequate monomer combination for polyimide with best balance of properties for the desired purpose. Several semiaromatic structures containing a common bis[4-(4-amino-phenoxy)phenyl] sulfone (BAPS) diamine sequence were analyzed through computer modeling evidencing the influence of the alicyclic dianhydride moiety on solubility, optical, and interfacial adhesion properties. The first set of imidic polymers was obtained from symmetric, rigid, and small dianhydride segments, while the second one is constructed based on nonsymmetric, semiflexible, and bulky dianhydride units. In the first case, the degree of freedom is mainly arising from the BAPS sequence, so the overall shape of the macromolecular coil is less affected by the small dianhydride units. In the second situation, the dianhydride substituents affect the chain packing—aspect reflected in the values of the lower values refractive index, dielectric constant, and molar polarization. Both macromolecular architecture and backbone polarizability had an impact on the refraction and polarization properties of polyimides. As the rigidity and size of the dianhydride part are smaller, the refractive index values were higher. For some samples, the refraction properties were validated by the experimental data. Furthermore, the dipole moment allowed calculation of the potential energy of interaction between the polyimide and certain nematics used in the manufacturing of the LCDs. The highest differences in the interaction energy with liquid crystal (LC) values are noted for the system of polyimides containing symmetric dianhydride segments. The energy of interaction could be also assessed from molecular modeling. Figure 5.3 presented an example of computed interaction between cyanobiphenyl nematic and a semiaromatic polyimide structure.

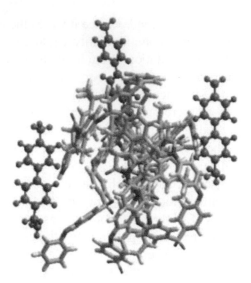

FIGURE 5.3   The molecular modeling interaction between cyanobiphenyl nematic LC (colored in purple) and a semiaromatic polyimide structure containing three SUs.

In another work, Albu et al.[38] analyzed the influence of the interaction among polyimide chain on the overall chain conformation. Semiaromatic polyimides were prepared from the same semiflexible alicyclic dianhydride and aromatic diamines with distinct rigidity. The length and the number of kinks in the diamine segment have a great influence on the resulting spatial arrangement of imide polymer chains. In order to understand the manner in which the SU configuration actions on the overall polymer conformation at macromolecular scale, the shape of two interacting polyimide chains containing three SUs was computed. The obtained spatial molecular conformations of the semialicyclic polyimides are presented in Figure 5.4. It was remarked that the macromolecule that contains the backbone 4,4′-(hexa-fluoroisopropylidene)bis(p-phenyleneoxy) dianiline (6FADE) units displays a higher number of bending points, which support its higher chain flexibility comparatively with the polyimide derived from 4,4′diaminodiphenylmethane (DDM). The conformational aspects are reflected in the polyimide film morphology. Atomic force microscopy reveals that such imidic polymers have a porous surface with distinct features. Combined with the good blood compatibility, this polyimide is a good candidate for membrane oxygenator applications.

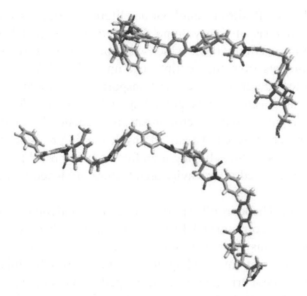

FIGURE 5.4 The conformation of two interacting macromolecular chains containing three SUs for (a) polyimide derived from DDM and (b) polyimide derived from 6FADE.

5.4.2 SMALL-MOLECULE PERMEATION

Chang et al.[39] employed some computing investigations on imidic polymers with intrinsic microporosity. They designed polyimide structures containing spiro-centers and reported their impact on the gas transport behavior by monitoring torsional angle, free volume, and cavity dimensions in membrane. It was revealed that the free volume did not affect so much the gas sorption; however, the type of the gas molecules was the determinant factor regarding gas solubility inside the polymer. The diffusion data indicated that the polyimide membranes with bigger free volume displayed more pathways for gas molecules traveling between cavities and thus enhancing the permeability. The quantitative information extracted from the molecular simulation was in accordance with the experimental data. The combined MD and MC methods showed to be very efficient tools for the analysis of the membrane structure and gas separation characteristics of polyimides containing spiro-centers.

The work of Luo et al.[40] has studied the microporous architecture and gas transport abilities of membranes designed on the basis of pentiptycene-containing polyimides. The simulated spatial conformation of the membranes revealed that the rigid H-shaped pentiptycene segments perturb the chain packing generating a big fraction of free volume and implicitly enhanced gas permeability. Positron annihilation lifetime spectroscopy allowed atomic-level analysis of free volume distribution. A bimodal microcavity size distribution was remarked, which was found to be susceptible to the type of substituents from the pentiptycene units. The diffusivity and solubility parameters describing the gas transport of membranes proved that the size sieving mechanism is prevailing, whereas bimodal microcavity dimension distribution with microporosity is accountable for the good H_2 gas separation performance. Advanced physical aging resistance was noted for the polyimide structures having high free volume probably as a result of configuration-generated microcavity architecture produced by pentiptycene sequences.

Vergadou and Theodorou[41] reported molecular modeling employment in the investigation of sorption and diffusion of molecules in poly(amide imide)s. The polymer atoms are able to move in some cases allowing local-chain motions during a penetrant traveling through the cavities. So, this aspect was included in the geometric approach developed by Greenfield[42] to evaluate accessible volume and its distribution in the polymer to attain a picture of sorption sites.

Narrow-necking zones linking regions of accessible volume for small penetrants are considered as the basic elements for determining transition states. The reaction path between the states was estimated using a subset of degrees of freedom that were important to the specific transition along the pathway. The approach was further developed[43,44] for macromolecules in Cartesian coordinates and in the polymer-penetrant multidimensional space utilizing an entirely flexible representation for carbon dioxide penetrant.

5.4.3   PIEZOELECTRIC PROPERTIES

Young et al.[45] analyzed through molecular modeling the piezoelectric characteristics of polyimides. Semiempiricial molecular orbital calculations were performed using MOPAC with the AM1 Hamiltonian attaining data concerning the potential energy surfaces and also the electron distributions of the chain segments. The torsional barriers were especially investigated since they influence the orientation of the dipoles by determining the flexibility of the polyimide main chain. The potential energy surfaces were afterwards utilized to adapt a force field for MD computations. The simulations were useful in studying the poling process. The used approach enabled the evaluation of the dielectric relaxation strength of an amorphous imidic polymer starting from its chemical structure. The model of an electrode polyimide imposed the use of periodic boundary conditions to mimic the bulk polymer and the use of plating atoms by which to introduce the electric field. It was reported that pendant nitrile dipole and the backbone dianhydride moiety dipole have essential role in the material's dielectric response. It was proved that the distinct aspects in the magnitude of the dielectric relaxations are ascribed to the nitrile dipole. Analysis of the relaxations shows the lack of the cooperative dipolar motions. All simulation data on relaxation strengths (11.5) led to close values to those obtained from experiments (17.8).

5.4.4   THERMAL AND MECHANICAL PROPERTIES

Li et al.[46] performed simulation to understand how the vitrification temperature (T_g) ranges from an isomeric polyimide to the corresponding symmetrical structure. MD method enabled evaluation of the T_g of the imide polymer derived from 3,3',4,4'-biphenyltetracarboxylic dianhydride and 4,4'-oxydianiline and its isomeric counterpart. Modeling data indicated that the nonbond energy is very important in glass transition process since it suddenly changes near T_g. The experimental results were in good correlation with theoretical data. The higher free volume fraction noticed for the isomeric structure can generate the macromolecular chains with more space to attain segmental motion. Based on the torsion angle distribution determinations, it is proved that the torsion angle of its biphenyl segment is constrained. In addition, from mean square displacement and vector autocorrelation function computations, this group is able to rotate against other groups in the glassy state, and significantly enhances the chain rigidity. As a result, the isomeric polyimide requires a larger relaxation time for the segment motion. Thus, the bigger T_g of isomeric imidic polymer is ascribed to the backbone rigidity for the time scale, and not the free volume for the space scale.

Lei et al.[34] reported thermo-mechanical characterization of six types of aromatic polyimides using hierarchical multiscale method. The Condensed-phase Optimized Molecular Potential for Atomistic Simulation Studies (COMPASS) force field[47,48] was employed in minimizing energy and in the MD calculations. The computed total potential energy contains information on bonded energy (stretching, angle bending, and torsion), nonbonded energy, and energy of cross terms among either two of the bonded energy. After energy optimization of polyimide SU, the chain length of the models was assessed. In the next step, the SU number was set to 20 and quantitative structure–activity relationship (QSPR) was applied to predict the size of Kuhn segments. When chain geometry is optimized, a cubic simulation cell containing 10 macromolecular chains for a polyimide structure was built. Then, a series of MD runs were performed based on details reported in Refs. [34,49]. The model validity[50] and equilibrium check[51] have been done by specifying their density, energy, and gyration radius fluctuations when the MD runs. For determination of the thermal properties, the samples were equilibrated at several temperatures in the 500-ps NPT-MD run, and the fluctuations in the density and gyration radius were monitored. The T_g was achieved from the inflection point of the fitting curves. It is well known that T_g is linked to the chain rigidity and mobility. The rigidity of a linkage can be diminished when torsion angle is present. The estimated T_g values were slightly higher than those experimentally recorded.[52] This is due to the faster cooling rate, but when high molecular weight result the measured, T_g value overcome the theoretical one. Regarding the mechanical performance, Lei et al.[34] showed that the tensile modulus (E) can be achieved from the "Stress–Strain" scripting in the simulation software. The estimated values of E are close to experiment, revealing certain contradictions with polyimides theoretical linkage rigidity as noticed from chemical structures. The tensile modulus is influenced by the deformation rate as the T_g is affected by the cooling rate, remarked in experiments and simulation,[53] which could be a starting point to explain the differences in the magnitude of E or T_g of the same system.

Mo et al.[54] analyzed the mechanical performance, including Young modulus, shear modulus, bulk modulus, and Poisson ratio, of pure polyimide and their SiO_2 nanocomposites using the MD simulation.

The unit cell models of the systems were constructed based on specific conditions reported in Mo et al.[54] The mechanical properties were computed after MD optimization via Forcite module. The elastic stiffness matrix of unit cell can be achieved. More polymer chains in cell are randomly placed maintaining the initial density requirement. For the composite, the phenomenon of nanoparticles centroid deviates from the unit cell body-centered position, influencing the stiffness of the system. So, the unit cell model of pristine polyimide and that corresponding to the composite were independently computed. It was noticed that Young modulus, shear modulus, and bulk modulus are enhanced by adding SiO_2 in the imidic polymer matrix, whereas no important change was observed for Poisson's ratio at 5% filler. These data prove that the mechanical characteristics of material can be enhanced by being filled embedding nanosilica.

MD simulations were employed by Nazarychev et al.[55] for examination of a thermoplastic aromatic polyetherimide. In the simulation protocol, it used 27 polymeric chains with degree of polymerization of 8. To avoid the residual stresses, a cyclic annealing was applied in temperature interval 300–600 K. The 11 instantaneous states were further chosen at starting states for cooling procedure to 290 K. The impact of the simulated cooling velocity on the mechanical features was computed. After the cooling stage, each of the prepared 11 systems has been subjected to mechanical deformation at different but fixed stretching velocities. Before the deformation, it disabled the algorithm that maintains the length of chemical bonds constant and they were instead described by the harmonic potential from the same force field. The uniaxial deformation modifies the periodic cell dimension at a certain rate. In the direction of the deformation, the systems are incompressible, namely the compressibility was fixed to zero, while in the transverse direction the compressibility was chosen to be 4.5×10^{-10}/Pa. Thus, during computing, the cell elongates along the deformation and compresses in orthogonally to deformation in response to the external pressure. The elasticity modulus was very little influenced by the degree of equilibration, molecular mass, and the size of a simulation box. This could be explained based on the assumption that the major contribution to the elasticity modulus arises from processes occurring at scales lower than the entanglement length. From the temperature dependence of the elasticity modulus, it was extracted the flow temperature of the polyimide, namely 580 K, which agrees with the experimental data. In the presence of high external pressure, the imidic polymer subjected to uniaxial deformation displays better mechanical properties. Even if MD simulations are useful approach to evaluate properties of heterocyclic polymers, it did not show a clear impact of the thermal history or the load application rate.

5.4.5 MISCIBILITY AND INTERFACIAL ADHESION

The molecular modeling can be used to evaluate the miscibility of polyimide with other polymers. The study of Chen et al.[56] employed MD to assess the radial distribution functions and Flory–Huggins parameters of the polyimide/polythene/compatibilizer systems. The computed data indicate that the two polymers are miscible up to a point after which intervene the positive effects of compatibilizer. DPD approach reveals that microphase separation takes place in the polyimide/polythene mixing systems. Moreover, effective interfaces are formed between polyimide and polythene phases after insertion of the compatibilizer. The results of the simulations emphasize that the capacity of mixing systems to resist compression, stretching, and shear deformation is enhanced upon addition of the compatibilizer. Such simulations are useful for predicting the compatibility of polyimide blends with other materials before actual system preparation and attempt to achieve an adequate compatibilizer for both counterparts.

Kumar et al.[57] examined the adhesion between the rutile substrate and an aromatic polyimide (PMR-15) through a coarse grained MD investigation. The initial step involves dividing the polyimide molecules into four segments, and an all-atom nonbonded force field ruling the interaction between polymer and rutile material is developed. The parameters of Buckingham potential for interaction among all the atoms in the polymer fragments and the rutile surface are estimated, assuring that the sum of nonbonded and electrostatic interactions between the substrate and a huge number of configurations of the polymer segments, determined from the QM route and achieved from the fitted potential, is considerably matched. Two coarse grained models of the polyimide are proposed: the first one is related to the interaction between two polyimide molecules and the second one is for the forces acting between coarse grained PMR-15 and rutile. MD results evidenced a traction separation law – adequate for performing FEM simulations of polymer-rutile joints – ruling the separation of polyimide block from rutile layer. Deeper information concerning affinity polymer fragments to the substrate is reported.

Min et al.[58] analyzed the interfacial interaction of two polyimides with silica glass. Based on the framework of steered MD, several different techniques were applied for adhesion evaluation, namely pulling, peeling, and sliding. The approached computational method is useful to examine heterogeneous materials having different interfacial adhesion modes. A new hybridization of interatomic potential is formulated to investigate such interfaces. The modeling showed that pulling test displays a pronounced dependence on the pulling rate; meanwhile, the peeling test is less sensitive. Moreover, pulling also involves biggest force since the interfacial area requires being instantaneously detached. In the same time, the farthest distance at separation is remarked for peeling because it results in line by line adhesion. An aromatic and an aliphatic polyimide were tested to analyze the rigidity dependent adhesion features. The wholly aromatic structure, which has higher rigidity than the aliphatic one owing to the stronger CTC interactions among the chains, imposes a stronger force but a smaller distance at separation from silica in regard with the aliphatic polyimide. Such modeling data provide fruitful insights into the interfacial adhesion interactions of polyimide with SiO_2 glass and with other structures.

5.5 CONCLUDING REMARKS

The molecular modeling represents a remarkable tool for predicting a large category of phenomena occurring in polymer materials. The most important computing protocols and software are briefly presented. In particular, molecular modeling methods are often applied to polyimides in order to describe their conformational properties before preparation procedure. Moreover, the interaction of imidic polymer with other polymers or materials can be better understood using MD method. Transport phenomena like gas permeation can be accurately described through molecular modeling. Thermal, physical, or dielectric properties are better understood by examining the macromolecular architecture of polyimides achieved by simulations. In this way, polymer modeling remains an essential, reliable, and impactful tool for the present and future use in a wider macromolecular science community.

KEYWORDS

- **polyimide**
- **chemical structure**
- **molecular modeling**
- **conformation**
- **properties**
- **applications**

REFERENCES

1. Rigby, D.; Eichinger, B. E. Current Opinion in Solid State and Materials Science. *Poly. Modeling.* **2001**, *5*, 445–450.
2. Clarke, J. H. R. Molecular Modelling of Polymers. *Curr. Opin. Solid St. M.* **1998**, *3*, 596–599.
3. Satyanarayana, K. C.; Abildskov, J.; Gani, R.; Tsolou, G.; Mavrantzas, V. G. Computer Aided Polymer Design Using Multi-Scale Modeling. *Braz. J. Chem. Eng.* **2010**, *27*, 369–380.

4. Porter, D. *Group Interaction Modelling of Polymer Properties*; Marcel Dekker: New York, 1995.
5. Capasso, V. Mathematical Modelling for Polymer Processing: Polymerization. In *Crystallization, Manufacturing*; Springer: Italy, 2003.
6. Perpète, E.; Laso, M. *Multiscale Modelling of Polymer Properties*; Elsevier: Amsterdam, 2006.
7. Gujrati, P. D.; Leonov, A. I. *Modeling and Simulation in Polymers*; Wiley: Germany, 2010.
8. Andreassen, E.; Larsen, A.; Hinrichsen, E. L. *Computer Modelling of Polymer Processing*; Smithers Rapra Publishing: Norway, 1992.
9. Gartner III, T. E.; Jayaraman, A. Modeling and Simulations of Polymers: A Roadmap. *Macromolecules* **2019**, *52*, 755–786.
10. McCrackin, F. L. Configuration of Isolated Polymer Molecules Absorbed on Solid Surfaces Studied by Monte-Carlo Computer Simulation. *J. Chem. Phys.* **1967**, *47*, 1980–1986.
11. Zuilhof, H.; Yu, S. H.; Sholl, D. S. Writing Theory and Modeling Papers for Langmuir: The Good, the Bad, and the Ugly. *Langmuir* **2018**, *34*, 1817–1818.
12. Doi, M. Material Modeling Platform. *J. Comput. Appl. Math.* **2002**, *149*, 13–25.
13. Frenkel, D.; Smit, B. *Understanding Molecular Simulation: From Algorithms to Applications*, 2nd ed.; Academic Press: San Diego, 2002.
14. Rapaport, D. C. *The Art of Molecular Dynamics Simulation*, 2nd ed.; Cambridge University Press: Cambridge, 2004.
15. Wöhr, M.; Bolwin, K.; Schnurnberger, W.; Fischer, M.; Neubrand, W.; Eigenberger, G.; Dynamic Modelling and Simulation of a Polymer Membrane Fuel Cell Including Mass Transport Limitation. *Int. J. Hydrog. Energ.* **1998**, *23*, 213–218.
16. Fakhroleslam, M.; Samimi, A.; Mousavi, S. A.; Rezaei, R. Prediction of the Effect of Polymer Membrane Composition in a Dry Air Humidification Process Via Neural Network Modeling. *Iran. J. Chem. Eng.* 13, **2016**, 73–83.
17. Hulubei, C.; Albu, R. M.; Lisa, G.; Nicolescu, A.; Hamciuc, E.; Hamciuc, C.; Barzic, A. I. Antagonistic Effects in Structural Design of Sulfur-based Polyimides as Shielding Layers for Solar Cells. *Sol. Energ. Mater. Sol. Cells.* **2019**, *193*, 219–230.
18. Zhang, M.; Niu, H.; Wu, D. Polyimide Fibers with High Strength and High Modulus: Preparation, Structures, Properties, and Applications. *Macromol. Rapid Commun.* **2018**, *39*, e1800141.
19. Ha, C.-S.; Mathews, S. Polyimides and High Performance Organic Polymers. In *Advanced Functional Materials*; Woo, H.-G. et al., Eds.; Springer-Verlag Berlin Heidelberg: Korea, 2011.
20. Barzic, A. I.; Hulubei, C.; Stoica, I.; Albu, R. M. Insights on Light Dispersion in Semialicyclic Polyimide Alignment Layers to Reduce Optical Losses in Display Devices. *Macromol. Mater. Eng.* **2018**, *303*, 1800235.
21. Damaceanu, M.-D.; Rusu, R.-D.; Olaru, M.; Stoica, I.; Bruma, M. Nanostructured Polyimide Films by UV Excimer Laser Irradiation. *Rom. J. Inform. Sci. Technol.* **2010**, *13*, 368–377.
22. Mathews, A. S.; Kim, I.; Ha; C.-S.; Synthesis, Characterization, and Properties of Fully Aliphatic Polyimide and their Derivatives for Microelectronics and Optoelectronics Applications. *Macromol. Res.* **2007**, *15*, 114–128.
23. Sroog, C. E. Polyimide. *Prog. Polym. Sci.* **1991**, *16*, 561–694.

24. Ghosh, M. K.; Mittal, K. L. *Polyimides: Fundamental and Applications*; Marcel Dekker: New York, 1996.
25. Bakar, M.; Hausnerova, B.; Kostrzewa, M. Effect of Diisocyanates on the Properties and Morphology of Epoxy/Polyurethane Interpenetrating Polymer Networks. *J. Thermoplast. Compos. Mater.* **2013**, *26*, 1364–1376.
26. Yang, S. Y.; Yuan, L. L. Advanced Polyimide Films. In *Advanced Polyimide Materials: Synthesis, Characterization, and Applications*; Yang, S.-Y., Ed.; Elsevier: China, 2018; pp 1–66.
27. Ioan, S.; Hulubei, C.; Popovici, D.; Musteata, V. E. Origin of Dielectric Response and Conductivity of Some Alicyclic Polyimides. *Polym. Eng. Sci.* 2013, *53*, 1430–1447.
28. Nam, K. H.; Choi, H. K.; Yeo, H.; You, N. H.; Ku, B. C. Yu, J. Molecular Design and Property Prediction of Sterically Confined Polyimides for Thermally Stable and Transparent Materials. *Polymers* **2018**, *10*, 630.
29. Hesse, L.; Sadowski, G. Modeling Liquid-liquid Equilibria of Polyimide Solutions. *Ind. Eng. Chem. Res.* **2012**, *51*, 539–546.
30. Haile, J. M. *Molecular Dynamics Simulation: Elementary Methods*; John Wiley and Sons: New York, 1992.
31. Yamamoto, T. Computer Modeling of Polymer Crystallization—Toward Computer-assisted Materials' Design. *Polymer* **2009**, *50*, 1975–1985.
32. Karatrantos, A.; Clarke, N.; Kroger, M. Modeling of Polymer Structure and Conformations in Polymer Nanocomposites from Atomistic to Mesoscale: A Review. *Polym. Rev.* **2016**, *56*, 385–428.
33. Gooneie, A.; Schuschnigg, S.; Holzer, C. A Review of Multiscale Computational Methods in Polymeric Materials. *Polymers* **2017**, *9*, 16.
34. Lei, H.; Qi, S.; Wu, D. Hierarchical Multiscale Analysis of Polyimide Films by Molecular Dynamics Simulation: Investigation of Thermo-mechanical Properties. *Polymer* **2019**, *179*, 121645.
35. Sava, I.; Hurduc, N.; Sacarescu, L.; Apostol, I.; Damian, V. Study of the Nanostructuration Capacity of Some Azopolymers with Rigid or Flexible Chains. *High Perform. Polym.* **2012**, *25*, 13–24.
36. Albu, R. M.; Nica, S. L.; Barzic, A. I. Refraction and Polarization Properties of Some Fluorinated Imidic Polymers. *Polym. Bull.* **2018**, *75*, 1535–1546.
37. Barzic, A. I.; Albu, R. M.; Ioanid, E. G.; Hulubei, C. Molecular Design of Some Semi-alicyclic Polyimides as a Route to Improve Refraction and Dielectric Properties for Liquid Crystal Display Applications. *High Perfom. Polym.* **2018**, *30*, 776–786.
38. Albu, R. M.; Hulubei, C.; Stoica, I.; Barzic, A. I. Semi-alicyclic Polyimides as Potential Membrane Oxygenators: Rheological Implications on Film Processing, Morphology and Blood Compatibility. *eXPRESS Polym. Lett.* **2019**, *13*, 349–364.
39. Chang, K.-S.; Tung, K.-L.; Linb, Y.-F.; Lin, H.-Y. Molecular Modelling of Polyimides with Intrinsic Microporosity: From Structural Characteristics to Transport Behavior. *RSC Adv.*, **2013**, *3*, 10403–10413.
40. Luo, S.; Wiegand, J. R.; Gao, P.; Doherty, C. M.; Hill, A. J. Molecular Origins of Fast and Selective Gas Transport in Pentiptycene-containing Polyimide Membranes and their Physical Aging Behavior. *J. Membr. Sci.* **2016**, *518*, 100–109.
41. Vergadou, N.; Theodorou, D. N. Molecular Modeling Investigations of Sorption and Diffusion of Small Molecules in Glassy Polymers. *Membranes* **2019**, *9*, 98.
42. Greenfield, M. L.; Theodorou, D. N. Geometric Analysis of Diffusion Pathways in Glassy and Melt Atacticpolypropylene. *Macromolecules* **1993**, *26*, 5461–5472.

43. Vergadou, N. Prediction of Gas Permeability of Inflexible Amorphous Polymers Via Molecular Simulation. Ph.D. Thesis, University of Athens, Greece, 2006.
44. Theodorou, D. N. Hierarchical Modelling of Polymeric Materials. *Chem. Eng. Sci.* **2007,** *62,* 5697–5714.
45. Young, J. A.; Farmer, B. L.; Hinkley, J. A. Molecular Modeling of the Poling of Piezoelectric Polyimides. *Polymer* **1999,** *40,* 2787–2795.
46. Li, M.; Liu, X. Y.; Qin, J. Q.; Gu, Y. Molecular Dynamics Simulation on Glass Transition Temperature of Isomeric Polyimide. *eXPRESS Polym. Lett.* **2009,** *3,* 665–675.
47. Sun, H. COMPASS: An ab Initio Force-field Optimized for Condensed-phase Applicationss Overview with Details on Alkane and Benzene Compounds. *J. Phys. Chem. B.* **1998,** *102,* 7338–7364.
48. Chang, K. S.; Tung, C. C.; Wang, K. S.; Tung, K. L. Free Volume Analysis and Gas Transport Mechanisms of Aromatic Polyimide Membranes: A Molecular Simulation Study. *J. Phys. Chem. B.* **2009,** *113,* 9821–9830.
49. Heuchel, M.; Hofmann, D. Molecular Modelling of Polyimide Membranes for Gas Separation. *Desalination* **2002,** *144,* 67–72.
50. Park, C. H.; Tocci, E.; Kim, S.; Kumar, A.; Lee, Y. M.; Drioli, E. A Simulation Study on OH-Containing Polyimide (HPI) and Thermally Rearranged Polybenzoxazoles (TR-PBO): Relationship Between Gas Transport Properties and Free Volume Morphology. *J. Phys. Chem. B.* **2014,** *118,* 2746–2757.
51. Nazarychev, V. M.; Larin, S. V.; Yakimansky, A. V.; Lukasheva, N. V.; Gurtovenko, A. A.; Gofman, I. V.; Yudin, V. E.; Svetlichnyi, V. M.; Kenny, J. M.; Lyulin, S. V. Parameterization of Electrostatic Interactions for Molecular Dynamics Simulations of Heterocyclic Polymers. *J. Polym. Sci. B Polym. Phys.* **2015,** *53,* 912–923.
52. Vollmayr, K.; Kob, W.; Binder, K. How do the Properties of a Glass Depend on the Cooling Rate? A Computer Simulation Study of a Lennard-Jones System. *J. Chem. Phys.* **1996,** *105,* 4714–4728.
53. Nazarychev, V. M.; Lyulin, A. V.; Larin, S. V.; Gurtovenko, A. A.; Kenny, J. M.; Lyulin, S. V. Molecular Dynamics Simulations of Uniaxial Deformation of Thermoplastic Polyimides. *Soft Matter.* **2016,** *12,* 3972–3981.
54. Mo, Y.; Zhang, H.; Xu, J. Molecular Dynamic Simulation of the Mechanical Properties of PI/SiO$_2$ Nanocomposite Based on Materials Studio. *J. Chem. Pharm. Res.* **2014,** *6,* 1534–1539.
55. Nazarychev, V. M.; Lyulin, A. V.; Larin, S. V.; Gurtovenko, A. A.; Kenny, J. M.; Lyulin, S. V. Molecular Dynamics Simulations of the Uniaxial Deformation of Thermoplastic Polyimides. *Soft Matter* **2016,** *12,* 3972–3981.
56. Chen, C.; Li, J.; Jin, Y.; Khosla, T.; Xiao, J.; Cheng, B. A Molecular Modeling Study for Miscibility of Polyimide/Polythene Mixing Systems with/without Compatibilizer. *J. Polym. Eng.* **2018,** *38,* 891–898.
57. Kumar, A.; Sudarkod, V.; Parandekar, P. V.; Sinha, N. K.; Prakash, O.; Nair, N. N.; Basu, S. Adhesion Between a Rutile Surface and a Polyimide: A Coarse Grained Molecular Dynamics Study. *Model. Simul. Mater. Sci.* **2018,** *26,* 035012.
58. Min, K.; Rammohan, A. R.; Lee, H. S.; Shin, J.; Lee, S. H.; Goyal, S.; Park, H.; Mauro, J. C.; Stewart, R.; Botu, V.; Kim, H.; Cho, E. Computational Approaches for Investigating Interfacial Adhesion Phenomena of Polyimide on Silica Glass. *Sci. Rep.* 2017. doi:10.1038/s41598-017-10994-8.

CHAPTER 6

Patterning Polyimide Films at Nanoscale Using Dynamic Plowing Lithography

IULIANA STOICA*

"Petru Poni" Institute of Macromolecular Chemistry, Laboratory of Polymeric Materials Physics, 41A Grigore Ghica Voda Alley, 700487, Iasi, Romania

*E-mail: stoica_iuliana@icmpp.ro

ABSTRACT

Over the time, atomic force microscopy (AFM) turned out to be besides a powerful technique used to visualize and to measure nanometric surface structures, an innovative technique that can be successfully applied for various nanometer scale modifications of polymer samples. Among the AFM-based lithographic techniques, dynamic plowing lithography (DPL) offers the possibility of surface modeling by indenting it with a vibrating tip using semicontact mode after a predefined pattern with a spacing even of a few nanometers without leading to edge irregularities. The advantage of the method lies in the fact that both the lithography and the imaging of the lithographed surface can be made without changing the scanning tool or the cantilever with no undesirable modifications. Among the polymers that can be patterned at nanoscale with DPL procedure, polyimides represent a special class due to their attractive mechanical, thermal, and electrical properties. A complex characterization of the obtained morphology through the surface texture and functional parameters can reveal the influence of the diamine and dianhydride moieties on the characteristics of the pattern in the same nanolithography conditions, this being explained based on the aliphatic/aromatic nature of the monomers and the backbone flexibility. This procedure has paved the way for new applications where the surface anisotropy is mandatory, such as fabrication of bio-analytical assays, guided cell growth, micro-fluidics, and liquid crystal alignment layers.

6.1   INTRODUCTION

Polymer nanostructures, due to their optical, electrical, and mechanical properties, are widely used in diverse research fields, such as optoelectronics, microelectronics, or biomedicine. There are several methods to obtain ordered polymeric nanostructures of different sizes and appearances. These include, for example, rubbing with textile materials through specially designed devices, the size of the oriented rubbing-induced grooves being strongly related not only to the microstructure and flexibility of textile fibers but also with polymer structural organization.[1-4] Another method is UV laser irradiation diffraction through a phase mask, the dimensions of the obtained periodic tridimensional nanogrooves depending on the phase mask pitch, incident fluence, the number of irradiation pulses, and the structure of the polymers.[5-10] Plasma treatment through grid masks can also be used to create rectangular 3D-patterned microstructures with alternating hydrophilic and hydrophobic surface chemistries, depending on the plasma exposure time and power.[11,12] The introduction in the structure of the polymer of a lyotropic liquid crystal template was also used to obtain a certain pattern depending on the template concentration.[13,14]

Recently discovered, the low-cost atomic force microscopy (AFM) tip-based nanomachining method, namely dynamic plowing lithography (DPL) represents a potential manner to fabricate polymer nanostructures by direct modification of the polymer surface by means of AFM cantilever tips.[15] Research into AFM nanolithography, this unique tool for materials structuring, and patterning with nanometer precision is fast growing, and it continues to attract increasing interest in nanotechnology. This combined fabrication and characterization function in AFM nanolithography allows convenient in situ and in-line pattern creation and characterization with nanometer resolution.

In this context, the chapter will first present a brief description of this AFM powerful innovative technique used both for visualization, measurement, and for modification at nanometer scale of the polymer samples. Also, a classification of the lithographic techniques based on AFM with a short description of each force-assisted nanolithography and bias-assisted nanolithography is presented. Among the force-assisted AFM nanolithographies, the AFM scratching techniques, namely the static and dynamic plowing, performed in two different modes: vector and raster, along with their advantages and limitations were emphasized. Among the polymers that can be patterned at nanoscale with these procedures, polyimides represent a special class due to their attractive mechanical, thermal, and electrical properties. By comparison, it presented the influence of the DPL and static plowing lithography (SPL) in vector mode not only on the AFM tip sharpness, contamination, and degradation but also on the aspect of the polyimide surface pattern, concluding once again that the DPL is more adequate to be used to pattern a surface in order to obtain well-defined nanochannels in contradistinction with the SPL. Further, a complete characterization through the surface texture and functional parameters of the obtained nanostructured morphology was done revealing the influence of the dianhydride and diamine moieties on the characteristics of the pattern in the same nanolithography conditions. Two factors were taken into consideration to explain the phenomena: the backbone flexibility and the aliphatic/aromatic nature of the monomers. This technique, further developed, can successfully be used in new applications where the surface anisotropy is necessary. Among them are the fabrication of bioanalytical assays, microfluidics, guided cell growth, and liquid crystal alignment layers.

6.2　DPL AS A TECHNIQUE FOR NANOSCALE PATTERNING

Scanning probe microscopy is a branch of microscopy including techniques such as: scanning tunneling microscopy,[16] AFM,[17] and its additional methods: Kelvin probe force microscopy,[18] magnetic force microscopy,[19] scanning capacitance microscopy,[20] piezoresponse force microscopy,[21] and so on, scanning probe electrochemistry,[22] fluidic force microscopy,[23] scanning thermal microscopy,[24] scanning near-field optical microscopy,[25] and other. In these techniques, the images are formed by using a probe through which a sample is scanned. Among these, AFM, invented in 1986 by Binnig and its collaborators[17] can be used to obtain quantitative topographic images of both insulating and conducting materials surfaces with molecular and atomic resolution in liquid, air, or vacuum by recording the forces (either attractive or repulsive) sensed by a sharp tip (with a small radius of curvature of about 10 nm) as it scans a surface of a specimen and maintaining a certain parameter constant.

　　Depending on the nature of the tip motion, AFM can be operated in three different modes, namely contact mode (which is a static mode), tapping mode/intermittent contact/vibrating mode/semicontact mode, and noncontact mode (which are dynamic modes).

In contact mode, the tip of a low spring constant/stiffness cantilever track the surface of the sample at a depth where the global force is repulsive, the image being obtained directly by utilization of the deflection signal of the cantilever or by utilization of the feedback signal that is necessary to preserve the constant position of the cantilever.

In tapping mode, the most frequently used AFM mode in ambient conditions or in liquids, the cantilever oscillates by means of a small piezo element at or near its resonance frequency with a certain amplitude, kept constant, as long as there is no drift or interaction with the surface. When the tip reaches the surface, the interaction forces acting on the cantilever's tip such as: dipole–dipole interactions, Van der Waals forces, electrostatic forces, produce a change in the amplitude of the cantilever's oscillation. The electronics use the amplitude as an input parameter and supervise the height of the cantilever over the sample. By adjusting this height, cantilever oscillation amplitude is preserved at the set value during scanning. Thus, the image is obtained by recording the force acting for a short duration on the sample surface wherewith the AFM tip apex gets in intermittent contacts. In this way, the damage produced by the tip to the surface is reduced because in this mode the tip-surface lateral forces are notably lower.

Another mode in AFM is the one in noncontact, in which the oscillating cantilever's tip at or near its resonant frequency (where the oscillation amplitude is less than 10 nm) does not touch the surface. The cantilever's resonance frequency or oscillation amplitude decreased by the van der Waals forces or any other long-range force are always maintained constant through the feedback loop system by measuring in each data point the distance between the tip and the sample, and adjusting it as needed. This is also the procedure through which the topographic image of the sample surface is constructed by the software. Thus, the sample or tip degradation is avoided, the method being very useful to investigate soft samples.

Besides collecting images of the sample surface at nanoscale, AFM can also be used to create different patterns by nanolithography in various media on materials, such as polymers, biological molecules, semiconductors, and metals and offers the possibility of immediate analysis of their morphological and physical properties by adjusting the proper configuration and integrating additional measurement modules. The AFM nanolithography technique is based on the interaction, either of physical or chemical nature between the probe and surface of the substrate generating nanostructures due to the physical modifications induced by exerted forces and/or to the chemical reactions induced by applied external fields. Therefore, according to the working principle, the AFM-based lithographic techniques can be divided into two categories, namely force-assisted AFM nanolithography (scratching, mechanical indentation, thermomechanical writing, dip-pen nanolithography, and nanomanipulation,) and bias-assisted AFM nanolithography (anodic oxidation, electrostatic attraction, electrochemical deposition, shock wave propagation, and nanoscale explosion).[26] Scheme 6.1 presents the classification of the lithographic techniques based on AFM and a short description of each one of the two major classes, force-assisted AFM nanolithography and bias-assisted AFM nanolithography.

Among the AFM nanolithographic techniques assisted by force, AFM scratching techniques (static plowing and dynamic plowing) are the most handy to use in order to obtain deep trenches with the characteristic shape of the used probe, applying strong loading forces to remove the substrate. The scratching of the polymer is more complex than scratching of the metal or semiconductors owing to the effects of the adhesion force.[27] Scheme 6.2 represents the classification of the two main AFM scratching techniques together with a short description of each of them.

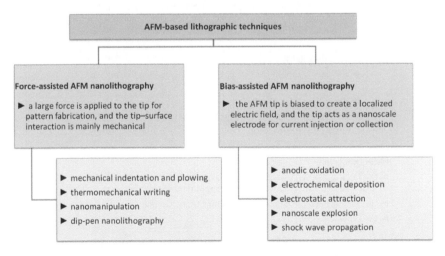

SCHEME 6.1 Classification of the lithographic techniques based on AFM.

Static plowing and DPL are two representative methods used in AFM to fabricate different nanostructured surfaces through the mechanical interactions of the cantilever tip and substrates.[26] In SPL, employed in contact mode, strong loading forces are applied to the tip in order to remove material from the substrate (e.g., to create an etch mask from a single resist layer) or to shape up only the substrate in a well-defined way leaving behind deep trenches, features in the shape of lines, rectangles, and squares, with the characteristic aspect of the used probe. Although low-cost and low-effort SPL has advantages, such as the precision of alignment, the absence of additional processing steps, and the possibility to be used to characterize microwear processes on materials of technological interest[28]; however, this method presents some disadvantages too because it is employed in contact mode and the twist of the cantilever may induce edge irregularities. In this case, additional modifications may be produced while imaging the surface before or after the lithography due to dragging of the surface depending on the local stiffness of the sample. Scheme 6.3 summarizes the advantages and limitations of the SPL.

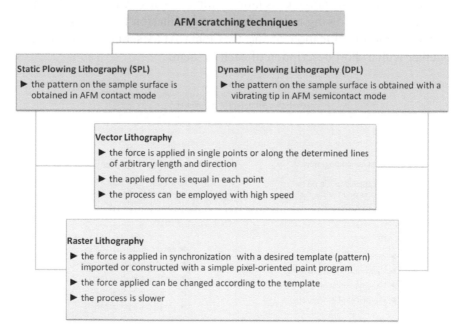

SCHEME 6.2   Classification of the main AFM scratching techniques.

Instead, DPL (using tapping mode) implies the modification of the surface by scratching it with a piezo drive vibrating cantilever's tip near its resonance frequency and its oscillation amplitude depending on the distance between the tip and the sample. Thus, the force acting on the tip is modulated by adjusting the amplitude of the cantilever oscillations. As the tip approaches the sample surface, structures are created by elastic and plastic deformations of the substrate.[29] The method has multiple advantages. Among the AFM scratching techniques, dynamic plowing is a type of lithography where the pattern irregularities caused by cantilever torsion are reduced because the soft polymers are modified by means of a low force in tapping mode without changing the whole scanning field because of dragging. Also, this method permits to image the lithographed surface with the absence of supplementary modification or any alteration without changing the scanning tool or the cantilever.[30] As a limitation, DPL sometimes leads to the formation of undesired remarkably high border walls surrounding the carved structures.[30] The advantages and limitations of the DPL are briefly presented in Scheme 6.3.

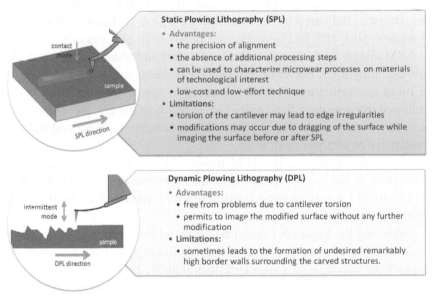

SCHEME 6.3   Advantages and limitations of the force-assisted AFM lithographic techniques.

Both static and DPL can be performed in two different modes: vector and raster (Scheme 6.2). In the case of vector lithography the software provides a set of commands that allow the AFM specialist to apply a force through the tip in single points or along the determined lines of arbitrary length and direction with defined scan speed/oscillation amplitude.[31] Thus, the vector lithography is convenient because it can be achieved with high speed, but meanwhile, a drawback could be that the force is equal in each point. Instead, the raster lithography can be made using a desired template, imported or constructed with a simple pixel-oriented paint program; the applied force can be changed according to the template by changing the beam bending, either by setting of the scanner displacement on defined distance along Z-axis or by setting of the SetPoint value. As presented also in Scheme 6.2, a disadvantage of the raster lithography is the low working speed, but this time an advantage can be considered the ability to change the applied force according to the pattern image.[32]

In our group it was studied, by comparison, the influence of the DPL and SPL in vector mode on the AFM tip sharpness, contamination, and degradation.[29] In order to visualize the shape of the scanning tip attached to a rectangular NSG10 cantilever (length = 95 ± 5 μm, width of 30 ± 3 μm, resonant frequency = 254 kHz, and nominal spring constant = 13 N/m), a test grating TGT1 having a multitude of sharp tips set on a silicon wafer surface with the characteristics delivered by the manufactures, namely: height = 0.3–0.5 μm, tip angle = 50 ± 10°, and tip curvature radius = 10 nm was used. Based on the AFM images of this grating obtained before and after dynamic and static plowing lithographs, the AFM probes were subsequently rebuilt by means of the blind reconstruction method presented in specialized literature[33–36] and employed using a specialized software. In this way the tips reconstructed by the deconvolution of the TGT1 scanned images were characterized and some quantitative results were extracted.

As seen in Figure 6.1, the AFM image of the TGT1 test grating and 3D perspective of the reconstructed tip obtained before the experiments reveal that the AFM tip was in good condition with a radius close to 10 nm, the same guaranteed by the manufacturer.

Analyzing the Figure 6.1 it can be observed that, although some sample materials adhered onto the apex of the cantilever's tip used to create very well-defined nanochannels with DPL technique by pressing and lifting off in many vibration cycles, still its conical shape was preserved undisturbed. The measurements indicate a tip radius of 12–13 cm. When the cantilever was used to perform nanochannels on polyimide film using SPL, its tip deconvoluted from the presented TGT1 test grating AFM image (Fig. 6.1) had on top some protuberances that influenced preferentially the tip radius (14 nm) and cone angle (109°) on the scratching direction. Because the scratching process happens in contact mode, the polymeric material tends to be moved in the direction of the scratch and on each side of the formed nanochannel (Fig. 6.1). Thereby, the created pattern will have not only uneven walls but also supplementary polymeric material which would prevent any subsequent application of it. Our conclusion was that, in the study of soft polymers of both force-assisted AFM lithographic techniques, DPL is more adequate to be used to pattern a surface in order to obtain well-defined nanochannels in contradistinction with the SPL.

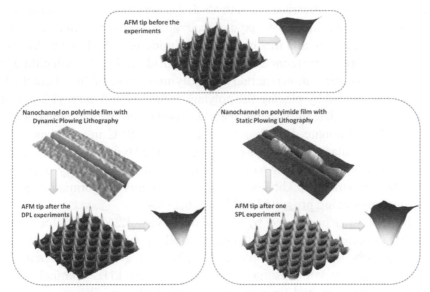

FIGURE 6.1 AFM images of the TGT1 test grating and 3D perspectives of the reconstructed tips obtained before and after dynamic and SPL were performed on a polyimide film.

6.3 DEVELOPMENT OF PERIODIC NANOSTRUCTURES ON POLYIMIDE FILMS DURING DPL

DPL was successfully used to perform lithography on polymer surfaces. Over the time, the studies were frequently reported on poly(methyl methacrylate),[15,37–44] polystyrene (PS),[38,44] polycarbonate,[44,45] poly(ethylene terephthalate),[46] and PS-*block*-poly(ethylene/butylenes)-*block*-PS triblock copolymer.[47] Thus, interesting nanoscale periodical structures such as, pits, nanodots, nanolines, nanogrooves, 2D nanostructures, 3D nanostructures, on flat or even curved surfaces were obtained.

Among the polymers that can be patterned at nanoscale with DPL procedure, polyimides represent a special class due to their attractive mechanical, thermal, and electrical properties. But studies regarding polyimides patterning via DPL are not very common in the specialized literature.

Consequently, in our group we used DPL to induce anisotropy of morphology on three different polyimide films, namely poly(DOCDA-ODA), poly(DOCDA-PDOA), and poly(BTDA-ODA) in order to adapt the obtained size-controlled nanochannels at specific applications.[29] The two semialiphatic polyimides poly(DOCDA-ODA) and poly(DOCDA-PDOA), respectively were synthesized from 5-(2,5-dioxotetrahydrofuryl)-3-methyl-3-cyclohexenyl-1,2-dicarboxylic anhydride (generically called epiclon) and two aromatic diamines: 4,4'-oxydianiline and 1,4-bis(p-aminophenoxy)-benzene, respectively and the aromatic polyimide poly(BTDA-ODA) from 3,3',4,4'-benzophenone tetracarboxylic dianhydride and 4,4'-oxydianiline by two-step polycondensation reactions.[48,49] Polyimide films of 40 ± 5 μm thickness were achieved by placing the polymer solutions in N-methyl-2-pyrrolidone on glass slides, introducing them in a preheated oven at temperatures 80–150°C to eliminate the solvent and drying at 120°C in a vacuum oven for 24 h. Molecular modeling was performed using Hyperchem professional for each polyimide in order to get an idea about the flexibility degree of the studied polymers backbones, the minimum energy conformations for 11 repeating units being presented in Figures 6.2–6.4.

The DPL was performed by means of Nova v.1.26.0.1443 for Solver software in vector mode—the set-point method on pre-scanned areas using a Scanning Probe Microscope Solver Pro-M platform, in air with a rectangular NSG10 cantilever with resonant frequency of 254 kHz, nominal spring constant of 13 N/m, and tip curvature radius of 10 nm. The nanochannels were set to be formed at 400 nm from each other by indenting the samples with the vibrating tip in moderate tapping mode. In this case, r_{sp}, the ratio between the free oscillation of the cantilever A_0 and the set-point amplitude A_{sp} was kept around 0.5. During DPL, a pressing force of 1200 nN was continually applied with a set velocity of 0.6 μm/s. Before and after DPL, 2D AFM morphology images, corresponding Fourier transform images (FFT) (Figs. 6.2–6.4), height profiles (Figs. 6.2–6.4), height histograms (Figs. 6.5–6.7), and bearing ratio curves (Abbott–Firestone curve) used to calculate the functional indices: surface bearing index (Sbi), core fluid retention index (Sci), valley fluid retention index (Svi), and the functional volume parameters: peak material volume (Vmp), core material volume (Vmc), core void volume (Vvc), and valley void volume (Vvv)[50] (Figs. 6.5–6.7) were obtained for each sample.

The representative 2D tapping mode AFM images revealed a grainy morphology for pristine polyimide films poly(DOCDA-ODA) and poly(BTDA-ODA) containing 4,4'-oxydianiline residues (Figs. 6.2 and 6.4) and, in addition, small circular pores with irregular contour for poly(DOCDA-PDOA) probably owed to higher number of ether linkages in the proximity of the phenylene units in polyimide backbone (Fig. 6.3). The cross-section profiles of the pristine polyimides indicate that the surface roughness was of a few nanometers influenced by the morphological features (Figs. 6.2–6.4). Also, the surface height distributions were bell shaped, with the bell top, indicated by an arrow in Figures 6.5–6.7 at the mean height in the image.

After the DPL was realized with the vibrating tip in vector mode moderate tapping, due to supplementary cantilever torsion, well-defined nanochannels, uniform along their length, of different characteristics depending on the structure and flexibility of the polyimide, were obtained and imaged without any further modification. The bimodal shape of the height histograms of the lithographed samples, having two data peaks, indicated in Figures 6.5–6.7 by the two arrows, confirmed as well the appearance of these formations on the surface. Thus, on the lithographed surfaces there are two distinct entities layout on two distinct levels, compared to the unmodified samples, in which case only one entity can be noticed.

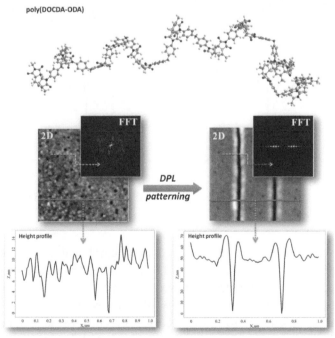

FIGURE 6.2   Minimum energy conformation for 11 repeating units and 2D AFM morphology image and corresponding fast FFT and height profile, before and after DPL, obtained for poly(DOCDA-ODA).

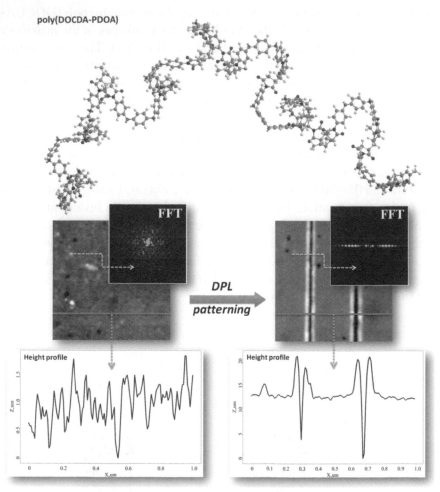

FIGURE 6.3   Minimum energy conformation for 11 repeating units and 2D AFM morphology image and corresponding fast FFT and height profile, before and after DPL, obtained for poly(DOCDA-PDOA).

For poly(DOCDA-ODA), obtained from epiclon dianhydride and the aromatic diamine containing an ether bridge in the structural unit, the intra/intermolecular charge-transfer complex interactions through the chains were reduced leading to the smallest ability of chain packing and by default to a less wrapped architecture, allowing the formation of nanochannels with great depth of about 60 ± 5 nm as it can be observed from the height profile from Figure 6.2. For this sample, after DPL, the root mean square roughness (Sq) increases even 10 times compared to pristine sample, namely from

1.7 to 11.2 nm. Also, the increase of the surface morphology complexity was highlighted by the increase of the surface area ratio (Sdr), a hybrid 3D parameter which is defined as the ratio of the interfacial area increment over the projected area increased from 1.878 to 20.409%.

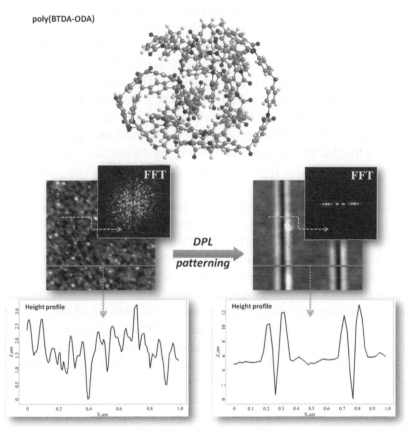

FIGURE 6.4 Minimum energy conformation for 11 repeating units and 2D AFM morphology image and corresponding fast FFT and height profile, before and after DPL, obtained for poly(BTDA-ODA).

In the case of poly(DOCDA-PDOA), achieved from epiclon and the aromatic diamine with two ether bridges close to phenyl rings in the structural unit, the nanochannels depth of about 19 ± 6 nm was much bigger than that of the pores from the surface, but smaller than that obtained in the poly(DOCDA-ODA) case, probably due to the supplementary aromatic rings from diamine residues. According to minimum energy conformation for 11 repeating units presented in Figure 6.3, these aromatic rings slightly intensify the intra/intermolecular charge-transfer complex interactions bringing the chains closer creating a folded conformation. Subsequent to the DPL, the Sq increases about four times compared to pristine sample from 0.8 to 3.2 nm leading to a moderate increase of the surface morphology complexity, illustrated by increasing Sdr parameter from 0.283 to 3.330%.

Poly(BTDA-ODA) had the most packed structure as it can be seen in Figure 6.4 from minimum energy conformation for 11 repeating units due to its fully aromatic nature, the original epiclon dianhydride met in previous two cases being replaced with BTDA which induces the greatest probability of inter and intramolecular CTC interactions. This makes the obtained film more rigid and this is how it explained the occurrence of smaller nanochannels of about 10 ± 3 nm after DPL. Even if the same force was applied as in previous cases, the penetration power of the cantilever was reduced. Also, the Sq and the Sdr (morphology complexity of the surface) slightly increase from 0.6 to 2.1 nm and from 0.212 to 0.653%, respectively, compared to pristine sample.

Since the application sought by the creation of these nanochannels was nanostructured substrates for liquid crystal orientation, other spatial and functional texture parameters which reveal important information on surface orientation, bearing and fluid retention abilities were calculated for a more detailed description of the interface where the polyimide/nematic molecules interaction occurs. The texture direction index of the surface (Stdi), defined as the average amplitude sum divided by the amplitude sum of the dominating direction was used to found the degree of surface orientation before and after DPL. Initially, the pristine polyimide surfaces had no particular orientation, the texture direction index being in all cases close to 1 as follows: for poly(DOCDA-ODA), Stdi = 0.820; for poly(DOCDA-PDOA), Stdi = 0.820; and for poly(BTDA-ODA), Stdi = 0.783. After DPL, due to nanochannels disposing, Stdi values close to zero, namely 0.114 for poly(DOCDA-ODA), 0.112 for poly(DOCDA-PDOA), and 0.151 for poly(BTDA-ODA) indicate a dominant direction of the morphology.

Based on Abbott–Firestone curves from Figures 6.5–6.7, functional indices displayed in Table 6.1 were used to describe the bearing and fluid retention properties of the samples before and after DPL. Thus, it was found that all the patterned films present better bearing properties and rising fluid retention in the core and valley zones. The poly(DOCDA-ODA)/DPL sample proving to be the best candidate for the pursued application, meeting all the demands regarding the bearing properties, such as the highest Sbi, Svi, and the lowest Sci.

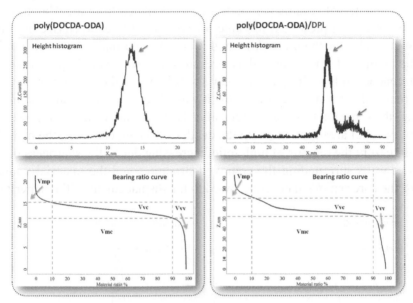

FIGURE 6.5 Height histogram and bearing ratio curve (Abbott–Firestone curve) used to calculate the functional volume parameters Vmp, Vmc, Vvc, and Vvv, before and after DPL, obtained for poly(DOCDA-ODA).

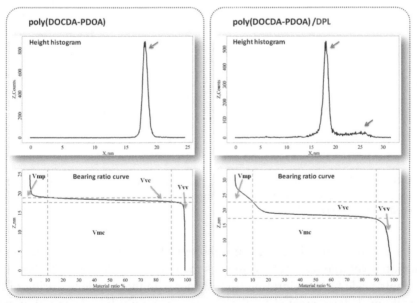

FIGURE 6.6 Height histogram and bearing ratio curve (Abbott–Firestone curve) used to calculate the functional volume parameters Vmp, Vmc, Vvc, and Vvv, before and after DPL, obtained for poly(DOCDA-PDOA).

The functional volume parameters also calculated from the Abbott–Firestone curves and presented in Table 6.1 were used to characterize both the new created nanochannels (Vvc and Vvv) and the undesired high border walls (Vmp and Vmc), formed as a result of the accelerated tip oscillations that conduct to high-frequency shear stress between the cantilevers tip and the material from the surface. Thus, it was observed that for all polyimide films, these parameters were larger after DPL, the air volume in the center and low relief area growing with the depth of the nanochannels, the material volume from the regions of high relief increasing due to the sharpness of the border walls surrounding the engraved structures and the material volume from the core regions increasing owed to the displacement of the material.

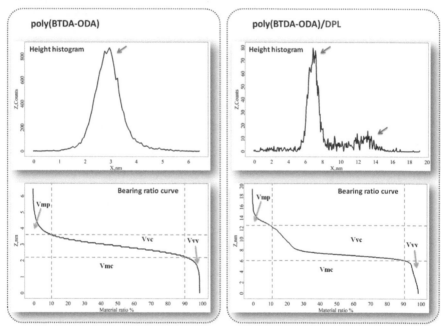

FIGURE 6.7   Height histogram and bearing ratio curve (Abbott–Firestone curve) used to calculate the functional volume parameters Vmp, Vmc, Vvc, and Vvv, before and after DPL, obtained for poly(BTDA-ODA).

Furthermore, the described morphological characteristics through these functional volume parameters, such as the trace from the walls of the nanochannels made by the vibrating tip together with the gentile texturing of the nanochannels base created by the tip apex can contribute to the alignment of the liquid crystals on patterned polyimide films, the best results being obtained once more for poly(DOCDA-ODA) sample.[29]

TABLE 6.1   Functional Indices and Functional Volume Parameters Calculated from the Abbott–Firestone Curves Obtained for Poly(DOCDA-ODA), Poly(DOCDA-PDOA), and Poly(BTDA-ODA) Films, Before and After DPL.

Sample	Functional index			Functional volume parameter			
	Sbi	Sci	Svi	Vmp (nm^3/nm^2)	Vmc (nm^3/nm^2)	Vvc (nm^3/nm^2)	Vvv (nm^3/nm^2)
poly(DOCDA-ODA)	0.310	1.289	0.151	0.077	1.231	1.665	0.259
poly(DOCDA-ODA)/DPL	0.629	1.361	0.173	0.430	5.214	12.204	1.944
poly(DOCDA-PDOA)	0.137	0.956	0.113	0.043	0.405	0.577	0.087
poly(DOCDA-PDOA)/DPL	0.504	2.001	0.132	0.229	1.280	4.328	0.417
poly(BTDA-ODA)	0.244	1.622	0.077	0.049	0.483	0.696	0.059
poly(BTDA-ODA)/DPL	0.503	2.181	0.097	0.116	1.294	3.562	0.164

6.4   HIGH-TECH APPLICATIONS OF DPL-INDUCED POLYIMIDE PERIODIC NANOSTRUCTURES

Taking into consideration their attractive thermal, mechanical, and electrical properties, polyimide films patterned at nanoscale using DPL can be employed in various domains, such as microelectronics and optoelectronics, biomedicine, in applications such as micro-fluidics, fabrication of bio-analytical assays, and liquid crystal alignment layers. For optimal orientation of liquid crystals on polyimide surfaces, the dimensions of the well-defined nanochannels can be controlled by the polymer structural organization and adjusting the loading force in DPL process. Therefore, custom nanostructures will be obtained using different polyimide structures depending on the type of the liquid crystal used in desired applications. Recently, Murray et al.[51] used AFM on a polyimide-coated substrate in order to create complex patterns consisting of two-dimensional topological defect arrays of arbitrary defect strength to serve as alignment templates for nematic liquid crystals. The atomic force microscope tip-based nanomechanical technique with the unique properties of low-cost with simple, highly accurate, and flexible control has been used to successfully fabricate nanodots, nanolines, and two-dimensional or three-dimensional nanostructures on flat or curved surfaces for applications in nanofluids, nanoelectronics, and nanosensors [52–54]. This manufacturing technique can be integrated with other conventional micro/nanofabrication methods including optical lithography, lift-off process, and wet etching. Thus, in order to achieve significant results, more attention is expected in the future to the association of this technique with other micro/ nanofabrication approaches. This definitely will advance AFM-based DPL closer to the industrial application.[27]

6.5   CONCLUSIONS

The chapter presented a short description, advantages, and limitations of the AFM as a method for visualization, measurement, and modification at nanometer scale of the polymer samples by means of the force-assisted lithographic techniques, such as static and dynamic plowing performed in two different modes, namely vector and raster. Among the polymers that can be patterned at nanoscale with AFM, polyimides were chosen due to their attractive mechanical, thermal, and electrical properties. In order to establish the proper conditions to obtain well-defined nanochannels, a preliminary study was made presenting the influence of the DPL and SPL in vector mode on the sharpness, contamination, and degradation of the AFM tip, and on the polyimide surface pattern aspect. The conclusion was that in contradistinction with the SPL, the DPL is more adequate to be used to pattern a surface to result even nanochannels with desired shape without any supplementary polymeric material moved in the direction of the lithography or on each side of it. Furthermore, in the same nanolithography conditions, nanostructured morphology was obtained on three different polyimide films. By applying a complete characterization through the surface texture and functional parameters, the influence of the dianhydride and diamine moieties (the backbone flexibility and the aliphatic/ aromatic nature of the monomers) on the characteristics of the pattern was revealed. By setting the optimum conditions, this technique can successfully be used to obtain custom nanostructures useful in new applications where the surface anisotropy is necessary, such as the fabrication of bio-analytical assays and liquid crystal alignment layers, or in micro-fluidics and guided cell growth.

KEYWORDS

- **polyimide**
- **atomic force microscopy**
- **dynamic plowing lithography**
- **patterning**
- **nanostructures**
- **high-tech applications**

REFERENCES

1. Stoica, I.; Barzic, A. I.; Hulubei, C. The Impact of Rubbing Fabric Type on the Surface Roughness and Tribological Properties of Some Semi-alicyclic Polyimides Evaluated from Atomic Force Measurements. *Appl. Surf. Sci.* **2013**, *268* (1), 442–449.
2. Stoica, I.; Barzic, A. I.; Popovici, D.; Vlad, S.; Cozan, V.; Hulubei, C. An Insight on the Effect of Rubbing Textile Fiber on Morphology of Some Semi-alicyclic Polyimides for Liquid Crystal Orientation. *Polym. Bull.* **2013**, *70* (5), 1553–1574.
3. Barzic, A. I.; Rusu, R. D.; Stoica, I.; Damaceanu, M. D. Chain Flexibility Versus Molecular Entanglement Response to Rubbing Deformation in Designing Poly(oxadiazolenaphthylimide)s as Liquid Crystal Orientation Layers. *J. Mater. Sci.* **2014**, 49, 3080–3098.
4. Barzic, A. I.; Hulubei, C.; Stoica, I.; Albu, R. M. Insights on Light Dispersion in Semi-alicyclic Polyimide Alignment Layers to Reduce Optical Losses in Display Devices. *Macromol. Mater. Eng.* **2018**, *303* (12), 1800235(1–11).
5. Sava, I.; Săcărescu, L.; Stoica, I.; Apostol, I.; Damian, V.; Hurduc, N. Photochromic Properties of Polyimide and Polysiloxane Azopolymers. *Polym. Int.* **2009**, *58*, 163–170.
6. Damaceanu, M.-D.; Rusu, R.-D.; Olaru, M.; Stoica, I.; Bruma, M. Nanostructured Polyimide Films by UV Excimer Laser Irradiation. *Rom. J. Inform. Sci. Techol.* **2010**, *13* (4), 368–377.
7. Stoica, I.; Epure, L.; Sava, I.; Damian, V.; Hurduc, N. An Atomic Force Microscopy Statistical Analysis of Laser-induced Azo-polyimide Periodic Tridimensional Nanogrooves. *Microsc. Res. Tech.* **2013**, *76* (9), 914–923.
8. Damian, V.; Resmerita, E.; Stoica, I.; Ibanescu, C.; Sacarescu, L.; Rocha, L.; Hurduc, N. Surface Relief Gratings Induced by Pulsed Laser Irradiation in Low Glass-transition Temperature Azopolysiloxanes. *J. Appl. Polym. Sci.* **2014**, *131* (24), 41015(1–10).
9. Sava, I.; Burescu, A.; Stoica, I.; Musteata, V.; Cristea, M.; Mihaila, I.; Pohoata, V.; Topala, I. Properties of Some Azo-copolyimide Thin Films Used in the Formation of Photo induced Surface Relief Gratings. *RSC Adv.* **2015**, *5*, 10125–10133.
10. Sava, I.; Stoica, I.; Mihaila, I.; Pohoata, V.; Topala, I.; Stoian, G.; Lupu, N. Nanoscale analysis of laser-induced surface relief gratings on azo-copolyimide films before and after gold coating. *Polym. Test.* **2018**, 72, 407–415.
11. Stoica, I.; Barzic, A. I.; Aflori, M.; Hulubei, C.; Harabagiu, V.; Vasilescu, D. S. Three-dimensional Nanostructures with Biocidal Activity Created on a Siloxane-containing Copolyimide Film. *Key Eng. Mat.* **2015**, *638*, 98–103.
12. Stoica, I.; Aflori, M.; Ioanid, E. G.; Hulubei, C. Effect of Oxygen Plasma Treatment and Gold Sputtering on Topographical and Local Mechanical Properties of Copolyimide/Gold Micropatterned Structures. *Surf. Interf. Anal.* **2018**, 50(2), 154–162.
13. Cosutchi, I.; Hulubei, C.; Stoica, I.; Ioan, S. A New Approach for Patterning Epiclon-based Polyimide Precursor Films Using a Lyotropic Liquid Crystal Template. *J. Polym. Res.* **2011**, *18* (6), 2389–2402.
14. Barzic, A. I.; Hulubei, C.; Avadanei, M. I.; Stoica, I.; Popovici, D. Polyimide Pecursor Pattern Induced by Banded Liquid Crystal Matrix: Effect of Dianhydride Moieties Flexibility. *J. Mater. Sci.* **2015**, *50* (3), 1358–1369.
15. Yan, Y. D.; Hu, Z. J.; Liu, W. T.; Zhao, X. S. Effects of Scratching Parameters on Fabrication of Polymer Nanostructures in Atomic Force Microscope Tapping Mode. *Procedia CIRP*. **2015**, *28*, 100–105.

16. Binnig, G.; Rohrer, H. Scanning Tunneling Microscopy. *IBMJ. Res. Dev.* **1986,** *30* (4), 355–69.
17. Binnig, G.; Qate, C. F.; Gerber, Ch. Atomic Force Microscope. *Phys. Rev. Lett.* **1986,** *56* (9), 930–933.
18. Nonnenmacher, M.; O'Boyle, M. P.; Wickramasinghe, H. K. Kelvin Probe Force Microscopy. *Appl. Phys. Lett.* **1991,** *58* (25), 2921–2923.
19. Hartmann, U. Magnetic Force Microscopy: Some Remarks from the Micromagnetic Point of View. *J. Appl. Phys.* **1988,** 64(3), 1561–1564.
20. Matey, J. R.; Blanc, J. Scanning Capacitance Microscopy. *J. Appl. Phys.* **1985,** *57* (5), 1437–1444.
21. Roelofs, A.; Bottger, U.; Waser, R.; Schlaphof, F.; Trogisch, S.; Eng, L. M. Differentiating 180° and 90° Switching of Ferroelectric Domains with Three-dimensional Piezoresponse Force Microscopy. *Appl. Phys. Lett.* **2000,** *77* (21), 3444–3446.
22. Lai, S. C. S.; Macpherson, J. V.; Unwin, P. R. In Situ Scanning Electrochemical Probe Microscopy for Energy Applications. *MRS Bull.* **2012,** *37* (7), 668–674.
23. Meister, A.; Gabi, M.; Behr, P.; Studer, P.; Vörös, J.; Niedermann, P.; Bitterli, J.; Polesel-Maris, J.; Liley, M.; Heinzelmann, H.; Zambelli, T. FluidFM: Combining Atomic Force Microscopy and Nanofluidics in a Universal Liquid Delivery System for Single Cell Applications and Beyond. *Nano Lett.* **2009,** 9(6): 2501–2507.
24. Majumdar, A. Scanning Thermal Microscopy. *Annu. Rev. Mater. Sci.* **1999,** 29(1), 505–585.
25. Heinzelmann, H.; Pohl, D. W. Scanning Near-field Optical Microscopy. *Appl. Phys. A* **1994,** *59* (2), 89–101.
26. Xie, X. N.; Chung, H. J.; Sow, C. H.; Wee, A. T. S. Nanoscale Materials Patterning and Engineering by Atomic Force Microscopy Nanolithography. *Mater. Sci. Eng.* **2006,** *R54,* 1–48.
27. Yan, Y.; Geng, Y.; Hu, Z. Recent Advances in AFM Tip-based Nanomechanical Machining. *Int. J. Mach. Tool. Manu.* **2015,** *99,* 1–18.
28. Carpick, R. W.; Salmeron, M. Scratching the Surface: Fundamental Investigations of Tribology with Atomic Force Microscopy. *Chem. Rev.* **1997,** *97*(4), 1163–1194.
29. Stoica, I.; Barzic, A. I.; Hulubei, C. Fabrication of Nanochannels on Polyimide Films Using Dynamic Plowing Lithography. *App. Surf. Sci.* **2017,** *426,* 307–314.
30. Cappella, B., Sturm, H. Comparison between Dynamic Plowing Lithography and Nanoindentation Methods. *J. Appl. Phys.* **2002,** *91,* 506–512.
31. Klehn, B.; Kunze, U. Nanolithography with an Atomic Force Microscope by Means of Vector-scan Controlled Dynamic Plowing. *J. Appl. Phys.* **1999,** *85,* 3897–3903.
32. Napolitano, S.; D'Acunto, M; Baschieri, P.; Gnecco, E.; Pingue, P. Ordered Rippling of Polymer Surfaces by Nanolithography: Influence of Scan Pattern and Boundary Effects. *Nanotechnol.* **2012,** *23* (47), 475301(1–6).
33. Villarrubia, J. S. Morphological Estimation of Tip Geometry for Scanned Probe Microscopy. *Surf. Sci.* **1994,** *321,* 287–300.
34. Williams, P. M.; Shakesheff, K. M.; Davies, M. C.; Jackson, D. E.; Roberts, C. J. Blind Reconstruction of Scanning Probe Image Data. *J. Vacuum Sci. Technol. B* **1996,** *14,* 1557–1562.
35. Villarrubia, J. S. Algorithms for Scanned Probe Microscope Image Simulation, Surface Reconstruction and Tip Estimation. *J. Nat. Inst. Stand. Technol.* **1997,** *102,* 435–454.

36. Villarrubia, J. S. *Strategy for Faster Blind Reconstruction of Tip Geometry for Scanned Probe Microscopy*. Proceedings of SPIE, Metrology, Inspection, and Process Control for Microlithography XII, Volume 3332, 1998; pp 10–18.

37. Heyde, M.; Rademann, K.; Cappella, B.; Geuss, M.; Sturm, H.; Spangenberg, T.; Niehus, H. Dynamic Plowing Nanolithography on Polymethylmethacrylate Using an Atomic Force Microscope. *Rev. Sci. Instrum.* **2001,** *72,* 136–141.

38. Cappella, B.; Sturm, H.; Weidner, S. M. Breaking Polymer Chains by Dynamic Plowing Lithography. *Polymer.* **2002,** *43,* 4461–4466.

39. He, Y.; Geng, Y.; Yan, Y.; Luo, X. Fabrication of Nanoscale Pits with High Throughput on Polymer Thin Film Using AFM Tip-based Dynamic Plowing Lithography. *Nanoscale Res. Lett.* **2017,** *12,* 544(1–11).

40. He, Y.; Yan, Y.; Geng, Y.; Hu, Z. Fabrication of None-ridge Nano grooves with Large-radius Probe on PMMA Thin-film Using AFM Tip-based Dynamic Plowing Lithography Approach. *J. Manuf. Process.* **2017,** *29,* 204–210.

41. Geng, Y.; Li, H.; Yan, Y.; He, Y.; Zhao, X. Study on Material Removal for Nano channels Fabrication Using Atomic Force Microscopy Tip-based Nano Milling Approach. *Proc. IMechE Part B: J. Eng. Manuf.* **2017,** *233* (2), 095440541774818 (1–9).

42. Geng, Y.; Brousseau, E. B.; Zhao, X.; Gensheimer, M.; Bowen, C. R. AFM Tip-based Nanomachining with Increased Cutting Speed at the Tool Work Piece Interface. *Prec. Eng.* **2018,** *51,* 536–544.

43. He, Y.; Yan, Y.; Geng, Fang, Z. Energy Dissipation Contributed on the Machined Depth Via Dynamic Plowing Lithography of Atomic Force Microscopy. *J. Vac. Sci. Technol. B* **2018,** *36* (4), 041802(1–6).

44. He, Y.; Yan, Y.; Wang, J.; Geng, Y.; Xue, B.; Zhao, X. Study on the Effects of the Machining Parameters on the Fabrication of Nanoscale Pits Using the Dynamic Plowing Lithography Approach.. *IEEE Trans. Nanotechnol.* **2019,** *18,* 351–357.

45. Iwata, F.; Yamaguchi, M.; Sasaki. A. Nanometer-scale Layer Modification of Polycarbonate Surface by Scratching with Tip Oscillation Using an Atomic Force Microscope. *Wear* **2003,** *254,* 1050–1055.

46. Kassavetis, S.; Mitsakakis, K.; Logothetidis, S. Nanoscale Patterning and Deformation of Soft Matter by Scanning Probe Microscopy. *Mater. Sci. Eng. C* **2007,** *27* (5–8), 1456–1460.

47. Wang, Y.; Hong, X. D.; Zeng, J.; Liu, B. Q.; Guo, B.; Yan, H. AFM Tip Hammering Nanolithography. *Small* **2009,** *5,* 477–483.

48. Hulubei, C.; Hamciuc, E.; Brumă, M. New Polyimides Based on Epiclon. *Rev. Roum. Chim.* **2007,** *52,* 1063–1069.

49. Song, Y. J.; Meng, S. H.; Wang, F. D.; Sun, C. X.; Tan, Z. C. A Study on the Thermodynamic properties of Polyimide BTDA-ODA by Adiabatic Calorimetry and Thermal Analysis. *J. Therm. Anal. Calorim.* **2002,** *69,* 617–625.

50. Carneiroa, K.; Jensena, C. P.; Jorgensena, J. F.; Garnoesa, J.; McKeown, P. A. Roughness Parameters of Surfaces by Atomic Force Microscopy. *CIRP, Ann. Manufact. Technol.* **1995,** *44,* 517–522.

51. Murray, B. S.; Pelcovits, R. A.; Rosenblatt, C. Creating Arbitrary Arrays of Two-dimensional Topological Defects. *Phys. Rev. E. Stat. Nonlin. Soft Mater. Phys.* **2014,** *90,* 052501(1–6).

52. Kunze, U. Nanoscale Devices Fabricated by Dynamic Ploughing with an Atomic Force Microscope. *Superlattices Microstruct.* **2002,** *31,* 3–17.

53. Wang, Z.; Wang, D.; Jiao, N.; Tung, S.; Dong, Z. Nanochannel System Fabricated by MEMS Microfabrication and Atomic Force Microscopy. *IET Nanobiotechnol.* **2011,** *5,* 108–113.
54. Hu, H.; Zhuo, Y.; Oruc, M. E.; Cunningham, B. T.; King, W. P. Nanofluidic Channels of Arbitrary Shapes Fabricated by Tip-based Nanofabrication. *Nanotechnology* **2014,** *25,* 055301(1–8).

CHAPTER 7

Polyimide Materials for Transistors and Biosensors Manufacturing

RALUCA MARINICA ALBU* and RAZVAN FLORIN BARZIC

*"Petru Poni" Institute of Macromolecular Chemistry,
Laboratory of Physical Chemistry of Polymers,
41A Grigore Ghica Voda Alley, 700487 Iasi, Romania*

Corresponding author. E-mail: albu.raluca@icmpp.ro

ABSTRACT

Polyimides are considered a special type of thermostable polymers, with excellent mechanical resistance combined with good dielectric, morphological, and optical features. For this reason, this category of high-performance plastics is widely exploited in electronic industry, particularly in transistor or biosensor manufacturing. Basic aspects regarding transistors or biosensors are shortly presented for better understanding of the polyimide role in these devices. The chapter also describes the current advances in transistors based on polyimides by highlighting their impact on the device performance when used as substrates or gate dielectrics. Moreover, the latest developments concerning the processing of polyimides for several types of biosensors are reviewed.

7.1 INTRODUCTION

The miniaturization of electronic devices concomitantly with increasing their performance imposed requirements regarding both material properties and processing technologies. Thermostable polymers are often used in manufacturing of high-performance electronic products like computation, mobiles, TV screens, human interactivity, energy generation/storage, and electronic textiles.[1–5]

Thin-film transistors (TFTs) have led to crucial developments in electronics industry, particularly in their application in the active-matrix organic light-emitting diodes (AMOLEDs), TFT-liquid crystal displays (TFT-LCDs), e-Papers, and several types of sensors.[6,7] Traditional TFTs were fabricated on glass support, whereas the latest trends involve utilization of flexible plastic substrates. Beside the advantages, there are some new process issues, such as deformation during heating/cooling generating misalignment in the photolithography process. Moreover, additional treatments like ion bombardment, high-temperature annealing, etchant corrosion, and UV irradiation in the TFTs manufacturing could create unwanted effects on the bonding ability. The stage of TFT annealing requires elevated temperatures to enhance the electrical stability. The interval of TFTs fabricating temperature is affected by both the vitrification temperature of the polymer flexible substrate and the thermal behavior of the functional layers used to optimize the bonding property. A less adequate handling procedure will lead to failure in fabrication stage or might deteriorate the electrical performance. In the same time, improper debonding procedure will induce mechanical, thermal, or optical issues in TFTs operation. In this context, bonding and debonding of the flexible substrate on/from a rigid carrier represent one of the main challenges in the field of flexible technology. Given the relatively small mobility of the organic semiconductor utilized in organic thin-film transistors (OTFTs), they cannot overcome the performance of field–effect transistors (FETs), which work using single-crystalline inorganic semiconductors with three times higher mobility.[8] This is the reason why OTFTs are not fitted in devices that demand large switching speeds. OTFTs are successfully competitive for available or new TFT applications that impose structural flexibility, processing at temperatures, large-area coverage, and also reduced costs.

Synthetic polymers used in electronic devices provide higher flexibility and higher reliability in conditions of high-mechanical load. Polymer substrates and organic semiconductors represent the key materials for a large variety of OTFTs that represent the present and future of hi-tech products.

7.2 GENERAL ASPECTS ON TRANSISTORS

There are two main categories of transistors, namely, bipolar junction transistors and FET. The first one operates based on two p-n (hole-electron) junction junctions, whereas the second one is working based upon voltage control. Transistors are considered active components in electronic circuits which are functioning as amplifiers and switches.

When introduced in digital circuits, the transistors operate as switches. The latter can be found either in an "on" or "off" state for high-power devices, like switched-mode power supplies and for low-power devices, including logic gates. The basic parameters for such applications are the voltage handled, the current switched, and the switching speed, related to the rise and fall times. In a grounded-emitter transistor circuit, it can be noted that the increase of the base voltage is accompanied by an exponential raise for the emitter and collector currents. The drop in collector voltage is due to the diminished resistance from collector to emitter. In conditions where there is a null voltage difference between the collector and emitter, it can be said that the collector current is affected mainly by the load resistance and the supply voltage. This is widely known as saturation since current is flowing from collector to emitter freely. In the saturated state, the switch is considered to be "on".[9] The main issue of bipolar transistors used as switches is to ensure sufficient base drive current. The transistor gives current gain, enabling a relatively high current in the collector to be switched by a lower current into the base terminal. The ratio of these currents is influenced by the type of transistor, and in some cases is affected by the collector current. In a switching circuit, the purpose is to simulate the ideal switch that presents analogous features to an open circuit when "off," short circuit in "on" state, concomitantly with an instantaneous transition both states. Parameters should be selected in a manner that "off" output is narrowed to leakage currents too low to influence the connected circuitry, the resistance of the transistor in the "on" state is too small to afflict the circuitry, and the transition between the two states is quite fast avoiding any detrimental effect.

When transistors are utilized as amplifiers, they are working in high- and low-level frequency stages, modulators, oscillators, and detectors. The common-emitter amplifier is projected so that a tiny change in voltage modifies the current through the base of the transistor. The enhancement of the transistor's current combined with the characteristics of the circuit shows that small swings in the base voltage create large changes in collector voltage. Several types of configurations of single-transistor amplifier are known, some of them giving current gain, voltage gain, and others both. From mobile phones to flat TV screens, a wide number of products have amplifiers for signal processing, sound reproduction, and radiotransmission. The performance of the transistors as the amplifier architecture evolved.

On the other hand, a large number of semiconductors producers exist on the market particularly for transistors. Therefore, there are low-, medium-, and high-power transistors for high/low frequencies or high current/high voltages.

A TFT is a particular category metal-oxide-semiconductor FET (MOSFET)[10] prepared by depositing thin films of an active semiconductor layer and also the dielectric coating and metallic contacts over a supporting (but insulating) substrate. The most encountered substrate is glass as a result of the main application in LCDs. This is distinct from that of bulk MOSFET devices where the semiconductors typically are the substrate. The materials from which the substrate and transistors terminals are made to determine its performance. In this context, the following section is devoted to the impact of polyimides as substrates or gate dielectrics on transistor reliability.

7.3 POLYIMIDE MATERIALS FOR TRANSISTOR MANUFACTURING

In a simplistic form, TFT devices have be constructed to contain at least the following basic components:

- the substrate
- the dielectric
- the semiconductor layer
- the source, drain, and gate electrodes

The source and drain contacts are placed in contact with the semiconductor, whereas the gate is in the vicinity of the dielectric layer instead of the semiconductor. TFTs can be constructed using a number of distinct structural layouts. Among the most common device architectures, these are worth mentioning:

- coplanar, top-gate,
- staggered, top-gate staggered,
- bottom-gate, and
- coplanar, bottom-gate.

Figure 7.1 illustrates an example of a TFT architecture, namely the staggered, top-gate. Polyimides are widely used in TFT manufacturing. They can be used as substrates, as gate dielectric and in some cases as both.

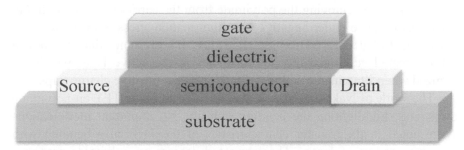

FIGURE 7.1 Schematic representation of staggered, top-gate configuration of a TFT.

7.3.1 POLYIMIDES AS SUBSTRATES

The use of polyimides substrates for TFTs is widely reported.[11–17] In the next paragraphs, we make a short presentation of the most recent breakthrough regarding this research topic.

Pecora et al.[11] describe a polysilicon TFT fabricated on polyimide substrates. The latter has good thermal stability and low coefficient of thermal expansion, that is, 3 ppm/°C. These aspects are helpful to avoid the problem of plastic shrinkage during processing at elevated temperatures, which in turn determine issues in the photolithographic steps. This is why the polyimide is spun on Si-support, which is used as rigid carrier. On the spun polyimide having a mean surface roughness around 3 nm, a silicon nitride layer is deposited not only to enable protection against moisture and liquid chemicals, but also to facilitate adhesion and to diminish mechanical stress. The TFT presents good electrical characteristics with field effect mobility up to 50 cm²/V s.

Xu et al.[12] developed a lanthanide-doped In–Zn–O TFT on polyimide support for AMOLED applications. A surface isolation layer is casted on the carrier before polyimide film deposition. This avoids chemical bonding between fluorinated imidic polymer and the carrier. The polyimide was easily debonded from the carrier layer without meaningful deterioration in electrical properties even at elevated temperatures (~350°C). A buffer coating SiN_x/SiO_2 was deposited on the polyimide to promote adhesion of metal, while keeping a good barrier against water/oxygen permeation. An inorganic film is casted on the glass prior polyimide coating to enable the debonding. The TFT displayed a subthreshold swing of 0.248 V 1/dec, a field-effect mobility of 6.97 cm²/V s, and an I_{on}/I_{off} ratio of 5.19 × 10⁷, which is enough for an organic light-emitting diode to function.[13]

Park et al.[13] reported the use of polyimide (PI) substrates in fabrication of In–Ga–Zn–O (IGZO) TFT and studied and analyzed the electrical features before and after removing the polyimide from the carrier glass. In addition, TFT performance under negative bias illumination stress is lowered as the polyimide foil is removed from the rigid glass. It is presumed that mechanical strain created additional oxygen vacancies in the active layer which are ionized when illuminated and they function as net positive charge traps.

The work of Nakata et al.[14] is also devoted to oxide TFT on polyimide substrates, highlighting the bending stability. No important modifications were noted in the transfer characteristics of the TFTs at a bending with curvature radius of 1 mm. The fabricated devices on polyimide present a mobility of 28 cm^2/V s at a channel length of 4.8 μm.

Lee et al.[15] examined the reliability of an IGZO TFT on a polyimide substrate subjected to extreme bending or folding. It was revealed that the formation of cracks could be lowered by reducing the widths of both Mo electrodes and IGZO layers. At a strain of 2.17%, attributed to a radius of 0.32 mm, no cracks are observed even after 60,000 bending cycles; thus, the robustness of the IGZO TFTs is still high. The mechanical stability of the device can be enhanced through the split of source/drain electrodes and semiconductor film.

Ruan and coworkers[16] fabricated an amorphous tungsten-doped indium-oxide TFT. The device was deposited on a transparent polyimide film in low-temperature conditions. Depending on the active layer thickness, the flexible TFT can reach high-carrier mobility and small subthreshold swing.

The efficiency of polyimide supports in comparison with those made of glass in TFT is reported by Han et al.[17] The manner in which the imidic polymer alters hydrogen concentration in the IGZO active zone affects the device operation and stability in bias-temperature–stress conditions. Hydrogen is responsible for the carrier density and electrically deactivates intrinsic defects. When the TFT is deposited on a polyimide layer, the hydrogen percent is 5% lower after annealing which enhances the hysteresis characteristics from 0.22 to 0.55 V. Also, the threshold voltage shift under positive bias temperature stress is two times higher in regard with TFT on a glass substrate. Therefore, the careful control of hydrogen amount is essential to preserve the good device performance and stability of IGZO TFTs.

7.3.2   POLYIMIDES AS GATE DIELECTRICS

The polyimides have suitable thermal and physical properties to be used as gate dielectrics in transistor manufacturing. Several investigations were devoted to this subject in the attempt to achieve transistors with upgraded performance.[18–25] The most recent advances in this area are further reviewed.

Kato and coworkers[18] reported organic FETs with polyimide as gate dielectric. The latter has been subjected to thermal curing at 180°C. The polyimide solution was spin coated on the base film with gate electrode. The polymer the film is placed into the clean oven in a constant temperature of 90°C, and then solvent is removed for 10 min or more. Subsequently, the spin-coated film is heated in the in nitrogen atmosphere at 180°C during 1 h and then cooled down naturally. Below 100°C the polyimide coating is taken from the oven. The gate dielectric has the surface roughness was around 0.2 nm. The pentacene FET with polyimide gate presents I_{on}/I_{off} ratio of 10^6 and mobility of 0.3 cm²/V s. The mobility is further enhanced up to 1 cm²/V s for the sample placed on 540 nm polyimide gate dielectric layers. The probability of device failure take place as a result of gate leakage and the initial yields are around 1% for the FETs with 540 nm polyimide gate dielectric films.

Sekitani et al.[19] describe the effects induced by annealing on pentacene FETs with polyimide gate dielectrics encapsulated with polychloro-*para*-xylylene parylene passivation layers. The annealing was produced in nitrogen medium at various substrate temperatures. It was shown that the annealed FETs are operational at a temperature of 160°C and do not present degradations concerning the performance after being exposed to a number of heat cycles between 25 and 160°C. The annealed device displays a variation of less than 5% in the source-drain currents. This happens even after the application of DC voltage biases. The charge carrier mobility increases from 0.52 to 0.56 cm²/V s, particularly when the FETs are cured at 140°C for 12 h in a nitrogen atmosphere. In these conditions, the I_{on}/I_{off} ratio also is enhanced to 10^6.

Zhen et al.[20] present fabrication of an organic TFT memory with polyimide gate dielectric reinforced with gold nanocrystals. The active layer used in this case is copper phthalocyanine (CuPc). Discrete gold nanocrystals are used to ensure the charge storage. In adequate gate bias conditions, gold nanocrystals are charged and discharged. In this way, the modulation of the channel conductance occurs. The registered current–voltage characteristics indicate that the developed device has a memory behavior at room temperature. The programming and erasing operations are affected by the can be better understood by analyzing the current–voltage curves and energy-band structures. The main advantages of the device are the low temperature of processing and reduced costs.

Chen et al.[21] prepared new gate dielectric by embedding tantalum pentoxide (Ta_2O_5) in polyimide matrix. The used metal oxide nanofiller is known for its good insulating characteristics and high dielectric constant. The polymer nanocomposites were prepared by in-situ synthesis. The spin-coated films have a smooth surface morphology in regard with the pristine polyimide films. The electrical capacitance increased almost two times for the composite containing 30% Ta_2O_5 compared to unfilled polyimide. This type of gate dielectric led to pentacene TFTs with considerably high-carrier mobility and current I_{on}/I_{off} ratio at small operation voltage.

Wang et al.[22] proposed to use polyimide electrets as gate dielectrics for organic FET memories. They have shown that there is a correlation between electrical characteristics and polarity of imidic polymeric gate electrets with quasi-permanent electric charges. The memory features and drain currents of the device could be essentially improved by using polyimide electrets which are able to trap the charge carriers in the vicinity of its polar groups by bias-induced dipole fields. Furthermore, these charges can be quickly released when the dipoles are reorienting in bias field. Thus, the electric dipoles from the polymer electrets are key factors in widening the memory window and improving the electrical characteristics of FET-based memory device.

Baek et al.[23] have used fluorinated polyimides to produce gate dielectrics with impact on electrical stability of transistors. The presence of fluorine atoms in a polymer backbone is known to increase surface hydrophobicity, chemical inertness, and reduce polarizability. Thus, it is expected to reduce the probability of trap formation, leading to good electrical properties and reliable device operation. The fluorinated imidic polymers contain 6 and 18 fluorine atoms per repeat unit, which induced smoother surface morphology and lower surface energy. These characteristics favored the crystalline morphology in the semiconductor layer deposited on the surface of the fluo-rinated polyimide. The field-effect mobility was enhanced, and the threshold voltage value moved toward positive values in the FET with pentacene and triethylsilylethynyl anthradithiophene. Also, the highly fluorinated poly-imide dielectric displayed negligible hysteresis and a considerable gate bias stability in air and nitrogen.

Kim et al.[24] modified the surface of polyimide gate insulators with self-assembled monolayers (SAMs) by inserting metal-oxide interlayers. To generate a site for surface treatment with SAM of the polyimide, a metal oxide layer (Al_2O_3) is spin-coated on the polyimide with a sol–gel reaction to achieve a template for preparation of SAM with phosphonic acid anchor groups. The SAM was spin-coated on the metal oxide-deposited polyimide film. The surface energy of the resulted system significantly decreased, so the compatibility with organic active layer is enhanced. The semiconductor molecules are uniformly grown in the modified polyimide gate. Using this fabrication procedure, the mobility almost doubled, while the threshold voltage was reduced by lowering the trap sites and enhancing the interfacial properties.

In a recent work, Zou et al.[25] prepared novel gate dielectrics by designing a polyimide containing a cross-linkable olefin group and a long alkyl chain with biphenyl via mild chemical synthesis, which do not involve thermal imidization. The resulted materials are thermostable having vitrification temperatures above ~150°C. The morphology of the film insulators is pinhole-free and the roughness is around 0.3–0.5 nm. The leakage current densities are small (2×10^7 A/cm² at 150 MV/m) while the dielectric loss values are also small (under 0.02, 10^3–10^7 Hz). These highlight the good insulating abilities. The *para*-hexaphenyl/vanadyl-phthalocyanine TFTs with the new polyimide dielectric gate layers present remarkable performance, having a charge carrier mobility of 1.11 and 1.23 cm²/V s, I_{on}/I_{off} ratio higher than 5×10^5, and small threshold voltage (3.6–10 V).

7.4 PROGRESSES IN POLYIMIDE MATERIALS FOR BIOSENSORS

Analytical devices that integrate biological elements are called biosensors. The essential elements that constitute a biosensor are illustrated in Figure 7.2.

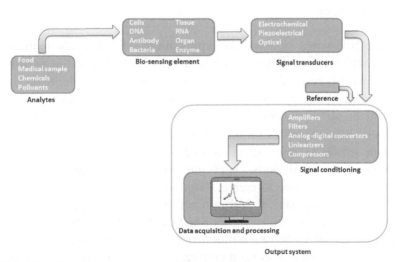

FIGURE 7.2 The schematic diagram illustrating the basic components and working principle of a biosensor.

The main idea residing in the operation principle of a biosensor is to transform a biological signal into an electrical one. Such devices work by coupling a biosensing element with a detector system using a transducer. In other words, the transduction of various signals produced upon analyte sample or biomaterial interaction into detectable signal takes place. Then, it is required to perform signal conditioning mainly by amplification, filtering, and converting from analog to digital. The final stage relies on data acquisition, processing, and storing using proper software.

There are several categories of biosensors and their classification is mainly based on the type of components. Therefore, the biosensors can be divided by considering:

- *the biological element*:
  - o protein-based biosensors: enzymes, purified protein (nonenzyme), antibodies
  - o whole-cell-based biosensor: natural, genetically engineered
- *the transducing method*:
  - o optical biosensors: fiber optics, Ramand and Fourier-transform infrared spectroscopy (FTIR), surface plasmon resonance, resonant mirror
  - o electrochemical biosensors: conductometric, amperometric, potentiometric, superficial charge detection, impedimetric
  - o mass-based biosensors: magnetoelectric, piezoelectric (quartz crystal microbalance, surface acoustic wave, surface transverse wave)

In the following sections, the implications of polyimide materials in biosensor fabrication are briefly presented by highlighting the most recent advances in this research domain. Figure 7.3 is depicting the main uses of this category of high-performance materials in biosensors field.

Mutlu et al.[26] developed a biosensor by the electrochemical etching of nickel films on gold. For this purpose, it was used a micromachined polyimide probe having a metal-coated tip (diameter under 200 nm) and incorporated electrical leads. Applying short electrical pulses (500 ns, 2 V) determines the probe to create 3 μm wide patterns on metal layer. The probe has small mechanical stiffness and thus avoids the need to preserve the proximity between the workpiece and the tool. The methodology allows probes to reach the substrate without making an ohmic contact. The small spring constant assures a reduced contact pressure of the probes and has less variation during rastering over minor topological features.[27] The structure is basically a cantilever in where two layers of imide polymer sandwich two Au leads of 250 nm thickness. At the end, this layer is somehow extended over a protruding pyramidal tip. The whole structure is connected to an insulated Si wafer on which metal pads enable electrical access to the embedded leads. Selective adsorption of alkanethiol SAMs to the patterned Au zones was noticed Ni was probed to be a suitable mask. Several kinds of biomolecules can be selectively binded to the SAMs and utilized for biosensing purposes.

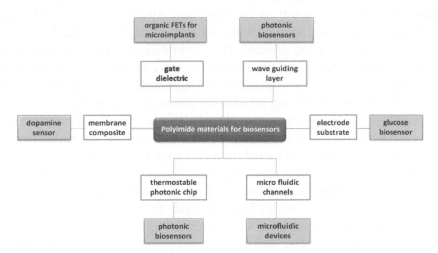

FIGURE 7.3   The main uses of polyimide materials in biosensors production.

Flexible organic FETs for microimplants having polyimide gate was investigated by Feili et al.[28] Biomedical microimplants are essential for restoring brain functions through functional electrical stimulation (FES), when used as neural prostheses. The used aromatic polyimide is of commercial nature, biocompatible, and biostable. It also has suitable features for flexible multichannel microelectrodes in treating certain neural affections.[29] Thus, it can be utilized as gate insulator. Polyimide was subjected to spin coating on Si support and afterward pentacene was deposited as active layer. Plasma activation of SAMs was suitable for advancing the electrical performance of the transistor. The surface treatments enhance the interfacial characteristics between the oxide and metal layers with organic molecules[30,31] and therefore upgrades the efficiency of the transistors. The designed device using polyimide as the insulator led to promising results.[28]

Melnik et al.[32] have used a complex procedure to modify polyimide film surface for photonic biosensors. The polyimide waveguiding layer was achieved by exposing the polymer to oxygen plasma, followed by binding of a 3-mercaptopropyl trimethoxysilane and then derivatization of the reactive thiol groups with maleimide-PEG$_2$-biotin and noncovalent attachment of streptavidin. The used approach led to a surface density of 144 fmol/mm^2 streptavidin with high reproducibility (13.9% RSD (relative standard deviation), $n = 10$) and had no inflicted damage to the surface. The described surface treatment was applied to polyimide-derived Mach-Zehnder interferometer sensors to accomplish a real-time assessment of streptavidin binding confirming the functionality of the biodevice. After this point, the streptavidin surface was used to react with biotinylated single-stranded DNA and applied to investigate the selective DNA hybridization. The polyimide was useful as guiding wave layer in evanescent wave biosensors. Such devices are helpful for monitoring relevant biomolecules and for the identification of kinetic parameters, which are relevant for biosensing applications.

Saftics et al.[33] studied thermostable polymer photonic chip fabrication for immobilization occurring in aqueous buffers. The samples are made of high-refractive-index polyimide casted on top of tetraethyl orthosilicate (TEOS) deposited on Si wafer. The in-situ spectroscopic ellipsometry using various aqueous buffers showing high drift in aqueous media which is ascribed to thickness and refractive index of the polyimide foil.

The polyethyleneimine layer was placed on the PI surface to allow biomolecule immobilization. The stability of imidic polymer in aqueous buffer solutions needs some improvements to meet the real biosensor application. Preliminary solution was to protect the polyimide by TiO$_2$ layers (obtained by atomic layer deposition) enabling the blocking of the waveguide from the aqueous media and also shielding effect for biosensor applications.

Zulfiqar and coworkers[34] created polyimide-based microfluidic channels which are crucial for biosensor applications. Many studies reported the suitability of polyimides microfluidic devices by employment of embossing with a silicon master mold,[35] layer transfer and lamination[36] or ion etching of nonphotosensitive polyimide.[37] Based on this, Zulfiqar[34] developed a procedure to pattern microfluidic channels, using partially cured polyimide, subjected to a dry etching process. The creation of closed microfluidic channel on wafer through joining two partially cured polyimides or a partially cured one to glass support has been proved. The uniformity and reproducibility of the channels' height were performed on the wafers with PI and showed low-standard deviations. The generated channels are appropriate for integrated microelectronics circuits which can control the biosensor devices.

Aksoy et al.[38] doped polyimide with boron nitride (BN) to achieve dopamine selective membranes.

It was observed that embedding the BN particles led to higher porosity, selectivity, and enhanced heat-resistance of virgin polyimide. The selectivity of the composite membranes toward dopamine was evidenced by differential pulse voltammetry. Electrochemical data confirmed that the polyimide membrane containing 5% BN present large linear ranges, high sensitivity, selectivity, reversibility, and small detection limits (4.10-8 M). Based on this approach, it was shown that the sample containing 5 % BN opens novel perspectives in fabricating dopamine sensor.

Shen et al.[39] produced an amperometric biosensor for glucose starting from AuPd-modified reduced graphene oxide/polyimide with glucose oxidase (PI/RGO). The method to obtain the bimetallic composite relied on electrodeposition. In this way, the conductivity and catalytic properties are increased. The resulted AuPd/polyimide/RGO electrode has good activity toward hydrogen peroxide. The noted electrochemical activity lies at the basis of the amperometric glucose biosensor which had glucose oxidase (GO_x) onto the AuPd/polyimide/RGO system. The device presented a fast amperometric response regarding glucose in 3 s, combined with a large linear range from 0.024 to 4.6 mM. The biosensor was tested for glucose detection in human blood serum leading to data to close those recorded with commercial instruments.

Yoon et al.[40] developed another type of biosensor with needle shape for glucose monitoring. Three electrochemical electrodes were microprocessed on a flexible polyimide layer using electroplating method. The electrodes were introduced in a medical polytetrafluoroethylene-based catheter using silicone adhesive. The fabricated device was shown high stability and reliability due to the used catheter as supporting structure of flexible sensor. The latter could be delivered in a facile manner to subcutaneous tissue through a stainless steel guide needle. The advantage is that the procedure does not involve surgery. Moreover, the catheter can be utilized in the same time as a tool for drug delivery. The manufactured biosensor displayed 415 nA/mM of sensitivity and 0.996 of R-square for glucose and high selectivity towards the interfering reagents. The in-vivo experiments performed in by placing the device in the back of a rabbit were showing the changes of glucose during 2 h, which correlated well with the values achieved by other methods.

In a recent study, Vanegas et al.[41] fabricated a biosensor for detection of biogenic amines (BA) in food. The BA could appear in such samples as a result of temperatures and storage conditions. Therefore, the BA amount is closely related to food safety and quality. To attain reagent-free food safety biosensing, laser-scribed graphene (LSG) electrodes, having locally sourced materials, were produced. The electrodes were obtained based upon a procedure reported by Teharani and Bavarian.[42] Briefly, the working, reference, and auxiliary electrodes were created using computer software. Kapton polyimide film was placed to photo-paper and the composite was set in the base of a UV laser engraver having bidimensional stepper motors. The fabrication was optimal at a laser-sample distance of 13 cm, applying a laser pulse rate of 21 ms and at the beam size of 1.3 mm. Afterward, silver-based ink was put to the reference electrode but also on the bonding pad of all three electrodes to avoid the mechanical damage of electrode after repeated use. In the end, a passivation layer consisting in nitrocellulose lacquer was put between the electrodes and the connectors to electrically isolate the nonactive features. Copper was electroplated over the working electrode which was supplied by a 9-V battery and the copper had the role of cathode. The sensors were fabricated with either locally sourced materials or standard materials. For the devices made with locally sourced materials, a commercial fertilizer was used together with copper as cathode. For the device with analytical grade materials, a solution of $CuSO_4$ and Na_2SO_4 was utilized with a copper mesh as the cathode. After metallization, the electrodes were washed with deionized water, to take out the excess materials, and then dried at room temperature. To render a notable selectivity toward biogenic amines, LSG electrodes were biofunctionalized with diamine oxidaze encapsulated with cellulose and then cross-linked. An enzyme solution was place on the working electrode and in presence of glutaraldehyde cross-linked to avoid the protein leaching. The diamine oxidase enzyme triggers the deamination of primary amines, diamines, and substituted amines through oxidation, to produce aldehyde, ammonia, and hydrogen peroxide.[43] Hydrogen peroxide is thus decomposed at a working electrode (+500 mV) creating an electrical response that can be further linked to the amount of BA. The reported biosensors made with locally sourced materials are efficient in screening fermented fish samples in the acceptable range (100–200 mg histamine/kg sample) imposed by the majority of regulatory agencies.[44,45]

This domain is in continuous development as the transistors tend to evolve[46] in the same trend with the polymer materials.[47–49] The deep knowledge and progress in metallization techniques[50–52] of polymers might also contribute to higher reliability and functionality of the biosensors.

7.5 CONCLUDING REMARKS

The imidic polymers are useful materials for producing flexible substrates or gate dielectrics for several kinds of transistors. Their mechanical strength, chemical resistance, and excellent dielectric properties lead to devices with higher performance opening new perspectives in flexible electronics. The dielectric behavior can be enhanced by embedding metal or metal oxide fillers, by processing them in the presence of strong electrical fields, or by specials methods, like SAMs.

Another important research area is that of biosensor-based polyimides. These materials can be used as available on the market or they can be modified to produce gate insulators, selective membranes, wave-guiding layers, electrode substrates, or microfluidic channels.

The worldwide demands for clean water, quality food, energy, and proper health infrastructure are increasing and with them the need for upgraded biosensors. The convergence of nanotechnology, polymer science, and human health knowledge imposes a fast technology transfer in various scenarios.

In this context, biosensors can be viewed as tools essential for maintaining the safety and quality of food to consumers and implicitly to their health. The current innovations will help to recreate the technology of tomorrow toward novel and more accessible platforms extending the use of biosensors to a larger area of consumers.

KEYWORDS

- **imidic polymer**
- **gate dielectrics**
- **flexible substrates**
- **processability**
- **transistors**
- **biosensors**

REFERENCES

1. Yakimets, I.; MacKerron, D.; Giesen, P.; Kilmartin, K. J.; Goorhuis, M.; Meinders, E.; MacDonald, W. A. Polymer Substrates for Flexible Electronics: Achievements and Challenges. *Adv. Mater. Res.* **2010**, *93–94*, 5–8.

2. Malik, A.; Kandasubramanian, B. Flexible Polymeric Substrates for Electronic Applications. *Polym. Rev.* 2018, *58*, 630–667.
3. Chatterjee, K.; Tabor, J.; Ghosh, T. K. Electrically Conductive Coatings for Fiber-Based E-textiles. *Fibers* 2019, *7*, 51.
4. Thakur, S. Shape Memory Polymers for Smart Textile Applications. In *Textiles for Advanced Applications*; Kumar, B., Thakur, S., Eds.; InTech: Croatia, 2017.
5. Kim, S. D.; Lee, B.; Byun, T.; Chung, I. S.; Park, J.; Shin, I.; Ahn, N. Y.; Seo, M.; Lee, Y.; Kim, Y.; Kim, W. Y.; Kwon, H.; Moon, H.; Yoo, S.; Kim, S. Y. Poly(amide-imide) Materials for Transparent and Flexible Displays. *Sci. Adv.* 2018, *4*, eaau1956.
6. Rjoub, A.; Tarawneh, B.; Alghsoon, R. Active Matrix Organic Light Emitting Diode Displays (AMOLED) New Pixel Design. *Microelectron. Eng.* 2019, *212*, 42–52.
7. Martins, R.; Gaspar, D.; Mendes, M. J.; Pereira, L.; Martins, J.; Bahubalindruni, P.; Barquinha, P.; Fortunato, E. Papertronics: Multigate Paper Transistor for Multifunction Applications. *Appl. Mater. Today* 2018, *12*, 402–414.
8. Taur, Y.; Ning, T. H. *Fundamentals of Modern VLSI Devices*; Cambridge University Press: New York, 1998; p 11.
9. Kaplan, D. M.; Christopher, G. *White Hands-on Electronics*. Cambridge University Press: United Kingdom, 2003; p 226.
10. Kimizuka, N.; Yamazaki, S. *Physics and Technology of Crystalline Oxide Semiconductor CAAC-IGZO: Fundamentals*; John Wiley & Sons: Hoboken, NJ, 2016; p 217.
11. Pecora, A.; Maiolo, L.; Cuscuna, M.; Simeone, D.; Minotti, A.; Mariucci, L.; Fortunato, G. Low-Temperature Polysilicon Thin Film Transistors on Polyimide Substrates for Electronics on Plastic. *Solid-State Electron.* 2008, *52*, 348–352.
12. Xu, Z.; Li, M.; Xu, M.; Zou, J.; Gao, Z.; Pang, J.; Guo, Y.; Zhou, L.; Wang, C.; Fu, D.; Peng, J.; Wang, L.; Cao, Y. *Flexible Amorphous Oxide Thin-Film Transistors on Polyimide Substrate for AMOLED*. In Proceedings Volume 9270, Optoelectronic Devices and Integration V, Beijing, China, October 9–11, 2014.
13. Park, J.; Kim, C. S.; Ahn, B. D.; Ryu, H.; Kim, H. S. Flexible In–Ga–Zn–O Thin-Film Transistors Fabricated on Polyimide Substrates and Mechanically Induced Instability Under Negative Bias Illumination Stress. *J. Electroceram.* 2015, *35*, 106–110.
14. Nakata, M.; Tsuji, H.; Motomura, G.; Nakajima, Y.; Takei, T.; Fujisaki, Y.; Shimidzu, N.; Yamamoto, T. Oxide Thin-Film Transistors Fabricated on Polyimide Film: Bending Stability and Electrical Properties. *IEEE Trans. Ind. Appl.* 2016, *52*, 5213–5218.
15. Lee, S.; Jeong, D.; Mativenga, M.; Jang, J. Highly Robust Bendable Oxide Thin-Film Transistors on Polyimide Substrates via Mesh and Strip Patterning of Device Layers. *Adv. Funct. Mater.* 2017, *27*, 1700437.
16. Ruan, D. B.; Liu, P. T.; Chiu Y. C.; Yu, M. C.; Gan, K. J.; Chien, T. C.; Kuo, P. Y.; Sze, S. M. *High Mobility Tungsten-Doped Thin-Film Transistor on Polyimide Substrate with Low Temperature Process*. In Proceedings of 7th International Symposium on Next Generation Electronics (ISNE), Taipei, Taiwan, May 7–9, 2018.
17. Han, K. L.; Cho, H. S.; Ok, K. C.; Oh, S.; Park, J. S. Comparative Study on Hydrogen Behavior in ingazno Thin Film Transistors with a $SiO_2/SiN_x/SiO_2$ Buffer on Polyimide and Glass Substrates. *Electron. Mater. Lett.* 2018, *14*, 749–754.
18. Kato, Y.; Iba, S.; Teramoto, R.; Sekitani, T.; Someya, T.; Kawaguchi, H.; Sakurai, T.; High Mobility of Pentacene Field-Effect Transistors with Polyimide Gate Dielectric Layers. *Appl. Phys. Lett.* 2004, *84*, 3789–3791.

19. Sekitani, T.; Someya, T.; Sakurai, T. Effects of Annealing on Pentacene Field-Effect Transistors Using Polyimide Gate Dielectric Layers. *J. Appl. Phys.* 2006, *100*, 024513.

20. Zhen, L.; Guan, W.; Shang, L.; Liu, M.; Liu, G. Organic Thin-Film Transistor Memory with Gold Nanocrystals Embedded in Polyimide Gate Dielectric. *J. Phys. D: Appl. Phys.* 2008, *41*, 135111.

21. Chen, L. H.; Lin, P.; Ho, J. C. Lee, C. C.; Kim, C.; Chen, M. C. Polyimide/Ta_2O_5 Nanocomposite Gate Insulators for Enhanced Organic Thin-Film Transistor Performance. *Synth. Met.* 2011, *161*, 1527–1531.

22. Wang, Yu. F.; Lin, C. Y.; Tsai, M. R.; Cheng, H. L.; Chou, W. Y. *Towards the High Performance N Channel Organic Memory Transistors with Modified Polyimide Gate Dielectrics*. In Proceedings of 21st International Workshop on Active-Matrix Flatpanel Displays and Devices (AM-FPD), Kyoto, Japan, July 2–4, 2014.

23. Baek, Y.; Lim, S.; Yoo, E. J.; Kim, L. H.; Kim, H.; Lee, S. W.; Kim, S. H.; Park, C. E. Fluorinated Polyimide Gate Dielectrics for the Advancing the Electrical Stability of Organic Field-Effect Transistors. *ACS Appl. Mater. Inter.* 2014, *6*, 15209–15216.

24. Kim, S.; Ha, T.; Yoo, S.; Ka, J.; Kim, J.; Won, J. C.; Choi, D. H.; Jang, K.; Kim, Y. H. Metal-Oxide Assisted Surface Treatment of Polyimide Gate Insulators for High-Performance Organic Thin-Film Transistors. *Phys. Chem. Chem. Phys.* 2017, *19*, 15521–15529.

25. Zou, J.; Wang, H.; Zhang, X.; Wang, X.; Shi, Z.; Jiang, Y.; Cui, Z.; Yan, D. Polyimide-Based Gate Dielectrics for High Performance Organic Thin Film Transistors. *J. Mater. Chem. C* 2019, *7*, 7454 7459.

26. Mutlu, S.; Basu, A. S.; Gianchandani, Y. B. Maskless Electrochemical Patterning of Gold Films for Biosensors Using Micromachined Polyimide Probes. *Sensors* [Online] 2005, IEEE, https://ieeexplore.ieee.org/document/1597914 (accessed September 15, 2019).

27. McNamara, S.; Basu, A.; Lee, J.; Gianchandani, Y. B. Ultracompliant Thermal Probe Arrays for Scanning Non-planar Surfaces without Force Feedback. *J. Micromech. Microeng.* 2005, *15*, 237–243.

28. Feili, D.; Schuettler, M.; Doerge, T.; Kammer, S.; Hoffmann, K. P.; Stieglitz, T. Flexible Organic Field Effect Transistors for Biomedical Microimplants Using Polyimide and Parylene C as Substrate and Insulator Layers. *J. Micromech. Microeng.* 2006, *16*, 1555–1561.

29. Stieglitz, T.; Beutel, H.; Meyer, J. U. A Flexible Light-Weight Multichannel Sieve Electrode with Integrated Cables for Interfacing Regenerating Peripheral Nerves. *Sensors Actuat.* 1997, *60*, 240–243.

30. Swiggers, M. L.; Xia, G.; Slinker, J. D.; Gorodetsky, A. A.; Malliaras, G. G.; Headrick, R. L.; Weslowski, B. T.; Shashidhar, R. N.; Dulcey, C. S. Orientation of Pentacene Films Using Surface Alignment Layers and its Influence on Thin-Film Transistor Characteristics. *Appl. Phys. Lett.* 2001, *79*, 1300–1302.

31. Wang, J.; Gundlach, D. J.; Kuo, C. C., Jackson, T. N. *Improved Contacts for Organic Electronic Devices Using Self-Assembled Charge Transfer Materials*. In Proceedings of 41st Electronic Materials Conf. Digest, Santa Barbara, USA, 30 June–2 July, 1999.

32. Melnik, E.; Bruck, R.; Hainberger, R.; Lämmerhofer, M. Multi-Step Surface Functionalization of Polyimide Based Evanescent Wave Photonic Biosensors and Application for DNA Hybridization by Mach-Zehnder Interferometer. *Anal. Chim. Acta* 2011, *699*, 206–215.

33. Saftics, A.; Agócs, E.; Fodor, B.; Patkó, D.; Petrik, P.; Kolari, K.; Aalto, T.; Fürjes, P.; Horvath, R.; Kurunczi, S. Investigation of Thin Polymer Layers for Biosensor Applications. *Appl. Surf. Sci.* 2013, *281*, 66–72.

34. Zulfiqar, A.; Pfreundt, A.; Svendsen, W. E.; Dimaki, M. Fabrication of Polyimide Based Microfluidic Channels for Biosensor Devices. *J. Micromech. Microeng.* 2015, *25*, 035022.

35. Youn, S. W.; Noguchi, T.; Takahashi, M.; Maeda, R. Fabrication of Micro-Mold for Hot-Embossing of Polyimide Microfluidic Platform by Using Electron Beam Lithography Combined with Inductively Coupled Plasma. *Microelectron. Eng.* 2008, *85*, 918–921.

36. Metz, S.; Holzer, R.; Renaud, P. Polyimide-Based Microfluidic Devices. *Lab Chip* 2001, *1*, 29–34.

37. Nguyen, T. N. T.; Lee, N. E. Deep Reactive Ion Etching of Polyimide for Microfluidic Applications. *J. Korean Phys. Soc.* 2007, *51*, 984.

38. Aksoy, B.; Paşahan, A.; Güngör, Ö.; Köytepe, S.; Seçkin, T. A Novel Electrochemical Biosensor Based on Polyimide-Boron Nitride Composite Membranes. *Int. J. Polym. Mater. Polym. Biomater.* 2017, *66*, 202–212.

39. Shen, X.; Xia, X.; Du, Y.; Ye, W.; Wang, C. Amperometric Glucose Biosensor Based on AuPd Modified Reduced Graphene Oxide/Polyimide Film with Glucose Oxidase. *J. Electrochem. Soc.* 2017, *164*, B285–B291.

40. Yoon, H. S.; Jeong, S. K.; Xuan, X.; Park, J. Y. *Semi-implantable Polyimide/PTFE Needle-Shaped Biosensor for Continuous Glucose Monitoring.* In Proceedings of 19th International Conference on Solid-State Sensors, Actuators and Microsystems (TRANSDUCERS), Kaohsiung, TAIWAN, June 18–22, 2017.

41. Vanegas, D. C.; Patiño, L.; Mendez, C.; de Oliveira, D. A.; Torres, A. M.; Gomes, C. L., McLamore, E. S. Laser Scribed Graphene Biosensor for Detection of Biogenic Amines in Food Samples Using Locally Sourced Materials. *Biosensors* 2018, *8*, 42.

42. Tehrani, F.; Bavarian, B. Facile and Scalable Disposable Sensor Based on Laser Engraved Graphene for Electrochemical Detection of Glucose. *Sci. Rep.* 2016, *6*, 27975.

43. Bóka, B.; Adányi, N.; Virág, D.; Sebela, M.; Kiss, A. Spoilage Detection with Biogenic Amine Biosensors, Comparison of Different Enzyme Electrodes. *Electroanalysis* 2012, *24*, 181–186.

44. U.S. Food and Drug Administration (USFDA). *Fish and Fishery Products Hazards and Controls Guidance*, 4th ed.; Center for Food Safety and Applied Nutrition: College Park, MD, 2011.

45. Ministerio de la Protección Social, República de Colombia. Resolución No. 776. *Por la Cual se Establece el Reglamento Técnico Sobre Los Requisitos Fisicoquímicos y Microbiológicos Que Deben Cumplir Los Productos de la Pesca, en Particular Pescados, Moluscos y Crustáceos Para Consumo Humano*; Instituto Nacional de Vigilancia de Medicamentos (INVIMA): Bogotá, Colombia, 2008.

46. Barzic, A. I.; Barzic, R. F. Materials for Organic Transistor Applications. In *Composite Materials for Industry, Electronics, and the Environment: Research and Applications*; Mukbaniani, O. V., Balköse, D., Susanto, H., Haghi, A. K., Eds.; Apple Academic Press, Taylor & Francis, 2019.

47. Popovici, D.; Barzic, A. I.; Barzic, R. F.; Vasilescu, D. S.; Hulubei, C. Semi-alicyclic Polyimide Precursors: Structural, Optical and Biointerface Evaluations. *Polym. Bull.* 2016, *73*, 331–344.

48. Barzic, A. I.; Barzic, R. F. Thermal Conduction in Polystyrene/Carbon Nanotubes: Effects of Nanofiller Orientation and Percolation Process. *Rev. Roum. Chim.* 2015, *60*, 803–807.

49. Barzic, R. F.; Barzic, A. I.; Dumitrascu, G. Percolation Network Formation in Poly(4-vinylpyridine)/Aluminum Nitride Nanocomposites: Rheological, Dielectric, and Thermal Investigations. *Polym. Compos.* 2014, *35*, 1543–1552.

50. Albu, R. M.; Stoica, I.; Avram, E.; Ioanid, E. G.; Ioan, S. Gold Layers on Untreated and Plasma-Treated Substrates of quaternized polysulfones. *J. Solid State Electrochem.* 2014, *18*, 2803–2813.

51. Filimon, A.; Albu, R. M.; Stoica, I.; Avram, E. Blends Based on Ionic Polysulfones with Improved Conformational and Microstructural Characteristics: Perspectives for Biomedical Applications. *Compos. B: Eng.* 2016, *93*, 1–11.

52. Albu, R. M.; Avram, E.; Stoica, I.; Ioan S. Polysulfones with Chelating Groups for Heavy Metals Retention. *Polym. Compos.* 2012, *33*, 573–581.

CHAPTER 8

Natural Fibers: Modifications and Enhancements in Applications

ANANTHU PRASAD[1*], SABU THOMAS[2], and ATHIRA JOHN[3]

[1]*International and Inter University Centre for Nanosicence and Nanotechnology (IIUCNN), Mahatma Gandhi University, Kottayam, Kerala, India*

[2]*School of Chemical Sciences, Mahatma Gandhi University, Kottayam, Kerala, India*

[3]*Centre for Biopolymer Science and Technology (CBPST), CIPET, Kochi, Kerala, India*

Corresponding author. E-mail: ananthuprasad069@gmail.com

ABSTRACT

Natural fibers have been used by humans for centuries. The intervention of new synthetic materials replaced natural fibers in many applications. But due to ecological concerns and better awareness of environmental pollution, natural fibers are attaining attention now. Natural fibers and derivatives of natural fibers are eco-friendly, cheap, and renewable. Large industrial sectors like automobile industry are promoting the use of natural fibers as reinforcements. However, when compared with synthetic fibers, natural fibers have several disadvantages like water absorption, low compatibility with matrix, and so on. Extensive research is being carried out to overcome these limitations. In this chapter, we can study natural fibers, its limitations, the methods adapted to overcome these limitations, and various applications.

8.1   INTRODUCTION

Natural fibers are attaining attention from all around the world due to its environmental friendly properties and widening application possibilities. Due to the depletion of oil resources and increasing ecological concerns, large industrial sectors like automobile industries are looking into products from natural fibers as a replacement for the synthetic-fiber-based products. One of the most studied areas is natural fiber composites, which can be considered as an eco-friendly alternative for the widely used synthetic fibers such as glass fibers. Lignocellulosic natural fibers like jute, sisal, coir, and pineapple have been used as reinforcements in polymer matrices.[1] If proper interfacial adhesion is achieved, natural fiber composites can exhibit very good properties and could compete with conventional plastic materials.

Basically, natural fibers are water based and plastics are oil based. As we all know, oil cannot be mixed with water. The same problem arises in the preparation of natural fiber composites. Natural fibers exhibit polar nature while plastics are nonpolar. To make a composite out of these two components, some kind of surface modification is essential. Over the years, several attempts were made to effectively modify surface properties of natural fibers. It includes both chemical and physical methods. In this chapter, we can see the classification and properties of natural fibers, their interfacial problems, and the surface modification techniques used to overcome those issues. Also, potential applications of natural fiber-based composites will be discussed.

8.2   CLASSIFICATION AND COMPOSITION OF NATURAL FIBERS

Natural fibers can be classified based upon the source of its origin, namely, plant derived and animal derived. Most of the natural fibers are of plant origin. Wool and silk are examples of fibers of animal origin. Plant-derived fibers mainly consist of cellulose whereas animal-derived fibers have proteins as the main component. Plant-derived natural fibers are often referred to as lignocellulosic fibers and it consists of cellulose, hemicellulose, lignin, pectin, and small amounts of fats and waxes. A general classification of natural fibers and some pictures of different natural fibers in use today are shown in Figures 8.1 and 8.2.[2]

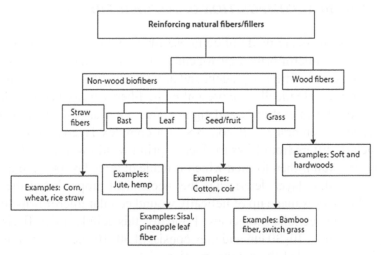

FIGURE 8.1 General classification of natural fibers.

FIGURE 8.2 Sources of natural fibers.

8.2.1 CHEMICAL COMPOSITION OF NATURAL FIBERS

Natural plant fibers are made of cellulose fibers, which consist of helically wound cellular microfibrils kept together by an amorphous lignin matrix. Lignin also acts as a stiffening agent and protects the fiber from gravitational and wind forces. Hemicellulose present in the fibers acts as a compatibilizer between cellular microfibrils and lignin. The cell wall of a fiber is not a homogeneous membrane.[3,4] It consists of a complex layered structure and this structure differs from fiber to fiber. A primary thin wall is first formed. A secondary wall is then formed which is made of three layers. Among these layers, the middle layer determines the mechanical properties of the fiber. The middle layer is made up of helically wound cellular microfibrils formed from long-chain cellular molecules. Typically, this cellular microfibris have diameters of about 10–30 nm and they consist of 30–100 cellulose molecules formed in an extended chain conformation.

The structure of hemicellulose, cellulose, and lignin is shown in Figures 8.3–8.5.

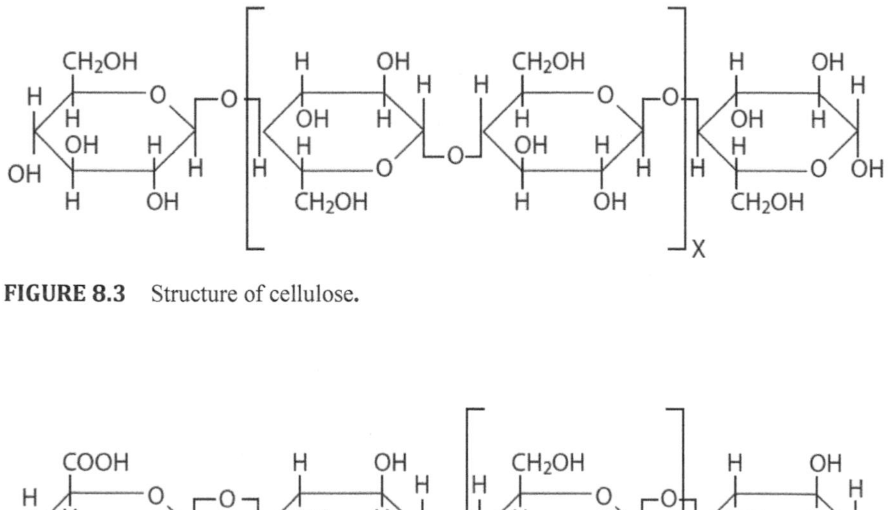

FIGURE 8.3 Structure of cellulose.

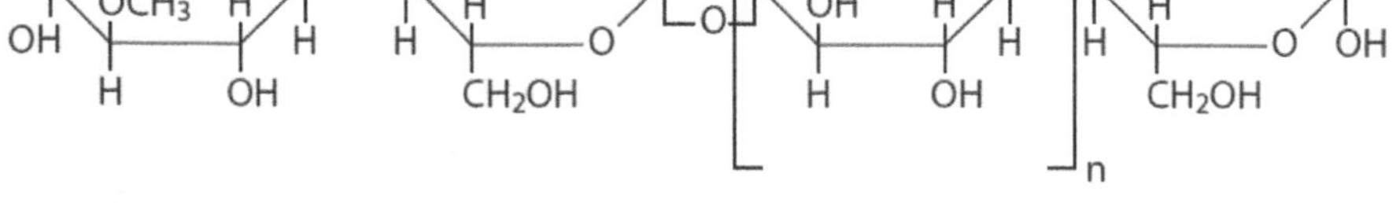

FIGURE 8.4 Structure of hemicellulose.

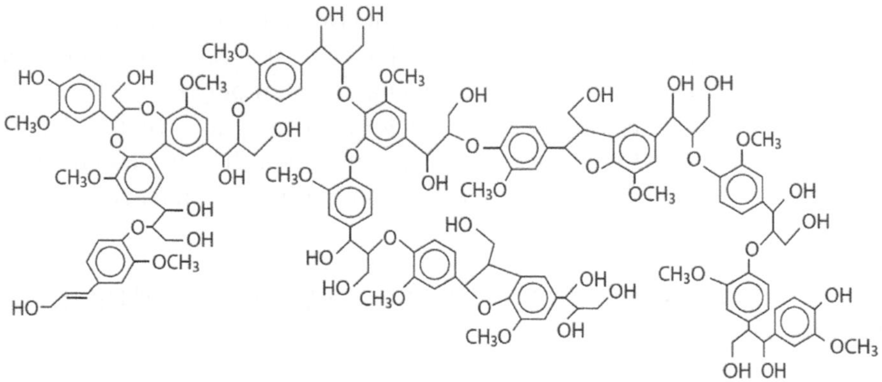

FIGURE 8.5   Structure of lignin.

8.3   FIBER SURFACES AND INTERFACIAL PROBLEMS

The interfacial bonding between fiber and matrix is the most important factor which affects the strength of a composite. Strong interfacial adhesion enables smooth stress transfer from matrix to the reinforcing fibers. But increased interfacial adhesion can result in crack propagation also. So a balanced optimum adhesion is preferred. As we mentioned before, the main hurdle in achieving this interfacial adhesion using natural fiber reinforcements is the difference in polarity. The hydrophilic natural fibers show limited interaction with the hydrophobic matrices which lead to decreased interfacial adhesion and low mechanical properties. For bonding to occur, the fibers must be kept in close contact with the matrix. In that regard, the wettability of fibers plays a key role.[5] Fiber wettability affects the toughness, tensile, and flexural strength of composites.[6] Physical treatment and chemical treatment can improve the wettability of the fiber and thus improve the interfacial strength.[7,8]

Interfacial bonding can be brought about by means of mechanisms of mechanical interlocking, electrostatic bonding, chemical bonding, and interdiffusion bonding.[9] Mechanical interlocking occurs due to the surface roughness of the fibers. The rough surface results in increased shear strength. Electrostatic bonding occurs mainly at metallic interfaces. Chemical bonding occurs due to the reaction between surface functional groups of fibers and matrix. The reactions result in the formation of chemical bonds. The interfacial strength of the composite depends on the type and density of these bonds. Chemical bonding can be bought about by the addition of a coupling agent. Coupling agent essentially acts as a bridge between fiber and matrix by forming bonds between them. Interdiffusion bonding happens at polymeric interfaces due to the chain entanglement of polymers. The extent of interfacial strength depends upon the length of polymer chains, degree of entanglement, and number of chains per unit area. It is to be noted that multiple bonding mechanisms can occur at an interface simultaneously.[10]

8.4   SURFACE MODIFICATION TECHNIQUES USED IN NATURAL FIBERS

To enhance the interfacial adhesion of the fibers and matrix, several surface modification techniques are used which can be broadly classified into chemical and physical treatments. These treatments result in the physical and chemical changes in the surface layer of the fiber without altering the bulk properties.[11]

8.4.1   CHEMICAL SURFACE MODIFICATION

The main chemical techniques used to surface modify natural fibers are mercerization (alkali treatment), acetylation, acrylation treatment, isocyanate treatment, maleic anhydride treatment.

8.4.1.1   MERCERIZATION

Mercerization is the alkali treatment of natural fibers.[12] Alkali treatment increases the surface roughness of the fibers which enables better mechanical interlocking at the interface and the amount of cellulose exposed on the surface. The following reaction takes place during alkali treatment:

$$\text{Fiber} - \text{OH} + \text{NaOH} \rightarrow \text{Fiber} - \text{O} - \text{Na} + \text{H}_2\text{O}$$

The addition of aqueous sodium hydroxide to the natural fibers results in the ionization of hydroxyl group to alkoxide. Several studies conducted on mercerized natural fibers reveal that alkali treatment results in increased amount of amorphous cellulose at the expense of crystalline cellulose in the network structure.[13]

8.4.1.2 ACETYLATION

Acetylation of natural fibers is esterification method known for its plasticizing effect on the fibers.[14] It is based on the reaction between hydroxyl groups present in the cell wall of lignocellulosic fibers with acetic or propionic anhydride, which takes place at elevated temperatures.

$$\text{Fiber} - \text{OH} + (\text{H}_3\text{C–CO})_2\text{O} \rightarrow \text{Fiber} - \text{O} - \text{CO} - \text{CH}_3 + \text{CH}_3\text{COOH}$$

It has been shown that acetylation enhances the dispersion and dimensional stability of natural fibers in the polymeric matrix.[13–15]

8.4.1.3 ACRYLATION TREATMENT

Acrylation treatment, maleated polypropylene/maleic anhydride treatment, and titanate treatment of cellulosic fibers are also reported.[16] Through these treatments, the surface energy of the fibers is increased, thereby providing better wettability and high interfacial adhesion.[17]

8.4.1.4 ISOCYANATE TREATMENT

Isocyanate consists of a –N=C=O group which could react with the surface hydroxyl groups to form strong covalent bonds, which enhances the compatibility of the fibers with the polymeric matrix. Thermoplastic–natural fiber composites show superior properties when isocyanates are used. Isocyanate can act as a promoter or as an inhibitor of interaction.[18]

$$\text{Fiber} - \text{OH} + \text{R–N=C=O} \rightarrow \text{Fiber} - \text{O–CO–NH–R}$$

8.4.1.5 MALEIC ANHYDRIDE TREATMENT

Maleic anhydride can be used as a grafting monomer for biopolymer matrices. The highly reactive anhydride group reacts with hydroxyl groups to form grafted biopolymers. Maleic anhydride is also used as an adhesion promoter in biocomposite applications.[19]

8.4.2   PHYSICAL SURFACE MODIFICATION TECHNIQUES

Physical surface modification techniques used are corona, plasma, ultraviolet, heat treatments.

8.4.2.1   CORONA TREATMENT

Corona treatment uses plasma generated by the application of high-voltage difference between two sharp electrode tips using quartz at a low temperature and atmospheric pressure. Oxygen-containing species are commonly used.[20] It introduces surface roughness and increased surface polarity which is due to the introduction of new functional groups (carboxyl and hydroxyl groups) on the fiber surface. But application of corona treatment to a three-dimensional surface is not feasible.[21,22]

8.4.2.2   PLASMA TREATMENT

Plasma treatment is similar to corona treatment, but a vacuum chamber is used while the treatment is carried out. Continuous gas supply is enabled which enables the appropriate pressure and desired gas composition for the introduction of specific functionalities on the fiber surface.[20] Plasma treatment results in increased fiber surface roughness and hydrophobicity of fibers, thereby increasing the interfacial adhesion.[23] Also, interlaminar shear strength and flexural strength have been increased up to 35% and 30%, respectively, using plasma treatment.[24]

8.4.2.3   HEAT TREATMENT

Heat treatment is the heating of fibers at temperatures close to its degradation temperatures which brings about changes in physical, chemical, and mechanical properties of the fibers including water content, chemistry, cellulose crystallinity, degree of polymerization, and strength. Some of the chemical changes include chain scission, free radical production, and formation of carbonyl, carboxyl, and peroxide groups.[20] The result of heat treatment depends on the time, temperature, and composition of gases present during the heat treatment.

8.5 APPLICATIONS OF NATURAL FIBERS

8.5.1 AUTOMOBILE INDUSTRY

Different natural fibers are extensively studied to explore their potential to be used as effective reinforcements for polymeric matrices in order to develop automobile components.[25] Polylactic acid (PLA) is a biopolymer which could provide sufficient reinforcement along with natural fibers. These composites have been used for developing different several automobile components by using different types of natural fibers like cotton, hemp, kenaf, and man-made cellulose fibers (Lyocell) as reinforcements. Flax/Hemp fibers dispersed in an epoxy matrix exhibit the highest strength among the natural fiber composites. It also shows resistance to environmental degradation. Many automobile parts can be developed using these composites due to these properties. Cotton fibers have high-impact strength but exhibit low-tensile strengths. So, it is used as reinforcements in interior parts in cars and in safety helmets.

8.5.2 CONDUCTING COMPOSITES

Conducting cellulose composites have increasingly attracted attention not only due to their environment-friendly nature but also because of their flexibility, low weight, and ease of processing overpiezoceramic and magnetostrictive materials. The basic function of the electroactive paper in such applications is to convert energy between electric and mechanical forms. Conducting polymers and their cellulose-based composites can be used in different devices, such as a polyaniline-based filter paper that acts as a sensor for acids, bases, and as endpoint indicators,[26] heating devices,[27] and actuators.[28]

Carboxy methyl cellulose, hydroxypropylcellulose (HPC), and acetoxypropylcellulose (APC) are being used for the fabrication of electro-optical sensors from cross-linked combination of HPC/APC[29,30] and liquid crystalline solutions with and without a low-molecular-weight nematic liquid crystal mixture. The use of cellulose in lithium batteries has also been elaborated to replace the polyolefin-based separator in rechargeable lithium ion batteries. A thin composite of cellulosic material (39–85 mm) used as a separator. These cellulosic separators exhibited good initial discharge capacity and capacity retention over 41 charge/discharge cycles.[31]

8.5.3 BARRIER APPLICATIONS

The impermeable crystalline domains present in the natural fibers generate a tortuous path which leads to slower diffusion rates and low permeability. From 1952, the filters used in cigarettes were made out of cellulose derivatives.[32] A greater specific surface area and barrier properties of cellulose fibers enables 44% reduction in tar and a 35% reduction of the nicotine in smoke.[25]

8.5.4 DRUG-DELIVERY SYSTEMS

The limitation of poor mechanical properties of biodegradable polymers used for drug-delivery systems can be resolved by blending with different fillers, especially layered silicates like hydroxyl apatite. The chemical structures, molecular weight, composition, and synthesis conditions are parameters that influence the final morphology and drug-delivery nature of the polymer and its composites. Cellulose alone is frequently used as adhesive for drug-delivery systems, whereas its composites are gaining considerable interest in this area where the cellulose derivative, ethylcellulose, with a synthetic polymer such as polystyrene may be used for the release of water-soluble drugs, such as phenobarbital.[33]

Cellulose nitrate and cellulose acetate monolayer membranes containing thermotropic liquid crystals have been developed as thermoresponsive barriers using methimazole and paracetamol as hydrophilic and hydrophobic drugs, respectively.[34]

8.5.5 BUILDING AND CONSTRUCTION INDUSTRY

Biobased structural composites in construction industry are of significant importance for fencing, decking, siding, doors, windows, bridges, fiber cement, and so on. Advantages associated with the use of natural fibers to reinforce cement, known as fiber cement, include the availability of raw material from renewable sources, high-fiber-tensile strength, high modulus of elasticity, relatively low cost, and well-developed technology for fiber processing.[35] Fiber cement presents improved toughness, ductility, flexural capacity, and crack resistance compared to nonfiber-reinforced, cement-based materials.[25]

Fiber reinforcements have two functions. Their primary function is to transfer the stresses and strengthen the composites. It also increases the toughness of the cement composites by means of energy-absorbing mechanisms. Different methods are available to make fiber–cement composites. In general, all processes utilize fiber concentrations of 8–12%.[36-38]

8.6 CONCLUSION

Natural fibers are evolving to be a material of superior characteristics and wide range of applications. It has broken the limits of its traditional applications and is competing with novel synthetic materials. The lack of interfacial adhesion in natural fiber composites is effectively countered with chemical and physical treatments. Better interfacial interaction between fibers and polymer matrices is observed by grafting of natural fibers. The hydrophilic natural fibers have been effectively converted to hydrophobic by the introduction of functionalities in the fiber surface by means of plasma treatment. Natural fibers find extensive applications in the automobile industry. The composites of natural fibers and PLA give a fully compostable biopolymer composite as a replacement for conventional composites. Flax/hemp fibers dispersed in epoxy matrices give high-strength composites with resistance to environmental degradation. Natural fibers are also used as filters due to its barrier properties, reinforcements in drug-delivery systems, and as reinforcement for fiber–cement composites. Further research in natural fibers is encouraged due to its positive impact on environment. Existing challenges are moisture absorption and low fire retardancy of fibers. The growth of natural fiber consumption is rapid and there would be a bright future ahead for their applications.

KEYWORDS

- **natural fibers**
- **composites**
- **surface modification**
- **cellulose**

REFERENCES

1. Hassan, M. M.; Wagner, M. H. Surface Modification of Natural Fibers for Reinforced Polymer Composites: A Critical Review. *Rev. Adhes. Adhesives* **2016,** *4* (1), 1–46.
2. Franck, R. R., Ed. *Bast and Other Plant Fibres*; CRC Press: Boca Raton, FL, 2005; Vol 39.
3. Rong, M. Z.; Zhang, M. Q.; Liu, Y.; Yang, G. C.; Zeng, H. M. The Effect of Fiber Treatment on the Mechanical Properties of Unidirectional Sisal-Reinforced Epoxy Composites. *Composites Sci. Technol.* **2001,** *61* (10), 1437–1447.
4. Nevell, T. P.; Zeronian, S. H. *Cellulose Chemistry and Its Applications*; Halsted Press, John Wiley: New York, 1985.
5. Chen, P.; Lu, C.; Yu, Q.; Gao, Y.; Li, J.; Li, X. Influence of Fiber Wettability on the Interfacial Adhesion of Continuous Fiber-Reinforced PPESK Composite. *J. Appl. Polym. Sci.* **2006,** *102* (3), 2544–2551.
6. Wu, X. F.; Dzenis, Y. A. Droplet on a Fiber: Geometrical Shape and Contact Angle. *Acta Mech.* **2006,** *185* (3–4), 215–225.
7. Bénard, Q.; Fois, M.; Grisel, M. Roughness and Fibre Reinforcement Effect onto Wettability of Composite Surfaces. *Appl. Surf. Sci.* **2007,** *253* (10), 4753–4758.
8. Sinha, E.; Panigrahi, S. Effect of Plasma Treatment on Structure, Wettability of Jute Fiber and Flexural Strength of Its Composite. *J. Composite Mater.* **2009,** *43* (17), 1791–1802.
9. Matthews, F. L.; Rawlings, R. D. *Composite Materials: Engineering and Science*; CRC Press: Boca Raton, FL, 1999.
10. Beckermann, G. Performance of Hemp-Fibre Reinforced Polypropylene Composite Materials. Doctoral Dissertation, The University of Waikato, 2007.
11. Acda, M. N.; Devera, E. E.; Cabangon, R. J.; Ramos, H. J. Effects of Plasma Modification on Adhesion Properties of Wood. *Int. J. Adhes. Adhesives* **2012,** *32,* 70–75.
12. Heap, S. A. Fundamental Aspects of the Mercerizing Process. *Int. Dyer* **1975,** *154,* 374–375.
13. Mohanty, A. K.; Misra, M.; Drzal, L. T. Surface Modifications of Natural Fibers and Performance of the Resulting Biocomposites: An Overview. *Composite Interfaces* **2001,** *8,* 313–343.
14. Misra, S.; Misra, M.; Tripathy, S. S.; Nayak, S. K.; Mohanty, A. K. The Influence of Chemical Surface Modification on the Performance of Sisal-Polyester Biocomposites. *Polym. Composites* **2002,** *23* (2), 164–170.
15. Bogoeva-Gaceva, G.; Avella, M.; Malinconico, M.; Buzarovska, A.; Grozdanov, A.; Gentile, G.; Errico, M. E. Natural Fiber Eco-composites. *Polym. Composites* **2007,** *28* (1), 98–107.
16. Rowell, R. M. Property Enhanced Natural Fiber Composite Materials Based on Chemical Modification. In *Science and Technology of Polymers and Advanced Materials*; Springer: Boston, MA, 1998; pp 717–732.
17. Mittal, K. L. The Role of the Interface in Adhesion Phenomena. *Polym. Eng. Sci.* **1977,** *17* (7), 467–473.
18. Kalia, S.; Kaith, B. S.; Kaur, I. Pretreatments of Natural Fibers and Their Application as Reinforcing Material in Polymer Composites—A Review. *Polym. Eng. Sci.* **2009,** *49* (7), 1253–1272.

19. Joseph, K.; Mattoso, L. H. C.; Toledo, R. D.; Thomas, S.; de Carvalho, L. H.; Pothen, L.; James, B. In *Natural Polymers and Agrofibers Composites*; Frollini, E., Leao, A. L., Mattoso, L. H. C., 2000, pp 159–201.

20. Pickering, K., Ed. *Properties and Performance of Natural-Fibre Composites*; Elsevier: Amsterdam, 2008.

21. Ragoubi, M.; Bienaimé, D.; Molina, S.; George, B.; Merlin, A. Impact of Corona Treated Hemp Fibres onto Mechanical Properties of Polypropylene Composites Made Thereof. *Ind. Crops Prod.* **2010,** *31* (2), 344–349.

22. Gassan, J.; Gutowski, V. S. Effects of Corona Discharge and UV Treatment on the Properties of Jute-Fibre Epoxy Composites. *Composites Sci. Technol.* **2000,** *60* (15), 2857–2863.

23. Sinha, E.; Panigrahi, S. Effect of Plasma Treatment on Structure, Wettability of Jute Fiber and Flexural Strength of Its Composite. *J. Composite Mater.* **2009,** *43* (17), 1791–1802.

24. Seki, Y.; Sever, K.; Sarikanat, M.; Güleç, H. A.; Tavman, I. H. The *Influence of Oxygen Plasma Treatment of Jute Fibers on Mechanical Properties of Jute Fiber Reinforced Thermoplastic Composites.* In Proceedings of the 5th International Advanced Technologies Symposium, 2009, May (IATS'09).

25. Pandey, J. K.; Ahn, S. H.; Lee, C. S.; Mohanty, A. K.; Misra, M. Recent Advances in the Application of Natural Fiber Based Composites. *Macromol. Mater. Eng.* **2010,** *295* (11), 975–989.

26. Bhat, N. V.; Seshadri, D. T.; Nate, M. M.; Gore, A. V. Development of Conductive Cotton Fabrics for Heating Devices. *J. Appl. Polym. Sci.* **2006,** *102*(5), 4690–4695.

27. Nalwa, H. S. *Handbook of Organic Conductive Molecules and Polymers*; Wiley: Hoboken, NJ, 1997.

28. Deshpande, S. D.; Kim, J.; Yun, S. R. Studies on Conducting Polymer Electroactive Paper Actuators: Effect of Humidity and Electrode Thickness. *Smart Mater. Struct.* **2005,** *14* (4), 876.

29. Costa, I.; Filip, D.; Figueirinhas, J. L.; Godinho, M. H. New Cellulose Derivatives Composites for Electro-Optical Sensors. *Carbohydr. Polym.* **2007,** *68* (1), 159–165.

30. Lukasiewicz, M.; Ptaszek, A.; Koziel, L.; Achremowicz, B.; Grzesik, M. Carboxymethylcellulose/Polyaniline Blends. Synthesis and Properties. *Polym. Bull.* **2007,** *58* (1), 281–288.

31. Kuribayashi, I. Characterization of Composite Cellulosic Separators for Rechargeable Lithium-Ion Batteries. *J. Power Sources* **1996,** *63* (1), 87–91.

32. Glasser, W. G. *Prospects for Future Applications of Cellulose Acetate*. In *Macromolecular Symposia*; Wiley-VCH Verlag: Weinheim, 2004; Vol 208, No 1, pp 371–394.

33. Ohara, Y.; Arakawa, M.; Kondo, T.; Lee, K. B. Preparation of Ethylcellulose/Polystyrene Composite Microcapsules of Two-Phase Structure and Permeability of the Microcapsule Membranes Towards Phenobarbital. *J. Membr. Sci.* **1985,** *23* (1), 1–9.

34. Atyabi, F.; Khodaverdi, E.; Dinarvand, R. Temperature Modulated Drug Permeation Through Liquid Crystal Embedded Cellulose Membranes. *Int. J. Pharm.* **2007,** *339* (1–2), 213–221.

35. http://www.cement.org.

36. Puitel, A. C.; Tofanica, B. M.; Gavrilescu, D. Environmentally Friendly Vegetal Fiber Based Materials. *Environ. Eng. Manage. J.* **2012,** *11* (3), 651–659.

37. Shao, Y.; Moras, S. Strength, Toughness and Durability of Extruded Cement Boards with Unbleached Kraft Pulp. *Int. Concrete, Spl. Publ.* **2002,** 206, 439–452.
38. Shao, Y.; Marikunte, S.; Shah, S. P. Extruded Fiber-Reinforced Composites. *Concrete Int.* **1995,** *17* (4), 48–53.

CHAPTER 9

Polyethylene Wax: Uses, Characterization, and Identification

SENEM YETGIN[1], MEHMET GONEN[2], SEVDIYE ATAKUL SAVRIK[3], and DEVRIM BALKOSE[4*]

[1]*Food Engineering Department, Kastamonu University, Kastomonu, Turkey*

[2]*Chemical Engineering Department, Suleyman Demirel University, Isparta, Turkey*

[3]*A.Ş Gaziemir, Akzo Nobel Boya, Izmir, Turkey*

[4]*Department of Chemical Engineering Gulbahce, Izmir Institute of Technology, Urla, Izmir, Turkey*

Corresponding author. E-mail: devrimbalkose@gmail.com

ABSTRACT

The characterization of a commercial particulate sample labeled as "polyethylene (PE) wax" used in pigment masterbatch preparation was performed in the present study by comparing its properties with similar materials; a paraffin wax, sorbitan monostearate (Span 60), and myristic acid. The samples were characterized by Fourier transform infrared (FTIR) spectroscopy, x-ray diffraction, scanning electron microscopy (SEM), differential scanning calorimetry (DSC), and thermal gravimetric analysis (TGA). Infrared spectroscopy showed that there were vibrations of CH_2 and CH_3 groups in paraffin wax, CH_2, CH_3, C-O, and C=O groups in commercial wax and Span 60 and myristic acid. It was found that commercial wax has a melting point of 54.9°C and heat of fusion of 115 J g^{-1}. The particles were in spherical form with 170 μm average diameter and mixed with cylindrical fibers of 5.45 μm diameter and 75 μm in length. There were particles having 560 μm diameter with holes of 145 μm on their surface. In these large particles, there was a crust of 32 μm in thickness coating aglomerated small particles. It was found that the commercial wax was not "PE wax" as it was labeled. The melting point, functional groups, and the crystal structure of commercial wax had closer similarity to Span 60 than that of paraffin wax and myristic acid. However, its onset temperature of mass loss was much lower than that of Span 60 and the heat of fusion of Span 60 was much lower than that of commercial wax. Thus, it was concluded that further studies with more advanced methods are necessary to make a complete identification of commercial PE wax samples.

9.1   INTRODUCTION

Polyethylene (PE) wax is used in many industrial applications such as making water repellent hydrophobic surfaces, candles, external lubricant in thermoset and thermoplastic extrusions and other polymers, preparation of pigment masterbatches, as well as phase change materials (PCM) in building industry. PE wax is produced by mixing the paraffin wax and PE in a certain ratio. The properties of PE wax is determined by selected PE either high density or low density and the properties of paraffin wax used in the formulation. Paraffin wax and PE are melted together in a stirred reactor which is heated by superheated steam through the jacket. The temperature of the mixture is held above the melting point of PE (120°C). The melted mixture is crystallized on a chilled drum on which flakes are removed by a doctor blade or it is sprayed into a chamber through a cold air that is circulated to obtain smaller particles. PE wax is supplied to the market in various forms, such as bulk form, powder form, or microgranular form depending on where they are used.

9.1.1   PIGMENT MASTERBATCH

Dekmush et al.[9] reported that the role of PE in the preparation of pigment masterbatches. They showed that both wax (Low molecular weight PE) and polymer-based linear density PE pigment masterbatches had similar color properties. However, polymer-based predispersed pigment masterbatches showed superior dispersion property. PE wax particles were nanoencapsulated by in-situ emulsion polymerization technique for color toner by Zang et al.[30]

There are other carriers used in pigment masterbatches. They are a combination of two groups. The copolymers of ethylene acrylic acid, copolymers of ethylene maleic anhydride, polypropylene maleic anhydride, and polycaprolactone are in the first group and copolymers of ethylene. In the second group copolymers of ethylene butyl acrylate, ethylene methacrylic acid, polypropylene grafted with polymethyl methacrylate, and polypropylene grafted with polystyrene are present. The mixtures from the first and the second group are used in pigment masterbatches.[6] Hyperdispersants are also used in the preparation of masterbatches. Polycarboxylic acid hyperdispersant containing fundamental chain, sulfonic groups, carboxyl, and polyoxyethylene branches were synthesized by Zhou et al.[31] to obtain efficient coal dispersion in water. 3-Pentadecylphenyl acrylate, methacrylic acid, and butyl acrylate were copolymerized in order to get hyperdispersant for dispersion of TiO_2 in organic media by Lui et al.[21] There are commercial hyperdispersants such as Lubrizol Solplus series and BYK series for preparation of pigment masterbatches. Hyperdispersants are two-component structures. These are anchoring group that provides strong adsorption onto the pigment surface and polymeric chains which are attached to the anchor group that provide the stabilization.[26] The nature of the polymeric chain is critical to the performance of hyperdispersants. If the chains are not sufficiently solvated, then they will collapse onto the pigment surface allowing the particles to aggregate or flocculate. The molecular weights of the Solplus/ Ircolplus hyperdispersant products are sufficient to provide polymer chains of optimum length to overcome Van der Waals forces of attraction between pigment particles. In order to provide good compatibility, several different types of polymer chains are utilized in the Solplus and Ircolplus hyperdispersants range, effectively covering the variety of types of solvents and binder resins encountered. There is generally an optimum chain-length over and above which the effectiveness of the stabilizing material ceases to increase; indeed in some cases, molecules with longer than optimum chains can be less effective. Ideally, the chains should be free to move in the dispersing medium. To achieve this, chains with anchor groups at one end only have shown to be the most effective in providing steric stabilization.[26] Yetgin and Balkose[29] prepared pigment masterbatches by mixing pigments with molten wax and then by cooling the mixture by mixing in a porcelain mortar. The pigment masterbatches were used in coloring polyvinyl chloride plastisols in their study.

9.1.2 PCM

PCM are microencapsulated materials with low melting temperature and high heat of fusion. The core could be paraffin wax, low melting organic acids such as myristic acid and palmitic acid,[2] and the coating could be inorganic origin like silica, titania[13] and/or organic origin like ethyl cellulose[20] and amino formaldehyde.[16]

9.1.3 GRANULATION

Granulation is the process of substance forming into small particles (granules). Spherical wax particles could be obtained by dripping molten wax to cold ultrasonic bath,[14] melt emulsification, spray freezing, and granulating in a high-speed granulator. Melt emulsification is a top-down approach that so far has been applied in the food industries (cf. homogenization of milk) and in pharmaceutical applications. The applicability of the process for the production of spherical polymer microparticles was investigated by Fanselow et al.[12] A certain amount of the emulsifier and PE wax was heated until PE wax was completely melted. A certain amount of the stabilizer, dispersants, and water was stirred until complete dissolving, and then the mixture solution was heated to the temperature of the PE wax melting point. At a certain stirring speed, 10 g mixture solution was dripped into the melted PE wax, and the stirring time was 2 min. Then, the rest of the mixture solution was added to the melted PE wax and the stirring time was 20 min. The products were cooled at room temperature to obtained PE wax emulsion.[19] Size reduction of the polymer raw melt emulsion was realized in a rotor–stator device. The process was characterized or paraffin and PE wax in an aqueous environment using Tween 85 as an emulsifier. Process temperature and dispersed phase viscosity have the largest impact on the particle size distribution: decreasing temperature and viscosity leads to smaller particles. High-speed granulation unit disperses a solution, melt or a suspension of a loose material curtain in a moving air medium in rotating drum provided with internal blades.[27] A novel granulation technology for producing spherical wax particles that makes use of the insolubility of wax and water was developed by Fang et al.[11] The granulation was stably performed in a granulator with a water-cooling tower. The granulation process was analyzed and the influences of key factors on the process were studied. The experiments were carried out with semi-refined wax with 58°C melting point and crystalline wax with 70°C melting point in an experimental granulator which has 118 nozzles (D_i = 0.96 mm, D_o = 1.28 mm). The processing capacity and the corresponding particle diameter were 55–60 kg h^{-1} and 4–5 mm, respectively. This technology had the following advantages: good particle shape with excellent fluidity, high process capacity and heat-transfer efficiency, and a small space for the granulator. It is also suitable for producing spherical particles of other materials, which had similar physical properties to wax. It is possible self-assembly of the particles by controlling the concentration and the ratio, R of the neutralizing agent to myristic acid. Spherical particles were formed at high R values.[3] Spray freezing is a process where a hot melt is forced through a binary spray nozzle and atomized into very fine droplets. Air or nitrogen is supplied to the nozzle as an atomizing agent. The process temperature has to be above the wax melting point to prevent any form of nozzle blocking. Wax droplets solidify and form spherical particles by subsequent cooling.[17]

For prilling, the molten material is atomized using either a single-fluid or two-fluid nozzle system in a nitrogen-purged vertical chamber.[28] Typical particle sizes are in the 100–1000 μm range. For most materials, the optimal particle size is 200–300 μm. The final product is screened directly off the collection cyclone to remove any large chunks of wax.

The present study aims the characterization and identification of a commercial PE wax which is used for pigment masterbatch preparation by comparing with paraffin wax, sorbitan monostearate (Span 60), and myristic acid.

9.2 MATERIALS AND METHOD

9.2.1 MATERIALS

A commercial wax sample labeled as "PE wax" that was provided by Polymast Masterbatch Boya Company was characterized in the present study by comparing its properties with waxy materials such as paraffin wax (Snow) from Baykim Kimya, Span 60 from Sigma Aldrich and myristic acid. While the paraffin wax was in bulk form, the commercial wax had white micro granular particles.

9.2.2 CHARACTERIZATION METHODS

The Fourier transform infrared (FTIR) spectra of the samples were obtained with either Shimadzu 8601 or Shimadzu IR Prestige-21 using KBr disc method. X-Ray diffraction pattern of the samples were obtained by either Philips X'Pert Pro diffractometer or Bruker D8 Advance Twin-Twin diffractometer using CuK$_\alpha$ radiation. Scanning electron microscopic (SEM) microphotographs of the samples were taken with FEI Quanta 250 FEG scanning electron microscope. The thermal gravimetric analysis (TGA) of the samples were made by using Shimadzu TGA-51 at 10°C min^{-1} heating rate from 25 to 1300°C under 40 cm^3 min^{-1} N$_2$ flow. The differential scanning calorimetric (DSC) analysis was made with TA Instruments Q10 Differential calorimeter by heating the samples between 25 and 600°C at 10°C min^{-1} heating rate under 40 cm^3 min^{-1} N$_2$ flow.

9.3   RESULTS AND DISCUSSION

9.3.1   FTIR STUDY

The functional groups of compounds are identified by FTIR analysis.

The FTIR spectra of paraffin wax, commercial wax, and Span 60 are shown in Figures 9.1, 9.2 and 9.3, respectively. The main peaks observed in the spectra of the samples under consideration and in the spectrum of myristic acid,[25] are shown in Table 9.1. The paraffin wax had only peaks related to symmetric and asymmetric stretching vibrations at 2914 and 2847 and bending vibrations at 1470 cm^{-1} and wagging vibrations at 720 cm^{-1} of CH$_2$ group. On the other hand, the commercial wax had extra peaks related to hydrogen bonded OH groups at 3404 cm^{-1}, C=O stretching peak at 1737 cm^{-1}, ionized carboxylate group vibration peak at 1545 cm^{-1} in addition to CH$_2$ vibration peaks. Span 60 had peaks related to hydrogen bonded OH vibrations at 3400 cm^{-1}, stretching and bending and wagging vibrations of CH$_2$ group and C=O stretching vibration at 1730 cm^{-1}. Myristic acid had also peaks related to CH$_2$, C-H, OH, C-C, and C=O groups. At 2916 and 2848 cm^{-1} symmetrical and asymmetrical stretching of –CH$_2$ functional group and at 1701 cm^{-1} C=O stretching vibration were observed. The stretching peaks at 686, 721, and 939 cm^{-1} were due to –CH swinging or rocking mode, which were characteristics of aliphatic chain of myristic acid. The FTIR peaks at 1431 and 1464 cm^{-1} were due to –CH$_2$ bending vibration peaks. The other peaks were observed at 1286 and 1261 cm^{-1} represented the C-H and C-C bending vibrations, respectively.[23,25]

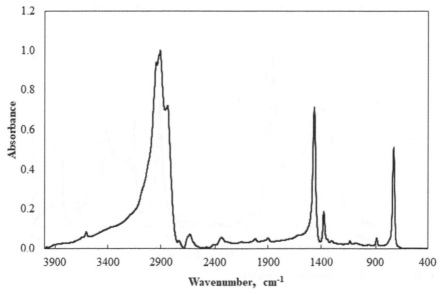

FIGURE 9.1   FTIR spectrum of paraffin wax.

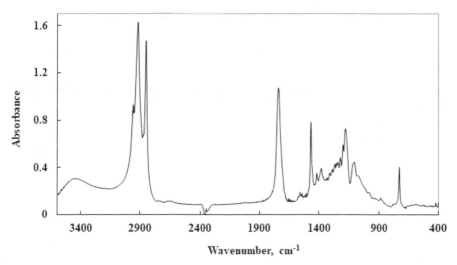

FIGURE 9.2   FTIR spectrum of commercial wax.

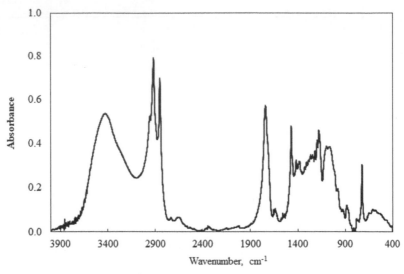

FIGURE 9.3   FTIR spectrum of Span 60.

The commercial wax is neither a paraffin wax nor a myristic acid since it has a different FTIR spectrum than the spectra of both substances. Its FTIR spectrum has features similar to that of Span 60, but the presence of carboxylate functional groups indicates that it is not Span 60 either.

TABLE 9.1   The Peaks Observed in FTIR Spectra of Paraffin Wax, PE Wax, Span 60, and Myristic Acid (cm⁻¹).

Vibration	Paraffin wax[8]	Commercial wax	Span 60[24]	Myristic acid[25]
OH strecthing		3404	3400	
CH₃ stretching		2953		
CH₂ symmetric stretching	2914	2916		2918
CH₂ asymmetric streching	2847	2818		2848
C=O stretching		1737	1730	1701
Ionized carboxylate streching C-O		1545		
CH₂ bending	1470	1467	1467	1481
				1464
C-H bending				1288
C-C bending				1261
		1182		
		1093		
CH rocking				935
CH₂ wagging	718	721	721	721

9.3.2 X-RAY DIFFRACTION

The crystalline order of substances is also an indicator of their identities. The X-ray diffraction diagram of paraffin wax was similar to that of PE.[10] In the morphological and thermal evaluation of PE wax and paraffin wax, the spectrum obtained by x-ray diffraction for the paraffin exhibited two intense and well-defined separate peaks, whereas the spectrum for the PE wax exhibited a single and broad peak. As the concentration of PE wax in the blends increased, the diffraction peaks became broader and less intense, and the crystallinity degree decreased because of increasing amounts of the lower-crystallinity PE wax.[1] The dimensions of the orthorhombic cell of linear PE at 23°C are $a = 0.74069$ nm, $b = 0.49491$ nm, and $c = 0.25511$ nm (chain axis). The monoclinic cell of PE is less stable and it is found in samples subjected to mechanical stress. The dimensions of the monoclinic cell of linear PE at 23°C are $a = 0.809$ nm, $b = 0.253$ nm (chain axis), and $c = 0.479$ nm. The rectangular face of the unit cell perpendicular to the chain direction is distorted into a parallelogram with one included angle, designated β, becoming 107.9°.[4] The X-ray diffraction diagram of paraffin wax in Figure 9.4 had the strongest peak at 21.4° for the 110 planes of the orthorhombic unit cells of the $-CH_2-CH_2-$ chains.[28] The peak at 19.4° is due to the presence of a small amount of monoclinic crystals in the paraffin wax sample.[5] The x-ray diffraction pattern of the commercial wax in Figure 9.5 had a very strong peak at 21.4° which could be attributed to 110 planes of the orthorhombic unit cells of $-CH_2-CH_2-$ chains. However, there was not an amorphous fraction in commercial wax since the broad peak at 19° did not exist in Figure 9.5. The peak observed at 5° could represent a layered C-chain structure with a layer distance of 1.77 nm. X-ray diffraction diagram of Span 60 is seen in Figure 9.6. Since only the crystallinity of the material encapsulated in Span 60 microspheres was investigated by Mady[22] and the crystallinity of Span 60 was not mentioned, it was thought that Span 60 did not have any crystal structure. Nevertheless, the present study showed that Span 60 is also crystalline with a maximum diffraction peak at 2θ value of 21.54° as seen in Figure 9.6. The long alkyl groups of Span 60 were crystallized in orthorhombic structure and 21.54° peak was due to 110 planes. A small shoulder that was observed at 2θ value of 19° indicated the presence of amorphous fraction The broad peaks at 2θ values of 5.1° and 13.7° pointed out the presence of bilayer order in Span 60. These peaks could be due to different orders diffraction of bilayer distance. The broad peak starting at 2θ value of 10° and ending with 16° could be due to the overlapping of the several diffraction peaks. Thus, the existence of a peak at 13.7° could be the maximum of the overlapped peaks. On the other hand, x-ray diffraction diagram of myristic acid demonstrated intense crystalline peaks at 2θ equals to 18.58°, 18.93°, 20.18°, 20.79°, 21.51°, 23.92°, and 37.80°.[25]

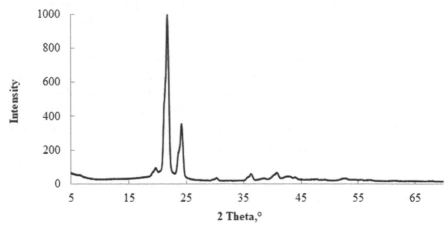

FIGURE 9.4   X-ray diffraction diagram of paraffin wax.

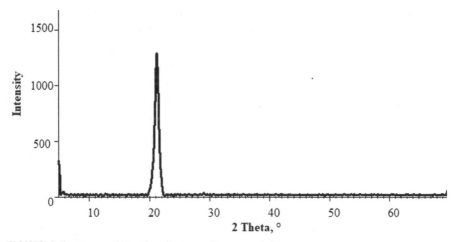

FIGURE 9.5   X-ray diffraction diagram of commercial wax.

Paraffin wax was in bulk form and the fracture surface of the paraffin wax was very smooth as seen in Figure 9.7. The SEM micrographs of the commercial wax are shown in Figure 9.8. The particles were in mainly spherical form with 170 µm average diameter and there were also cylindrical fibers of 5.45 μm diameter and 75 μm length. The range of spherical particle size was 32 nm–563 nm. There were particles having 560 μm diameter with holes of 145 μm on their surface. These large particles consisted of many small particles of 32 μm in size and had a thick coating with 53 μm thickness. The SEM micrographs in Figure 9.8 notified that these particles should have been formed by a melt freezing process.[17] The cylindrical particles were the ones freezing instantly when injected into the medium. The spherical particles should have formed from droplets formed from the liquid injection. The commercial wax is either one substance with different geometries or a mixture of two substances having spherical and fibrous shape. Since the x-ray diffraction showed the presence of one crystalline phase, it can be assumed that the spherical and fibrous particles are the same substance. The number distribution of the particle diameter of commercial wax is shown in Figure 9.9. Small particles were higher in number compared to larger ones.

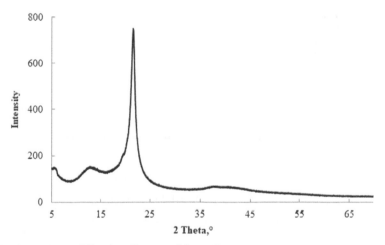

FIGURE 9.6 X-ray diffraction diagram of Span 60.

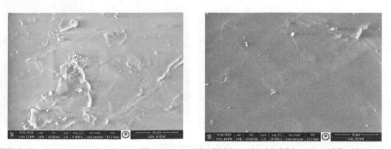

FIGURE 9.7 Micrograph of paraffin wax at (1) 5000x, (2) 10,000x magnification.

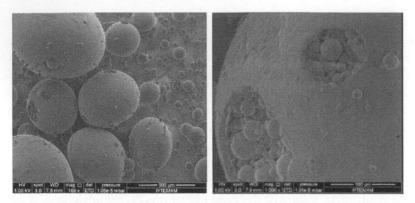

FIGURE 9.8   Micrographs of commercial wax at (1) 169x, (2) 1000x magnification.

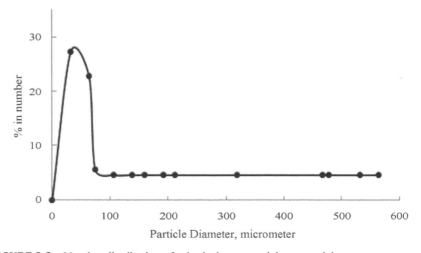

FIGURE 9.9   Number distribution of spherical commercial wax particles.

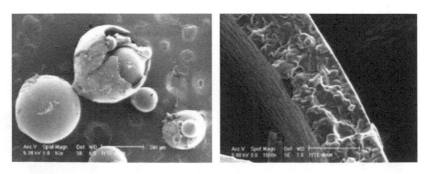

FIGURE 9.10

The SEM micrograph of Span 60 in Figure 9.10 showed that Span 60 had also spherical particles. The agglomerated small particles were encapsulated by a thick layer of material. The morphology of the Span 60 represented that the particles are spherical and they are one within the other and each large particle comprised of small particles. The particle size of the largest particle including agglomerated smaller ones was nearly 500 μm. The wall thickness of the largest particle was nearly 20 μm as demonstrated in Figure 9.9b.

9.3.3   TGA

Dynamic heating of the substances results in breaking of the chemical bonds and formation of volatile substances. In Figures 9.11–9.14 the thermogravimetry (TG) curves for paraffin wax, commercial wax, and Span 60 are seen. The onset temperature of the mass loss and remaining mass at 600°C for the samples examined and for myristic acid[1,2] are tabulated in Table 9.1. Paraffin wax had the highest onset of mass loss temperature, 250°C and the commercial wax had the lowest onset of the mass loss temperature 100°C among the four substances investigated. The onset of mass loss of Span 60 and myristic acid were very close to each other 148°C and 150°C, respectively. Paraffin wax degraded to 95% volatile substances up to 600°C, since it has 5% remaining mass at this temperature. Commercial wax, Span 60, and Myristic acid had higher remaining mass, 10–25%, indicating that less volatile substances were formed and carbonization to nonvolatile substances had also occurred during heating.

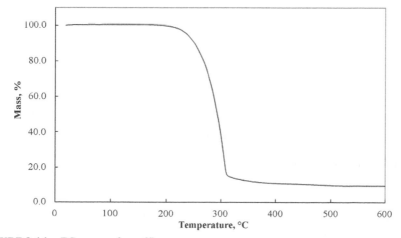

FIGURE 9.11   TG curve of paraffin wax.

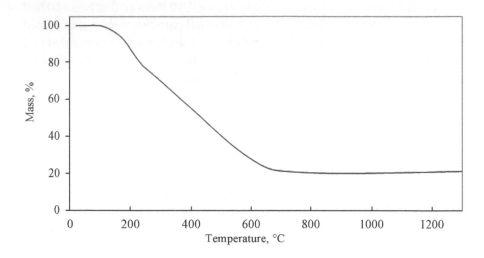

FIGURE 9.12   TG curve of commercial wax.

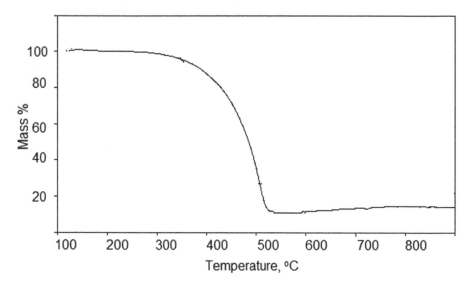

FIGURE 9.13   TG curve of Span 60.

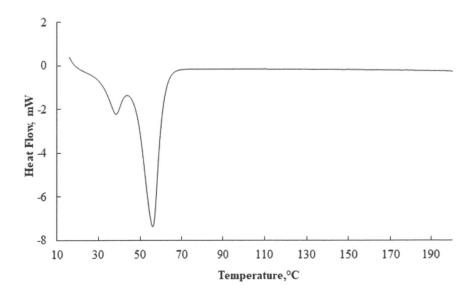

FIGURE 9.14   DSC curve of paraffin wax.

DSC analysis delivers the enthalpy changes occurring on heating the substances. The paraffin wax and PE wax could have different melting temperatures. For instance, the candle wax melting temperature was around 50°C, while PE wax had a melting temperature around 70°C.[7] In Figures 9.14–9.16 DSC curves of paraffin wax, commercial wax, and Span 60 are presented. The major endothermic peak occurring around 50°C is due to the melting process since there was no mass loss at this temperature as can be seen in TG curves in Figures 9.11–9.13. The peaks at 39 and 57°C in Figure 9.14 are belonging to orthorhombic to hexagonal transition and solid–liquid transition of paraffin wax, respectively.[13,18] Paraffin wax has 57°C melting point and 196 J g^{-1} heat of fusion. The peak temperatures of the melting and the heat of fusion of the substances under study are listed in Table 9.2. The melting temperature of commercial wax is 54°C and it has a lower heat of fusion than paraffin wax, 115 J g^{-1}. Span 60 has 50°C melting temperature but its heat of fusion is very low, 3.5 J g^{-1}. On the other hand, for myristic acid different melting points were reported as 54.10°C by Alve et al.[2] and 64°C by Genç and Genç.[13] Its heat of fusion was 196 J g^{-1} as reported by Alva et al.[2] and 178 J g^{-1} as reported by Genç and Genç.[13]

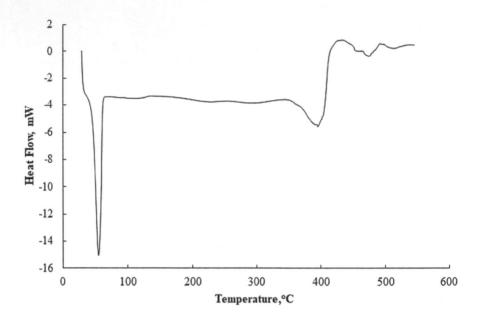

FIGURE 9.15   DSC curve of commercial wax.

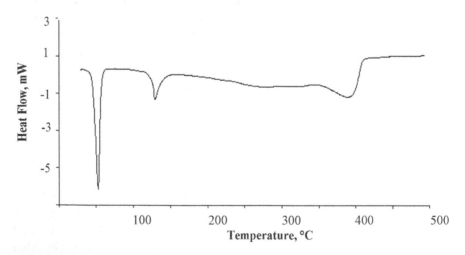

FIGURE 9.16   DSC curve of Span 60.

TABLE 9.2 Thermal Properties.

Material	Onset temperature of mass loss, °C	Remaining mass at 600, °C	Temperature of endothermic transition, °C	Enthalpy of transition, J g⁻¹
Paraffin wax	230	5	39 57	196
Commercial wax	100	20	54	115
Span 60	148	18	50	3.5
Myristic acid,	150[13]	25[13]	54.1[2] 64[13]	196[2] 178[13]

9.4 CONCLUSION

It is attempted to characterize a commercial spherical particulate sample labeled as "PE Wax" by FTIR, X-ray diffraction, SEM, TG, and DSC analyses, and it is found out that it was not a PE wax as it was labeled. Thus, the comparison of the results was performed with the analysis of paraffin wax, Span 60, and myristic acid which could also be used in the preparation of masterbatches. These substances can also be used in the preparation of PCM and as drug carriers for controlled release. All four substances had similar melting temperatures around 50°C. The high value of heat of fusion of paraffin wax, commercial wax, and myristic acid allows them to be used in the core of the PCM. The comparative investigation in the present study showed that the commercial sample had ester carbonyl groups and carboxylate groups in addition to -CH$_2$ and -CH$_3$ groups. It could be an oligomer of C-backbone and with ester and carboxylate side functional groups. It was found that commercial wax has 54.9°C melting point and heat of fusion of 115 J g⁻¹. The particles were in spherical form with 170 μm average diameter and mixed with cylindrical fibers of 5.45 μm average diameter and 75 μm average length. There are particles 560 μm diameter with holes of 145 μm on their surface. In these large particles, there was crust coating aglomerated small particles of 32 μm in size. The melting point, functional groups, and the crystal structure of commercial wax had closer similarity to Span 60 than that of paraffin wax and myristic acid. However, its onset temperature of mass loss was much lower than that of Span 60 and the heat of fusion of Span 60 was much lower than that of commercial wax. Further studies with more advanced methods are necessary to make a complete identification of commercial PE wax samples.

KEYWORDS

- **PE wax**
- **dispersant**
- **pigment masterbatch**
- **spherical particles**
- **phase change material**
- **microgranulation**
- **FTIR**

REFERENCES

1. Akishino, J. K.; Cerqueira, D. P. et al. Morphological and Thermal Evaluation of Blends of Polyethylene Wax and Paraffin. *Thermochim. Acta* **2016**, *626*, 9–12.
2. Alva, G.; Huang, X. et al. Synthesis and Characterization of Microencapsulated Myristic Acid-Palmitic Acid Eutectic Mixture as Phase Change Material for Thermal Energy Storage. *Appl. Energy* **2017**, *203*, 677–685.
3. Arnould, A.; F. Cousin et al. Impact of the Molar Ratio and the Nature of the Counter-Ion on the Self-Assembly of Myristic Acid. *J. Colloid, Interface Sci.* **2018**, *510*, 133–141.
4. Bilmayer, W.F. *Textbook of Polymer Science*; John Wiley and Sons: New York, 1984.
5. Caminiti, R.; Pandolfi, L. et al. Structure of Polyethylene from X-ray Powder Diffraction: Influence of the Amorphous Fraction on Data Analysis. *J. Macromol. Sci.-Phys.* **2000**, *B39* (4), 481–492.
6. Castanyer C. J. Pigment Preparation for Coloring Polymers. United States Patent 7935747, 2005.
7. Ciesinska, W.; Liszynska, B. et al. Selected Thermal Properties of Polyethylene Waxes. *J. Therm. Anal. Calorim.* **2016**, *125* (3), 1439–1443.
8. D'Amelia R. P. D.; Gentle, S.; Nirode, W .F.; Huand, L. Quantitative Analysis of Copolymers and Blends of Polyvinyl Acetate (PVAc) using Fourier Transform Infrared Spectroscopy (FTIR) and Elemental Analysis (EA). *World J. Chem. Educ.* **2016**, *4* (2), 25–31
9. Deshmukh, S. P.; Parmar, M. B. et al. Polymer- and Wax-Based Monoconcentrate Predispersed Pigments in the Colouring of Plastics. *Color. Technol.* 2010, *126* (4), 189–193.
10. Elnahas, H. H.; Abdou, S. M. et al. Structural, Morphological and Mechanical Properties of Gamma Irradiated Low Density Polyethylene/Paraffin Wax Blends. *Radiat. Phys. Chem.* **2018**, *151*, 217–224.
11. Fang, J. Y.; Xi, Y. Z. et al. A Novel Granulation Technology for Producing Spherical Wax Particles. *Chem. Eng. Technol.* **2004**, *27*(9), 1039–1047.
12. Fanselow, S.; Emamjomeh, S. E. et al. Production of Spherical Wax and Polyolefin Microparticles by Melt Emulsification for Additive Manufacturing. *Chem. Eng. Sci.* **2016**, *141*, 282–292.

13. Genc, M.; Genc, Z. K. Microencapsulated Myristic Acid-Fly Ash with TiO_2 Shell as a Novel Phase Change Material for Building Application. *J. Therm. Anal. Calorim.* **2018,** *131* (3), 2373–2380.
14. Ghobashy, M. M.; Elhady, M. A. et al. Pellets of Magnetized Polyethylene (Fe_3O_4/PE) Wax from Gamma Irradiated Polyethylene: Synthesis and Characterization. *Int. J. Plast. Technol.* **2018,** *22* (1), 1–9.
15. Gonen, M.; Balkose, D. et al. The Effect of Zinc Stearate on Thermal Degradation of Paraffin Wax. *J. Therm. Anal. Calorim.* **2008,** *94* (3): 737–742.
16. Karkri, M.; Lachheb, M. et al. Thermal Properties of Phase-Change Materials Based on High-Density Polyethylene Filled with Micro-Encapsulated Paraffin Wax for Thermal Energy Storage. *Energy Build.* **2015,** *88,* 144–152.
17. Koen, L. The Micronization of Synthetic Waxes. MS Thesis in Chemical Engineering, Stellenbosch University South Africa, 2003.
18. Kumar, S.; Agrawal, K. M. et al. Study of Phase Transition in Hard Microcrystalline Waxes and Wax Blends By Differential Scanning Calorimetry. *Petrol. Sci. Technol.* **2004,** *22* (3–4), 337–345.
19. Li, C.; Qin, C.; Li ,L.; Yan, F.; Wang, J.; Zhang, Y. Preparation and Performance of an Oil-Soluble Polyethylene Wax Particles Temporary Plugging Agent. *Hindawi J. Chem.* *2018,* Article ID 7086059, 7 pages, https://doi.org/10.1155/2018/7086059
20. Lin, Y. X.; Zhu, C. Q. et al. Microencapsulation and Thermal Properties of Myristic Acid with Ethyl Cellulose Shell for Thermal Energy Storage. *Appl. Energy* **2018,** *231,* 494–501.
21. Liu, W.; Cheng, L. et al. Synthesis and Properties of Novel Acrylic Polyester Hyper-Dispersant. *J. Coat. Technol. Res.* **2016,** *13* (5), 763–768.
22. Mady, O. Span 60 as a Microsphere Matrix: Preparation and in Vitro Characterization of Novel Ibuprofen-Span 60 Microspheres. *J. Surfactants Deterg.* **2017,** *20* (1), 219–232.
23. Miranda, J. R. S., de Castro, A. J. R. et al. Phase Transformation in the C Form of Myristic-Acid Crystals and DFT Calculations. *Spectrochim. Acta Part A-Mol. Biomol. Spectrosc.* **2019,** *208,* 97–108.
24. Savrik S.A. Enhancement of Tribological Properties of Mineral Oil by Addition of Sorbitan Monostearate and Zinc Borate, Ph.D. thesis, Izmir Institute of Technology, Izmir, 2010.
25. Trivedi, M.K.; Tallapragada, R.M.; Branton, A.; Trivedi, D.; Nayak, G. et al. Physical, Spectroscopic and Thermal Characterization of Biofield Treated Myristic Acid. *J. Fundam. Renewable Energy Appl.* **2015,** *5,* 180, DOI:10.4172/20904541.1000180.
26. WEB1: https://www.lubrizol.com/Coatings/Technologies/Surface-Modifiers (accessed Apr 24, 2019).
27. WEB 2: https://niik.ru/en/technology/technology-granulation-prilling/ (accessed Apr 24, 2019).
28. WEB3: https://www.aveka.com/processing/melt-spraying-prilling/ date of access (accessed Apr 24, 2019).
29. Yetgin, S.; Balköse, D. *Birmicrogranül pigment master beçinin karakterizasyonuve üretimi,* 1. Ulusal Ege Kompozit Malzemeler Sempozyumu, İzmir, 2011.
30. Zang, Y.; Ye, M. Q. et al. Preparation of Nano-Encapsulated Polyethylene Wax Particles for Color Toner by In Situ Emulsion Polymerization. *J. Appl. Polym. Sci.* **2017,** *134* (2), DOI: 10.1002/APP.44399

31. Zhou, M. S.; Huang, K. et al. Development and Evaluation of Polycarboxylic Acid Hyper-Dispersant used to Prepare High-Concentrated Coal-Water Slurry. *Powder Technol.* **2012,** *229,* 185–190.

CHAPTER 10

Supercapacitor: An Efficient Approach for Energy Storage Devices

SHAMA ISLAM*, HANA KHAN, and M. ZULFEQUAR

Department of Physics, Jamia Millia Islamia, New Delhi 110025, India

Corresponding author. E-mail: shamaphysics786@gmail.com

ABSTRACT

Supercapacitors are the most promising candidates in future generations for energy-storage devices. New energy devices are desirable because of the power and surrounding crisis is at hazard to growth the power source in squat exaggeration beat have skillful an incredible enthusiasm for electrochemical capacitors, additionally given as supercapacitors. In the present period, supercapacitors stipulate the high essence and comprehensive lastingness needful for a few current efficiency devices in thermoelectric vehicles, backup ascend for a few electrical devices and a constant desirable quality group. Novel challenging applications in usage counting micro autonomous robots and in mobile/portable energy storage, distributed sensors, and other devices to convene the supplies of high power density and long durability supercapacitor devices are responsible for a search of new and exciting materials. To extend superior electrode materials for supercapacitors an imperative approach and method is selected to fabricate materials. Graphene-based materials have been used for electric vehicles to provide improved resources for storing electricity and the remarkable enhancement of moveable electronics. To develop the electrode materials various carbon–metal oxide composite have been prepared by mixing of metal oxides in the matrix of carbon nanostructures counting zero-dimensional carbon nanoparticles, one-dimensional nanostructures (carbon nanotubes and carbon nanofibers), two-dimensional nanosheets (graphene and reduced graphene oxides) as well as three-dimensional porous carbon nanoarchitecture. In the present chapter, the efforts have been enthusiastic to achieved lightweight, thin supercapacitors with enhanced supercapacitance performance. In case, upright classify supercapacitors are a liability to establish second-hand generalship polymer to show starving reliable behavior among the way lump, exceptionally conducting polymers, for example, polyaniline, polypyrroles, and polyethenedioxythiophene with admirable electrical conductivity and moderate pseudo electrical capacity have a lively extensive interest as electrode materials for supercapacitors applications.

10.1　INTRODUCTION TO SUPERCAPACITORS

In order to solve the two challenging issues of exhaustion of fossils fuels and climate change, highly efficient and environmentally benign energy-storage devices with reasonable costs are highly desirable in our daily life. Supercapacitors have attracted a lot of attention in recent years as storage devices for applications in electrical vehicles, portable electronic devices, and power grids, etc. The electrical activity of supercapacitors is abundant via double-layer loading, faradic processes, or an aggregate of the two. Most of the stored activity is typically for small and can be instantly transmitted to authoritative supercapacitors, capable of supporting maximum pulse capacity rather than a large amount of energy. Various chords are accepted to designate the accessory, such as "double-layer capacitor," "supercapacitor," "ultra-capacitor," "power capacitor," "gold capacitor," "power mask" or "electrochemical double-layer capacitor". This deluge of names is confusing, and the term "double-layer capacitor" refers only to accessories that utilize a double-layer capability but do not include those that expect sufficient pseudocapacity not cover the wide range of devices. In this chapter, the term "supercapacitor" is used to avoid confusion.[1-3]

10.2　TYPES OF SUPERCAPACITORS

A double-layer supercapacitor[2] was made in 1746 in Leiden, Netherlands. This capacitor was found on the plates of the condenser now stored in the capacitor may be unfounded and untrue. However, if nothing of an indigenous origin is seen when a high it was clear that the width of the capacitor was described lasted until 1957. Supercapacitors can be divided into two categories depending on the mechanism of accumulation of activity: electric double-layer capacitors (EDLCs), whose potential capacity is generated as an interface between electrodes and electrolytes such as carbon materials and pseudo-capacitors. They accumulate in pure electrostatic charge, including oxidation/capacitance reduction (oxidation and reduction) or Faraday charges of optional particles due to a clear electrode, such as the identity of bulk metals and conductive polymers.[3-5] It is possible that both EDLC and pseudocapacitor are apparent phenomena during unloading operations; therefore, the achievement depends largely on the optical field. In general, a clear mutual presentation is required for overall efficiency. However, proper control of the specified apparent width and corresponding measurement of the electric tap hole is faster, because at higher current density, higher microprocessibility will increase to the amount of capacity affected. Regardless of how the charging accumulation mechanism works, electrochemical achievement lies mainly in the use of acceptable electrode materials. It is generally accepted that carbon-based annotations are a variety of electrode accepted annotations due to their background, including low cost, accessibility, manageability, environmental friendliness, and durability.[7-9] An EDLC mainly uses the charges accumulated on the surface of the electrode/interfacial electrode; the closure uses conductive polymers or metal oxides as electrode materials, which use faradic mechanisms to generate charges. It has been reported that all conductive resumes have superior collegial capabilities to EDLC-based materials, including polyaniline (PANI), polypyrrole and poly (3,4-ethylenedioxythiophene)[3]; these had low cycling stability due to a lowering of the structure during the charging/discharging process. The rupture ion battery (or ionic side and another electronic) occurs at the interface between the heat and debris solution and the unwieldy colloid, the semiconductor, and the metal electrode, giving acceleration as the double layer and what they call electrostatic capacitor in dual-layer nature qualification (ionic and electronic charge separation), and deposited in no chemical reactions. Therefore, loading and unloading is very lively and installed immediately. The first charge accumulates in the jar attributed to the supercapacitor capacity of the patented double layer. In addition to dual-layer, it is also possible to use a large pseudocapacitance ion metal and electrosorption associated with an invalid or in a redox process. Because of this thermodynamic transgression is said to be suitable for electrode process development, for example, pseudocapacitance. It is possible that both EDLC and pseudocapacitor are apparent phenomena during unloading operations; therefore, the achievement depends largely on the optical field. In general, a clear mutual presentation is required for overall efficiency.[10–15] However, proper control of the specified apparent width and corresponding measurement of the electric tap hole is faster, because at higher current density, higher microprocessibility will increase to the amount of capacity affected. Regardless of how the charging accumulation mechanism works, electrochemical achievement lies mainly in the use of acceptable electrode materials.[6] It is generally accepted that carbon-based annotations are a variety of electrode accepted annotations due to their background, including low cost, accessibility, manageability, environmental friendliness, and durability.[16–18]

10.3   MATERIALS USED IN SUPERCAPACITORS

Supercapacitors can be manufactured from a wide variety of materials. The option of which depends heavily on the type of capacitance to be utilized, as is evident from Figure 10.3. Carbon in its dispersion and the administration of anatomy is a huge amount of widely acclaimed advertisement supercapacitor material. These can be summaries obtained by full carbon activation pretreatment materials initially manufactured from thermal carbonization charcoal, pitch, wood, attic shells, or polymers.[19]. The most commonly used pretreatments are hot nitrogen,[20] hydrogen,[21] change of carbon dioxide or steam flux[22] maturity acceptable conductivity, wide specific area and stability of ion, carbon summaries display a wide and permanent double layer capacitance.

10.3.1   CARBON MATERIALS

Carbon structures that can be adapted to EDLC such as EDLC carbon (onion like carbon (OLC)), carbon nanotubes (CNTs), graphene, activated carbon (AC), carbon derived from carbon, and mold carbon. Each carbon anatomy symbol has its rules and disadvantages. For example, OLCs are fully achievable for ion uptake excellent conductivity, high performance but the binding capacity is 30 F g^{-1} ranked as 17 CNTs electrodes to reach the body due to their activity unique tubular and high energy pipette structures properties that facilitate the rapid transfer of ions and electrons. However, the supreme assembly deserves a full mass distribution application. Recently, graphene praises interest for EDLC applications.[23–25] As a 2D carbon nanostructure, graphene can add potential benefits maximum realization provides the maximum visual width and higher conductivity. However, the important question is how to avoid restacking of sheets while preparing the electrode. 3D carbon hierarchical extracts with the most obvious areas (e.g., AC, carbon-protected carbide, and carbon mold) a highly active staining structure for EDLC. Its specific capacity is unsatisfactory at the maximum current and density due to the presence of micropores and low conductivity. Therefore, it is important to compile a novel combine the benefits of modified carbon structure content. To further enhance their energy density later, density capacity sacrificed by high-energy innate carbon electroplating materials such as modified metal oxides polymers were conducted and extensively researched to accommodate battery activity requirements are high.[26–28] In context, abstract carbon atoms take double functions; they do not only according to the higher capacity of vehicles, and however, it can also act as a conductive current due to electron transport. Carbon materials further divided as follows:

10.3.1.1 0D CARBON NANOPARTICLES

0D carbon particles are based on round-shaped particles with an adjacent side arrangement of 1. The 0D carbon molecules cover AC, carbon nanosulfate, and mesopore carbon. They accept a width of width range (3000 m^2 g^{-1} for AC). More importantly, acceptor management can be designed and metrics can be introduced, making them acceptable for the recognition of metal oxides for electrode condenser electrodes.

10.3.1.2 1D CARBON NANOPARTICLES

1D nanoparticles are summed as fiber shape with an aspect ratio. CNTs, carbon nanotubes fiber, and carbon nanocoils in the category of 1D carbon nanostructures. They accept the upper-side arrangement and accept good electronic transport characteristics, which are accepted to facilitate.

10.3.1.3 2D NANOSTRUCTURES

2D nanostructures are paper summaries with high aspect ratios. Graphene, graphene oxide (GO) or reduced graphene oxide is a group of 2D carbon nanoparticles. Graphene and GO show abundant mechanical strength, and cater to electronic applications with specific area as the most obvious field, a filter capable of supercapacitor electrodes. For example, theoretical surface area of the individual band graphene is 2756 m^2 g^{-1} with a charge mobility 200,000 cm^2 v^{-1} s^{-1}.

10.3.1.4   3D DESIGN

3D designs are made up of low-dimensional architecture blocks. 3D absorption carbon nanostructures in the electrodes of supercapacitors are essentially carbon foams or sponges. When optimized for electrodes of supercapacitors, such foam has a high apparent area, electrolyte interface and sufficient amplitude of the electrode, and the path of associated electron transport.

10.3.2   METAL OXIDES

Metal oxides or hydroxides such as ruthenium, cobalt, nickel, and manganese are the intimates of oxide/hydroxide the material used in supercapacitor applications. These materials government or semi-operation and redox display active behavior, accelerating pseudocapacitance. Consideration of oxides (hydrated or anhydrous) as capacitive materials, ruthenium dioxide was widely suggested earlier reported as a new current electrode in 1971.[29] The application of RuO_2 in supercapacitors is based on metallic coat applications and capacitive redox reactions (Ru^{2+}, Ru^{3+}, and Ru^{4+} pairs) in an aqueous medium. Consideration of the oxide provides near wired and clear capacitance with specific capacitance above 1.4 V^{20} 600–1000 F g^{-1} depending on the preparation process, measurement conditions, use of support, etc.[30] Another compelling advantage is ultra better stability or continuation of eternal life. Thermally formed Ti RuO_2 films can be answered more than 10^5 times with TiO_2 or Ti_2O_5 between 0.02 and 1.2 V, or up to 1.4 V, with a slight drop. Checking for ruthenium alone, the maximum amount of oxide material is thus limited its use as supercapacitor material for the aerospace and military industries barter applications[31] are economical. Other small metal oxides also a capacitance for supercapacitor applications for example. CO_3O_4,[23,24] NiO,[32,33] and MnO_2.[34,35] Although these metal oxides are abundant in an effective excess, their specific capacitance is low, typically between 20 and 200 F g^{-1}; they are also no more commutative than ruthenium oxide.

10.4   SYNTHESIS TECHNIQUE OF CONDUCTING POLYMERS AND THIN FILMS USED AS ELECTRODE MATERIAL FOR SUPERCAPACITORS

The synthesis of the materials is required for the study of polymer composites. The development of systematic studies for the synthesis of polymer composites and polymer nanocomposites is a current challenge. A detailed description of the synthesis of conjugated polymers by various methods and the characterization techniques employed is given in this chapter. Characterization can take the form of actual materials testing or analysis. Although the techniques to be used depends upon the type of material and information one needs to know, usually one is interested in first knowing the structural properties and composition and then the chemical state, optical properties, DC conductivity, and other properties. A wide range of techniques are available in each of these areas and the systematic use of these tools leads to a complete understanding of the system. The information obtained from these techniques can be processed to yield images or spectra which reveal the structural, chemical, and physical details of the materials. To study the optoelectronic properties of synthesized semiconductor materials synthesized, a range of techniques such as UV-visible absorption spectroscopy and photoluminescence are used. The structural properties are studied using Fourier transform infrared spectroscopy, X-ray diffraction measurements.

10.4.1   CHEMICAL OXIDATION METHOD

Conducting polymers can be synthesized either chemically or electrochemically.[36] Chemical synthesis involves either condensation polymerization where the growth of polymer chains proceeds by a condensation reaction or addition polymerization where the growth is dependent on radical, anion, or cation formation at the end of the polymer chain. The conducting polymers produced chemically are in their undoped insulating states and the doping process can switch them into their conductive states. Chemical synthesis is preferred when it is necessary to synthesize or modify conducting polymers at large scale, or to obtain conducting polymers with defined structures.[37] In this work, the polymer has been synthesized by in situ oxidative polymerization of the monomer with hydrochloric acid (HCl), and by using ammonium persulfate ((NH_4) S_2O_8) as an oxidant. The monomer is doubly distilled prior to use. The oxidant to monomer ratio is (1:1). The monomer (1M) was dissolved in 10 mL of HCl (1M) taken in 200 mL round bottom flask and stirred well. Further, finely ground dopant material powder taken in different weight% with respect to monomer concentration was added to the previous mixture under vigorous stirring in order to keep dopant material suspended in solution. The reaction mixture was then cooled up to 5°C, and the precooled solution of $(NH_4)_2S_2O_8$ (1M) was added dropwise over a period of 30 min. The reaction was allowed to proceed for 6–8 h. The mixture was further cooled down to 4°C for 24–36 h. It was then filtered and washed until the filtrate was colorless. The dark-colored polymer powder so obtained was dried thoroughly in an oven at 100°C until a constant weight was attained, then grinded and sieved. Undoped polymer was synthesized in the same fashion with the absence of the dopant material. The chemical reaction for the synthesis of PANI is shown in Figure 10.1.

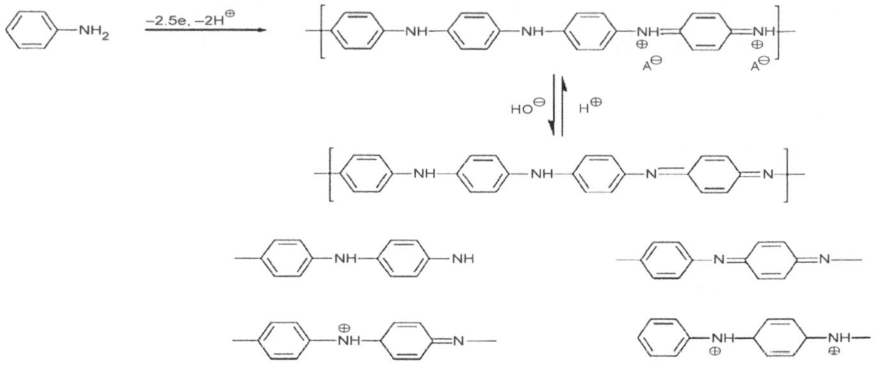

FIGURE 10.1 Schematic scheme proposed for the formation of PANI.

The same method was employed to synthesize poly(o-toluidine) and poly(m-toluidine) powders. The monomers o-toluidine and m-toluidine are the derivatives of aniline in which methyl group is attached at *ortho* (o-) and *meta* (m-) positions of aniline ring. The commercially available monomers were procured (Merck Chemicals) and the polymers were prepared for the present study. The monomers and polymers of o- and m- substituted anilines are shown in Figure 10.2.

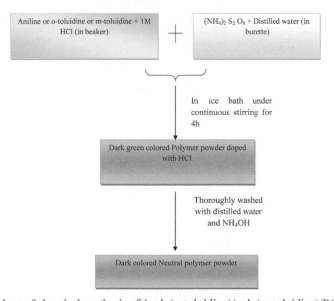

FIGURE 10.2   The *ortho-* and *meta-* substituted anilines and their polymers.

The polymer powders so obtained were found to be very less soluble in common organic solvents like THF (tetrahydrofuran), $CHCl_3$, etc.

Aniline or o-toluidine or m-toluidine + 1M HCl (in beaker)

$+$

$(NH_4)_2 S_2 O_8$ + Distilled water (in burette)

In ice bath under continuous stirring for 4h

Dark green colored Polymer powder doped with HCl

Thoroughly washed with distilled water and NH_4OH

Dark colored Neutral polymer powder

Flowchart of chemical synthesis of {poly(o-toluidine)/poly(m-toluidine)/PANI}.

10.4.2   RADIO FREQUENCY (RF) PLASMA POLYMERIZATION FOR THE PREPARATION OF POLYMER THIN FILMS

Plasma polymerized thin films from different monomer precursors were prepared by employing the RF plasma polymerization technique.[38,39] Plasma polymerized thin films of monomers on ultrasonically cleaned glass and silicon wafer substrates were obtained by polymerizing of monomers (99.9% purity) under RF plasma discharge in a home built set up. The setup consists of a custom-manufactured glass deposition chamber coupled to a vacuum system, RF amplifier, and a monomer feed-through set up. For maximum deposition, a novel setup has been designed. The position from where evacuation takes place and the position of monomer feed-through are crucial in this setup; a schematic of which shown in Figure 10.3. Feed-through setup consists of one on/off valve and one needle valve. The on/off valve is to create a quick vacuum in monomer pot and needle valve regulates the flow of monomer vapors into the chamber. This helps us to achieve the maximum deposition rate on substrates.[40]

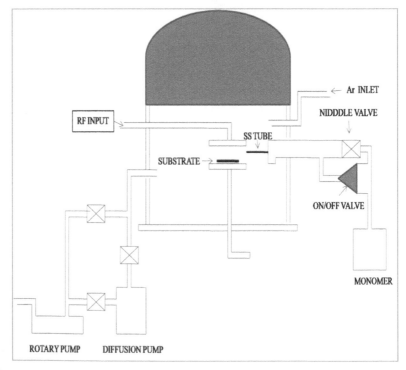

FIGURE 10.3   The RF-plasma polymerization setup.

The monomer vapors enter slowly into the chamber through the needle valve. During a typical experiment, the glass and silicon substrates are placed on the lower electrode. The system was evacuated to lower than 10^{-3} torr and argon gas was introduced into the chamber for plasma pretreatment for 20 s. The RF power applied was 15 W for about 1 h. Fragmentation of monomer takes place because of the argon plasma created between the electrodes. The deposition rate is estimated to 3.33 nm/min under constant deposition conditions.

10.5 FABRICATION OF SUPERCAPACITORS

For electrochemical measurements the capacitor anodes were made by mixing polyorthotoludine/multiwalled carbon nanotubes (POT/MWCNTs) nanocomposites powder, acetylene black, and PVDF-HFP (poly(vinylidine fluoride)-co-hexafluoropropylene) suitably in the weight ratio of 70:10:20. This mixture was then drop casted onto graphite sheets of dimensions 1 cm × 1 cm. These anodes were then vacuum dried. The capacitor is worked from anodes of nanocomposites isolated with a smooth electrolyte (H_2SO_4) dipped paper. Two anode cells were frequently connected yet a three cathode setup was additionally utilized. The electrochemical execution was performed in acidic (1 mol L^{-1} H_2SO_4) arrangement, utilizing galvanostatic, cyclic voltammetry, and impedance spectroscopy strategies. The capacitance was ascertained from the electrical current advance and inside the instance of a nonrectangular edge, the curves are enclosed.

10.6 PARAMETERS RESPONSIBLE FOR SUPERCAPACITORS PERFORMANCE

To check the electrochemical behaviors of the electrode active materials three-electrode configuration has been used frequently. A three-electrode configuration, in which working electrode, reference electrode, counter electrode, and electrolyte are dipped in the electrolyte materials. The materials which are under investigation are coated on the graphite sheet or directly used is called working electrode. Saturated calomel electrodes and platinum electrodes are often used as reference electrode and counter electrode, respectively. The electrolytes those mainly used are organic solvents and waste. Environmental pollution may increase due to organic solvents, so the most preferable electrolyte is aqueous solutions. Typically, the following methods for electrochemical performances of supercapacitors are characterized by: cyclic voltammetry (CV), galvanostatic charge-discharge, and electrochemical impedance spectroscopy. The parameters observed with these methods estimate the electrochemical performance of a supercapacitor, such as the cycle life, specific capacitance, charge-discharge rate, etc. The formula $C_{sp} = \int idv / s.\Delta v.m$ has been used to calculate specific capacitance from CV, whereby i is average current, s is scan rate, and m is the mass, while for galvanostatic estimations $C_d = 2i\ t/(m \times \Delta V)$ was appointed (Δt is the release time inside the range of voltage ΔV). It is important to word that in our counts the capacitance calculated from above-mentioned formulas is for one electrode only. Hence, the total capacitance of the cell turns out to $C_{sp} = 2C/m$ (m is that the mass of one electrode). The capacitance was conjointly anticipated from impedance spectroscopy investigation using the following formula: $C_{sp} = 2/(m \times Z'\omega)$ where ω is angular frequency, Z " is the imaginary impedance recorded at 10 mHz.

ACKNOWLEDGMENT

This work supported by the CSIR (file No: Pool No. 9056-A), Council of Scientific and Industrial Research, Human Resource Development Group (HRDG), Delhi, India.

KEYWORDS

- **supercapacitors**
- **energy devices**
- **electrodes materials**
- **graphene based materials**
- **carbon metal oxide**
- **conducting polymers**

REFERENCES

1. Zhai, Y.; Dou, Y.; Zhao, D.; Fulvio, P. F.; Mayes, R. T.; Dai, S. Carbon Materials for Chemical Capacitive Energy Storage. *Adv. Mater.* **2011**, *23* (42), 4828–4850.
2. Hu, N.; Zhang, L.; Yang, C.; Zhao, J.; Yang, Z.; Wei, H.; Liao, H. et al. Three-Dimensional Skeleton Networks of Graphene Wrapped Polyaniline Nanofibers: An

Excellent Structure for High-Performance Flexible Solid-State Supercapacitors. *Sci. Rep.* **2016,** *6,* 19777.

3. Meng, C.; Liu, C.; Chen, L.; Hu, C.; Fan, S. Highly Flexible and All-Solid-State Paper Like Polymer Supercapacitors. *Nano Lett.* **2010,** *10* (10), 4025–4031.

4. Zhang, C.; Peng, Z.; Lin, J.; Zhu, Y.; Ruan, G.; Hwang, C.-C.; Lu, W.; Hauge, R. H.; Tour, J. M. Splitting of a Vertical Multiwalled Carbon Nanotube Carpet to a Graphene Nanoribbon Carpet and its Use in Supercapacitors. *ACS Nano* **2013,** *7* (6), 5151–5159.

5. Arbizzani, C.; Mastragostino, M.; Soavi, F. New Trends in Electrochemical Supercapacitors. *J. Power Sources* **2001,** *100* (1–2), 164–170.

6. Li, Z. H.; Xia, Q. L.; Liu, L. L.; Lei, G. T.; Xiao, Q. Z.; Gao, D. S.; Zhou, X. D. Effect of Zwitterionic Salt on the Electrochemical Properties of a Solid Polymer Electrolyte with High Temperature Stability for Lithium Ion Batteries. *Electrochim. Acta* **2010,** *56* (2), 804–809.

7. Laforgue, A.; Simon, P.; Sarrazin, C.; Fauvarque, J.F. Polythiophene-based supercapacitors. *J. Power Sources* **1999,** *80* (1–2), 142–148.

8. Niu, Z.; Zhou, W.; Chen, J.; Feng, G.; Li, H.; Ma, W.; Li, J. et al. Compact-Designed Supercapacitors Using Free-Atanding Single-Walled Carbon Nanotube Films. *Energy Environ. Sci.* **2011,** *4* (4), 1440–1446.

9. Zhang, Z.; Deng, J.; Li, X.; Yang, Z.; He, S.; Chen, X.; Guan, G.; Ren, J.; Peng, H. Superelastic Supercapacitors with High Performances During Stretching. *Adv. Mater.* **2015,** *27* (2), 356–362.

10. Li, Y.; Wang, B.; Chen, H.; Feng, W. Improvement of the Electrochemical Properties via Poly (3, 4-ethylenedioxythiophene) Oriented Micro/Nanorods. *J. Power Sources* **2010,** *195* (9), 3025–3030.

11. Islam, S.; Lakshmi, G. B.; Zulfequar, M.; Husain, M.; Siddiqui, A. M. Comparative Studies of Chemically Synthesized and RF Plasma-polymerized poly (o-toluidine). *Pramana* **2015,** *84* (4), 653–665.

12. Niu, Z.; Zhou, W.; Chen, J.; Feng, G.; Li, H.; Ma, W.; Li, J. et al. Compact-Designed Supercapacitors Using Free-Standing Single-Walled Carbon Nanotube Films. *Energy Environ. Sci.* **2011,** *4* (4), 1440–1446.

13. Carlberg, J. C.; Inganäs, O. Poly (3, 4-ethylenedioxythiophene) as Electrode Material in Electrochemical Capacitors. *J. Electrochem. Soc.* **1997,** *144* (4), L61–L64.

14. Meng, C.; Liu, C.; Fan, S. Flexible Carbon Nanotube/Polyaniline Paper-Like Films and Their Enhanced Electrochemical Properties. *Electrochem. Commun.* **2009,** *11* (1), 186–189.

15. Islam, S.; Ganaie, M.; Ahmad, S.; Siddiqui, A. M.; Zulfequar, M. Dopant Effect and Characterization Of Poly (o-toluidine)/Vanadium Pentoxide Composites Prepared by In Situ Polymerization Process. *Int. J. Phys.* **2014,** *2,* 105–122.

16. Wu, Q.; Xu, Y.; Yao, Z.; Liu, A.; Shi, G. Supercapacitors Based on Flexible Graphene/Polyaniline Nanofiber Composite Films. *ACS Nano* **2010,** *4* (4), 1963–1970.

17. Ortega, P. F. R.; Trigueiro, J. P. C.; Silva, G. C.; Lavall, R. L. Improving Supercapacitor Capacitance by Using a Novel Gel Nanocomposite Polymer Electrolyte Based on Nanostructured SiO_2, PVDF and Imidazolium Ionic Liquid. *Electrochim. Acta* **2016,** *188*; 809–817.

18. Hu, N.; Zhang, L.; Yang, C.; Zhao, J.; Yang, Z.; Wei, H.; Liao, H. et al. Three-Dimensional Skeleton Networks of Graphene Wrapped Polyaniline Nanofibers: An

Excellent Structure for High-Performance Flexible Solid-State Supercapacitors. *Sci. Rep.* **2016,** *6,* 19777.

19. Patake, V. D.; Lokhande, C. D.; Joo, O. S. Electrodeposited Ruthenium Oxide Thin Films for Supercapacitor: Effect of Surface Treatments. *Appl. Surf. Sci.* **2009,** *255* (7), 4192–4196.

20. Pham, D. T.; Lee, T. H.; Luong, D. H.; Yao, F.; Ghosh, A.; Le, V. T.; Kim, T. H.; Li, B.; Chang, J.; Lee, Y. H. Carbon Nanotube-Bridged Graphene 3D Building Blocks for Ultrafast Compact Supercapacitors. *ACS Nano* **2015,** *9* (2), 2018–2027.

21. Liao, Y.; Li, X.-G.; Hoek, E. M. V.; Kaner, R. B. Carbon Nanotube/Polyaniline Nanofiber Ultrafiltration Membranes. *J. Mater. Chem. A* **2013,** *1* (48), 15390–15396.

22. Ge, D.; Yang, L.; Fan, L.; Zhang, C.; Xiao, X.; Gogotsi, Y.; Yang, S. Foldable Supercapacitors from Triple Networks of Macroporous Cellulose Fibers, Single-Walled Carbon Nanotubes and Polyaniline Nanoribbons. *Nano Energy* **2015,** *11,* 568–578.

23. Hyder, M. N.; Lee, S. W.; Cebeci, F. C.; Schmidt, D. J.; Shao-Horn, Y.; Hammond, P. T. Layer-by-Layer Assembled Polyaniline Nanofiber/Multiwall Carbon Nanotube Thin Film Electrodes for High-Power and High-Energy Storage Applications. *ACS Nano* **2011,** *5* (11), 8552–8561.

24. Yuan, L.; Xiao, X.; Ding, T.; Zhong, J.; Zhang, X.; Shen, Y.; Hu, B.; Huang, Y.; Zhou, J.; Wang, Z. L. Paper-Based Supercapacitors for Self-Powered Nanosystems. *Angewandte Chemie Int. Ed.* **2012,** *51* (20), 4934–4938.

25. Lin, H.; Li, L.; Ren, J.; Cai, Z.; Qiu, L.; Yang, Z.; Peng, H. Conducting Polymer Composite Film Incorporated with Aligned Carbon Nanotubes for Transparent, Flexible and Efficient Supercapacitor. *Sci. Rep.* **2013,** *3,* 1353.

26. Mastragostino, M.; Arbizzani, C.; Soavi, F. Polymer-Based Supercapacitors. *J. Power Sources* **2001,** *97,* 812–815.

27. Jost, K.; Dion, G.; Gogotsi, Y. Textile Energy Storage in Perspective. *J. Mater. Chem. A* **2014,** *2* (28), 10776–10787.

28. Stenger-Smith, J. D.; Webber, C. K.; Anderson, N.; Chafin, A. P.; Zong, K.; Reynolds. J. R. Poly (3, 4-alkylenedioxythiophene)-Based Supercapacitors Using Ionic Liquids as Supporting Electrolytes. *J. Electrochem. Soc.* **2002,** *149* (8), A973–A977.

29. Jayalakshmi, M.; Balasubramanian, K. Simple Capacitors to Supercapacitors-An Overview. *Int. J. Electrochem. Sci.* **2008,** *3* (11), 1196–1217.

30. Zhou, Y.; Xu, H.; Lachman, N.; Ghaffari, M.; Wu, S.; Liu, Y.; Ugur, A.; Gleason, K. K.; Wardle, B. L.; Zhang, Q. M. Advanced Asymmetric Supercapacitor Based on Conducting Polymer and Aligned Carbon Nanotubes with Controlled Nanomorphology. *Nano Energy* **2014,** *9,* 176–185.

31. Li, L.; Loveday, D. C.; Mudigonda, D. S. K.; Ferraris, J. P. Effect of Electrolytes on Performance of Electrochemical Capacitors Based on Poly [3-(3, 4-difluorophenyl) thiophene]. *J. Electrochem. Soc.* **2002,** *149* (9), A1201–A1207.

32. Frackowiak, E. Carbon Materials for Supercapacitor Application. *Phys. Chem. Chem. Phys.* **2007,** *9* (15), 1774–1785.

33. Belanger, D.; Ren, X.; Davey, J.; Uribe, F.; Gottesfeld, S. Characterization and Long-Term Performance of Polyaniline-Based Electrochemical Capacitors. *J. Electrochem. Soc.* **2000,** *147* (8), 2923–2929.

34. Kiamahalleh, M. V.; Sharifzein, S. H.; Najafpour G.; Abdsata, S.; Buniran, S. Multiwalled Carbon Nanotubes Based Nanocomposites for Supercapacitors: A Review of Electrode Materials. *Nano* **2012,** *7* (02), 1230002.

35. Bhattacharya, P.; Sahoo, S.; Das, C. K. Microwave Absorption Behaviour of MWCNT based Nanocomposites in X-band region. *eXPRESS Polymer Lett.* **2013,** *7* (3), 212–223.
36. Eftekhari, A.; Li, L.; Yang, Y. Polyaniline Supercapacitors. *J. Power Sources* **2017,** *347*, 86–107.
37. Ryu, K. S.; Kim, K. M.; Park, N. G.; Park, Y. J.; Chang, S. H. Symmetric Redox Supercapacitor with Conducting Polyaniline Electrodes. *J. Power Sources* **2002,** *103* (2), 305–309.
38. Ma, Y., Chang, H.; Zhang, M.; Chen, Y. Graphene-Based Materials for Lithium-Ion Hybrid Supercapacitors. *Adv.Mater.* **2015,** *27* (36), 5296–5308.
39. Yuan, L., Yao, B.; Hu, B.; Huo, K.; Chen, W.; Zhou, J. Polypyrrole-Coated Paper for Flexible Solid-State Energy Storage. *Energy Environ. Sci.* **2013,** *6* (2): 470–476.
40. Biswas, S.; Drzal, L. T. Multilayered Nanoarchitecture of Graphene Nanosheets and Polypyrrole Nanowires for High Performance Supercapacitor Electrodes. *Chem. Mater.* **2010,** *22* (20), 5667–5671.

CHAPTER 11

Green Perspectives of Imides toward Polymeric Membranes

DEEPAK PODDAR, ANKITA SINGH, SANJEEVE THAKUR, and
PURNIMA JAIN*

*Department of Chemistry, Netaji Subhas University of Technology
(Erstwhile Netaji Subhas Institute of Technology, University of Delhi)
Dwarka, New Delhi 110078, India*

Corresponding author. E-mail: prnm_j@yahoo.com

ABSTRACT

Polyimides (PIs) have found extensive use in the fabrication of polymeric membranes because of their high selectivity and permeability. Besides, these are advantageous as they have high-temperature resistance, flexibility, high chemical and mechanical stability, secure handling, and inexpensive nature compared to the inorganic material and ceramics used in filtration membranes and even readily accept the physical and chemical modification to improve its properties. PI has been used in various separation techniques, especially gas-phase separation processes. The chapter mainly focuses on the greener approach to synthesize the PI polymer and membrane fabrication at the same time. Various methods for synthesis have been discussed and elaborated with examples. The use of the ecofriendly solvent media has been discussed in detail with the particular emphasis on the selection of the solvent as a medium to be used in synthesis methods, and various parameters such as Hansen parameter have also been discussed to provide an insight to the students to choose solvent effectively.

The authors have tried to provide the content of the chapter in a concise yet understandable manner and those interested in apprehending more can go through the references provided at the end of the chapter.

11.1   INTRODUCTION

A membrane is a thin layer of semipermeable material that separates substances when a driving force is applied across the membrane.[1]

Polymeric membranes are fabricated using polymers as the matrix and are widely used in the separation process, energy production, fuel cells, and so on. These membranes can be both pressure driven or electricity driven. Microfiltration, ultrafiltration, nanofiltration, and reverse osmosis are the a few examples of pressure-driven polymeric membranes, and ion exchange membranes can be cited as an example of the electricity-driven membrane.

PI is a polymer of imide monomer and constitutes repeating units of –CONH– functional group. It has widely been used in the fabrication of polymeric membranes due to high selectivity and permeability especially for gas-phase separation processes, high-heat resistance, and excellent chemical and mechanical stability; these can be synthesized by the reaction of dianhydride with diamine or diisocyanate.

Current scenario demands for the greener synthesis approach to minimize the environmental impacts and reduce the waste and toxic by-products. This can be fractionally achieved by using water or nontoxic organic solvents instead of conventional solvent. There are few approaches for the greener approach for the fabrication of membrane shown in Figure 11.1.[2] In the case where the use of traditional toxic solvent could not be resisted, it can be used with the association of other solvents so that the mixture becomes less virulent. Also, chemicals and synthesis processes which might possess environmental threat should be avoided.[2] This chapter discusses about the use of PIs in the fabrication of polymer membranes; chemical modifications that can be used to improve the behavior of PIs, for example, cross-linking, which increases hydrophilicity, improved solvent stability, and suppressed plasticization; thermal annealing that enhances the selectivity for gas in the gas-phase separation process due to the introduction of inter- and intrastructure bond known as charge transfer complexes (CTCs) which can be confirmed by ^{13}C NMR spectra and fluorescence spectroscopy; thermal quenching which provides an increment in fractional free volume, thus decreasing density and increasing permeability; use of two or more PIs altogether and others. The chapter also discusses the purpose served by PIs when used in polymeric membranes.

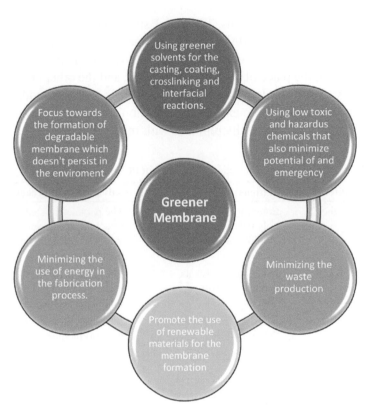

FIGURE 11.1 Few approaches to develop greener membranes following the green chemistry principle.

11.2 SOLVENTS USED IN POLYIMIDE MEMBRANES FABRICATION

In the current scenario where pollution is a global issue which needs to be addressed, sustainable development is the best way in the reduction on the impact on environmental and most importantly protection of resource consumption related to all human activities. These objects drive us toward the fresh renewable sources of energy, and there should be the main focus on the waste reduction and recovery of the material. The "greener" approach should be utilized for the fabrication of membrane compared to the corresponding conventional techniques without compromising the intrinsic feature like efficiency, straightforwardness in fabrication, stability, easily controllable, and can be scale-up under a broad spectrum of operational conditions. However, sustainable development for the membrane fabrication process should avoid the use of hazardous chemicals. Table 11.1 summarizes the basis of solvent selection and its possible effects on the environment and human health. For the sustainable course, there are two prospectives: one is the use of renewable material in the place of oil-derived material; and the other approach is to reduce the use of hazardous and harmful chemicals, solvents, additives, and modifiers involved in the membrane fabrication process which are toxic to the environment and optimally replaced with renewable source.

TABLE 11.1 Criteria for the Selection of Solvents in PI Membrane.

Selection criteria	Limitations	Advantages of greener solvent
Properties of the solvents	i. Solubility: Solubilization of polymer ii. Stability of the solvent iii. Volatility	The parameter listed helps to select the proper solvents for the synthesis and fabrication of PI membrane
Environmental hazard	i. Avoid toxic solvents ii. Biodegradability	The solvent used is the membrane system, create minimal environmental hazards.
Health hazards	i. Not possess acute toxicity ii. Organ failure iii. Carcinogenic effect	Since solvent plays a role from the start medium to filtration and thus not possess health hazard.

11.2.1   CHOICE OF SOLVENTS

The preliminary requirement for the choice of solvents includes the solvent being competent enough to solubilize the starting materials and should not in any case interfere with the synthesis process. For the exothermic processes, the solvent itself should not participate in heat production. Instead, it should act as a heat absorber. Table 11.2 shows the solubility parameter and T_g of the PI synthesis from the different sets of momers.[3]

TABLE 11.2 The Estimated Solubility Parameters (δ)[20] and Glass Transition Temperature (T_g) of Various Polyimides from Diamines and Dianhydrides.

Dianhydride	BTDA		PMDA		BPDA		ODPA		6FDA	
Characteristics diamine	T_g (°C)	δ	T_g (°C)	δ	T_g (°C)	δ	T_g (°C)	δ (°C)	T_g (°C)	δ
m-BAPS	252	30.6	286	31.2	235	30.2	244	30.3	236	27.1
APPS ($n = 11.1$)	–	16.3	–	15.6	–	16.0	–	16.1	–	15.7
m-BAPA	192	30.3	212	30.8	199	29.9	181	29.9	200	26.6

TABLE 11.2 *(Continued)*

Dianhydride	BTDA		PMDA		BPDA		ODPA		6FDA	
Characteristics diamine	T_g (°C)	δ	T_g (°C)	δ	T_g (°C)	δ	T_g (°C)	δ	T_g (°C)	δ
m-ODA	–	32.7	–	34.6	–	32.2	205	32.2	–	27.2
m-MDA	265	31.9	308	33.4	–	31.4	258	31.4	248	26.7
m-DABP	259	33.2	326	35.1	–	32.8	248	32.7	260	27.7
m-PDA	300	33.1	442	35.8	–	32.6	313	32.5	303	26.6
p-PDA	333	33.1	–	35.8	–	32.6	342	32.5	339	26.6
m-6F-diamine	239	27.7	–	27.7	267	27.1	224	27.2	250	24.1

11.2.2 CRITERIA FOR THE SELECTION OF SOLVENTS

The ability of the solvents to solubilize the polymer can be predicted by the Hansen solubility parameter (HSP) for solvents. Another essential requirement is that the solvent used must not be harmful and toxic to the environment and the same can be assessed at a glimpse by considering the information given on a Material Safety Data Sheets for solvents along with the physicochemical properties of the solvents.

11.2.2.1 TOXIC SOLVENTS

The role of solvents is versatile in the preparation of polymeric membranes. There are few techniques such as temperature-induced phase separation and nonsolvent-induced phase separation techniques. Both the techniques are used for the fabrication of highly porous microscopic membranes. The use of conventional chemicals gives the flexibility to the selection process of a solvent system which makes the fabrication easy. Table 11.3 shows the solubility parameter of the conventional solvents.[4]

TABLE 11.3 Hansen Solubility Parameters for the Traditional Solvent Used in the Membrane Fabrication.

Solvent	Hydrogen bonding (δ_h) $(MPa)^{1/2}$	Dispersion forces (δ_d) $(MPa)^{1/2}$	Polar forces (δ_d) $(MPa)^{1/2}$	Solubility parameters (δ) $(MPa)^{1/2}$
DMF	11.30	17.4	13.7	24.8
DMA	11.80	17.8	14.1	22.7
NMP	7.20	18.4	12.3	22.9
THF	3.70	19.0	10.2	22.5

TABLE 11.3   *(Continued)*

DBP	4.10	17.8	8.6	20.2
DOP	3.10	16.6	7.0	16.8
Acetone	6.90	15.5	10.4	19.9
Chloroform	5.50	17.8	3.1	19.0
1,4-Dioxane	8.00	16.8	5.7	18.5
Toluene	2.00	18.0	1.4	18.2
Methyl salicylate	12.30	16.0	8.0	21.7
Diphenyl ether	5.80	19.5	3.4	20.6

DMF, Dimethylformamide; DMA, dimethylacetamide; NMP, *N*-methyl-2-pyrrolidone; THF, tetrahydrofuran; DBP, dibutyl phthalate; DOP, dioctyl phthalate.

But at the end of most of the synthesis processes, the solvent used is either passed into the water bodies or goes into the atmosphere. In both cases, it interacts with the environment and possesses the threat if the solvent used is toxic. It is advisable to use the methods which either do not need the use of solvent and water or the solvents which can be filtered, degraded, or do not pose a challenge when they join the environment.

The long-term exposure to toxic solvents may also cause a damaging effect on the body organs, respiratory problems, or even disturb nervous system, for example, chlorinated solvents such as chloroform affects the liver, carbon disulfide may affect the heart, and some of them can also vandalize kidney.[5] Figure 11.2 shows the number of cited literature use of ecofriendly solvents for the fabrication of polymeric membranes.[4]

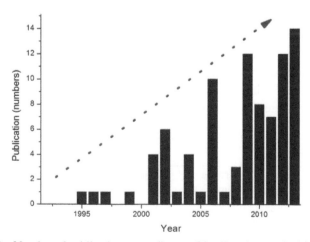

FIGURE 11.2   Number of publications regarding ecofriendly solvents cited.[4]
Note: Number of publications regarding harmless solvent cited in reference until 2014 May.[3]

11.2.2.2 ORGANIC SOLVENTS

The organic solvents conventionally used in the synthesis of PIs such as dimethylformamide (DMF), dimethylacetamide (DMAc), dimethyl sulfoxide (DMSO) are also harmful in ways like DMF causes formation of underground ozone and high exposure may cause liver problems and skin irritation through normal exposure does not have a major effect: sometimes it can be difficult task to substitute the organic solvent with the greener one without compromising the performance and properties of the membrane. In that case, depending on the HSP of the greener solvent mentioned in Table 11.4, various ratios of different solvent can be considered.[4,6] There are much living and environmental threat that can be caused by the toxic solvents so dry methods, aqueous solvent-based synthesis, or greener solvents such as supercritical liquids or gases such as CO_2 should be opted for wherever possible.

TABLE 11.4 HSP for the Nontoxic and Greener Solvents Used in the Membrane Fabrication.

Solvent	Hydrogen bonding (δ_b) $(MPa)^{1/2}$	Dispersion forces (δ_d) $(MPa)^{1/2}$	Polar forces (δ_d) $(MPa)^{1/2}$	Solubility parameters (δ) $(MPa)^{1/2}$
Methyl lactate	16.0	7.6	12.5	21.7
Ethyl lactate	15.8	6.5	10.2	19.9
TEP	9.2	16.8	11.5	22.3
DMSO	10.2	18.4	16.4	26.7
Water	15.5	16.0	42.3	47.8

TEP, Triethylphosphate, DMSO, dimethyl sulfoxide.

11.2.3 IONIC LIQUIDS

Ionic liquids (ILs) are the comparatively newer class of chemicals introduced in the field of chemistry, which freshly boomed as an alternative for conventional solvents as environmentally friendly solvents.[4] ILs are basically made up of ions like the cationic part can be one of pyridinium, imidazolium, quaternary ammonium, or quaternary phosphonium, while the counterion part (anionic part) can be a triflate, halogen, trifluoroborate, or hexafluorophosphate. The basic chemical structure of ILs is reported in Figure 11.3. It has been established that salt having a melting point (T_m) below of water is considered as an IL. But all the ionic liquids are not harmless; in fact, few of them are very toxic in nature and even products can be a difficulty. However, they are well thought out as a "greener" alternative for the conventional solvents, due to its alterable viscosity, T_m, density, and hydrophobicity by adjusting the respective ionic structures. It also is classified by the lack of detectable vapor pressure, nonflammable, noncombustibility, high thermal stability, and low viscosity.

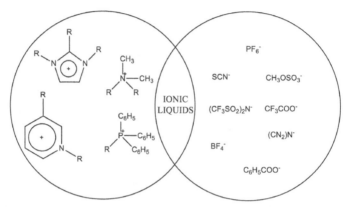

FIGURE 11.3 Conventional parts of cations and anions of ILS, R can be alkyl, alkenyl groups.

11.3 MEMBRANE FABRICATION: A GREEN APPROACH

Green method of synthesis helps to minimize or eliminate the environmentally hazardous substance in design, processability, and application of polymeric membranes. Since the functional and most important part of the filtration system is a membrane, solvent media and various methods, which can be used for the synthesis and fabrication, are summarized along with the precursors used.

There are many conventional solvents used in both the synthesis process of polymer and fabrication of membrane which is known to be hazardous. Also, the traditional method of synthesis create waste which is nonbiodegradable and thus pollute the environment. The following section discusses the green solvent or solvent systems which should consider replacing the so-used medium during the design and fabrication of the membranes. The solvent selection should be in a way that it compromises minimum with the efficacy of the product. Selecting precursors and methods for the synthesis process is also very important, and thus various methods in place of typical chemical methods have been discussed citing the examples for the fabrication.

11.3.1   GREEN SOLVENT IN THE PROCESS MEMBRANE FABRICATION

There are few examples which show the use of greener solvent in the fabrication of PI membranes. Soroko et al. showed the preparation of organic solvent nanofiltration (OSN) using PI by replacing the solvent mixture of DMF/1,4-dioxane (DX) with DMSO/acetone solvent, thus allowing to engineer PI OSN membrane to match the precise requirement for diverse application, and also greener membrane showed the alike performance in terms of rejection as compared with the PI OSN membrane prepared using DMF and DX. SEM results in Figure 11.4 show the formation of spongy like membrane and in Figure 11.5 show the rejection of PI membrane.[7] In addition with above, this author also successfully replaced the cross-linking agent for membrane, that is isopropanol with water, without affecting the performance of membrane.[8] Gao et al. fabricated the composite consisting of outerselective thin film (TFC) hollow fiber membrane. Authors focused on the removal of excess of *m*-phenylenediamine (MPD) and alkane solvent which was used with trimesoyl chloride in the process of synthesis of TFC and Figure 11.6 confirms the formation of hollow fiber.[9] For the removal of the solvents, initially TFC hollow fiber membrane substrate was modified with polyethyleneimine (PEI) for the cross-linking and then using by the small water-soluble molecules like glutaraldehyde and epichlorohydrin. The results obtained displayed the PEI cross-linking may help in the removal of the additional MPD during interfacial polymerization and introduced the additional functional sites for the functionalization of GA and ECH, but earlier alkane treatment was required for the site generation.

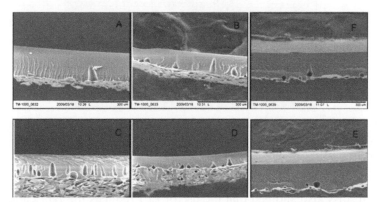

FIGURE 11.4   SEM micrograph of cross-section of PI membrane prepared from the varying ratio of DMSO/acetone. (a) M1 3/1, (b) M2 5/1, (c) M3 7/1, (d) M4 11/1, (e) M5 13/1, and (f) M6 15/1.

Source: Reprinted with permission from Ref. [7]. © 2011 Royal Society of Chemistry.

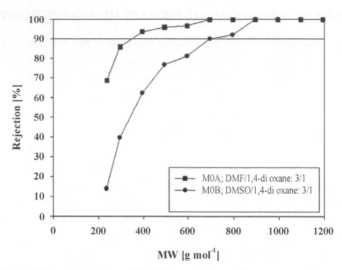

FIGURE 11.5   Effect on rejection of PI OSN membrane with respect to solvent system used in the polymer-dope solution at 30 bar.

Source: Reprinted with permission from Ref. [7]. © 2011 Royal Society of Chemistry.

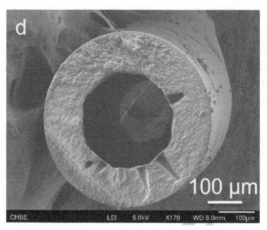

FIGURE 11.6   FESEM micrograph of the M-PEI and GA hollow fiber; membranes whole cross-section.

Source: Reprinted with permission from Ref. [9]. © 2016 Elsevier.

The results also showed that the use of PEI reduces the fabrication cost. Performance of the membrane was increased in a way that enhanced the salt rejection especially (Na_2SO_4) and NaCl, which can reach up to 90% with a pure water permeability (PWP) of 1.74 ± 0.01 L m^{-2} bar^{-1} h^{-1}. Modified membrane showed decent term stability up to 72 h of experiment with impressive thermal stability with broader range of operation temperature (5–60°C). Electrospinning is nowadays becoming an attractive technique used in membrane fabrication and its composite formation. Fabrication of the long-continued fiber with a diameter ranging in nanometer to few micrometers is one of the main advantages of the technique. Jiang et al. focused on the alternative of large amounts of toxic, flammable, and environmental unfriendly organic solvent used in electrospinning by replacing the solvents and solute system with water-soluble precursor solution of hydrophobic nanofibers membrane which showed the decrease in porosity and water permeability.[9] Instead of the use of DMF as a solvent for poly(amic acid) (PAA), author found the alternative by electrospinning of ammonium salt of PAA from water and in second step high temperature used for the imidization and removal of ammonia and the template polymer to prepare the PI nanofiber. The fiber diameter and the contact angle have been shown in Figure 11.7 of PAA-salt/polyethylene glycol (PEO) nanofiber and PI nanofibers imidized.[10] The unanticipated benefit for using the salt leads to an inherent increase in the conductivity of the solution. The membrane obtained from the greener route showed the better high-temperature liquid filtration (100°C), with better thermomechanical stability and good enough to accomplish quantitative filtration of iron oxide microparticles which could be of future interest for oil filtration in engines. Filtration test results were also comparable with the PI obtained from the DMF solution.[10]

FIGURE 11.7 SEM micrograph of electrospun PAA-salt/PEO nanofibers (a, b) and PI nanofibers imidized (d, e). Diameter distribution of electrospun PAA-salt/PEO and PI nanofibers.

Source: Reprinted with permission from Ref. [10]. © 2016 American Chemical Society.

11.3.2   SYNTHESIS OF POLYIMIDES

There are various green methods sought for the synthesis of PIs in various forms, for example, thermal condensation process where melamine and pyro metallic dianhydride are heated in semiclosed environment and result is highly crystalline PI.[11] Also, the directional freezing method can be used in the fabrication of PI aerogels. The PIs synthesized through this method were found to possess less density, a high degree of porosity and exhibited anisotropy, good dielectric properties, and found to be mechanically stable. Table 11.5 summarized the discussed method under the topic.

TABLE 11.5   Summary of the Various Methods Discussed above and Their Benefits over Conventional Techniques.

Methods Used	Monomer	Nature of the product	Advantages
Thermal condensation	Melamine & PMDA	Crystalline powder	• Aqueous medium • avoid organic solvent • easy to perform • products obtained is highly crystalline
Directional freezing	Ammonium salt of polyamic acid	Aerogels	• Use of non-toxic solvents
Solvothermal method	PDA & PMDA	Crystalline	• Aqueous medium • One-pot synthesis • Crystalline product obtained

11.3.2.1   THERMAL CONDENSATION OF MELAMINE AND PYROMETALLIC DIANHYDRIDE

PI is synthesized by a heating equimolar mixture of melamine and pyromellitic dianhydride (PMDA) at a high temperature of 325°C in a semiclosed system. Reaction mechanism, shown in Figure 11.8,[11] does not create hazardous waste and does not have requirement of sensitive environment. The water present if any is easily removed from the system. The process is easy to perform and does not create any hazardous waste which may pose a threat to the environment.

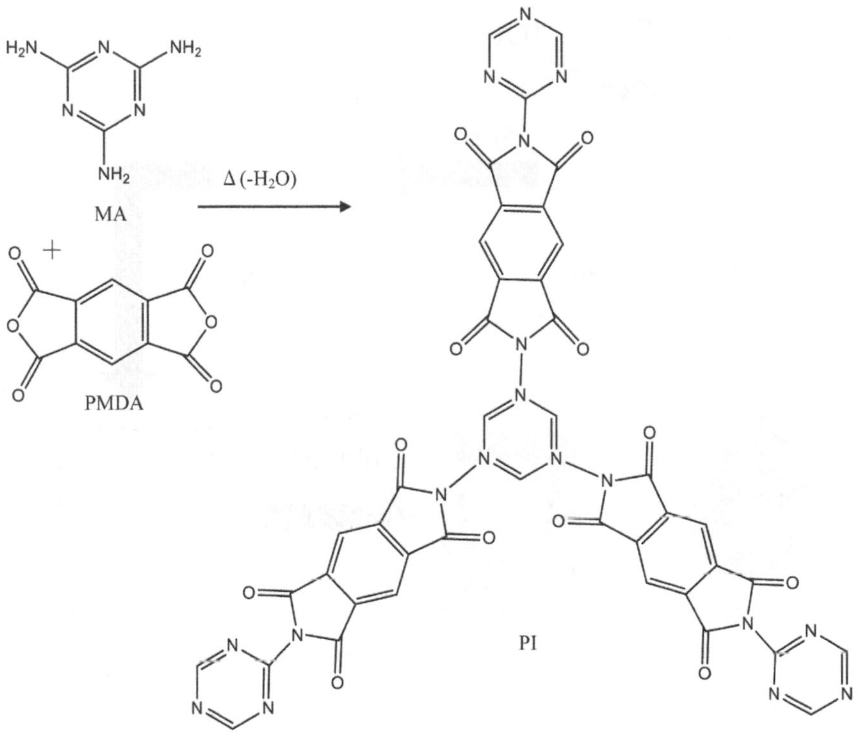

FIGURE 11.8 Scheme of reactions for the synthesis of PI.

11.3.2.2 DIRECTIONAL FREEZING METHOD

PAA, the monomer for PI synthesis, is soluble in toxic solvents. In order to make it water soluble, PAA ammonium salt is prepared and used for the fabrication of PI aerogels. For the purpose, PAA powder is first precipitated from the PAA solution of *N,N*-DMAc.[12] The synthesis process is simple and ecofriendly as it avoids the use of organic solvent in the method of preparation of fabricating PI aerogels from the ammonium salt of PAA as shown in Figure 11.9. The cylindrical mold was filled with the PAS solution at room temperature. An ice-ethyl alcohol mixture was prepared separately and kept in a beaker. The system's temperature was kept at −65°C. After the temperature is maintained, the PAS solution in the mold is kept in the container of freezing liquid in such a way that the level of freezing liquid is always above the PAS solution's level in the mold.

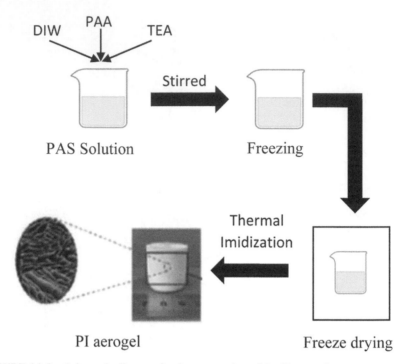

FIGURE 11.9 Schematic diagram for the preparation of the PI aerogel.

Due to the temperature difference between the PAS solution in the mold and the freezing liquid, there exists a temperature gradient which was aligned in a cylindrical way to the center of the axis of the mold. With the preferential growth of ice crystal along the radial direction to the mold, the PAS solution gets frozen. After the complete freezing of the PAS solution, it is transferred to freeze dryer.

11.3.2.3 SYNTHESIS OF POLYIMIDE THROUGH HYDROTHERMAL AUTOCLAVE METHOD

Autoclave reactor provides a controlled pressure environment to the reaction medium and can withstand high temperature depending on the material it is made up of. The hydrothermal method is adopted for the synthesis of aromatic imides, such as poly(*p*-phenylene pyromellitimide) (PPPI), one of the most challenging PIs with regards to synthesis and processability. The method gives highly crystalline environment, which is minimized. Interestingly, hydrothermal route for the synthesis of high performance polymer (HPPs) is inherently green, although the traditional approach depends on the harsh synthetic environments, longed-duration product, and the solvent, which is water which again is a green solvent, minimizes the threatening effects of environment. Intriguingly, hydrothermal synthesis of HPPs is inherently green, whereas the classical procedures rely on harsh synthetic environments, long reaction times, toxic solvents, and catalysts, hydrothermal (HT) synthesis uses only water and yields full conversion after only 1 h.[13] Two pathways for the synthesis of the PPPI shown in Figure 11.10 are as follows: a classical approach showing the transformation of the PAA into PPPI and hydrothermal approach showing the transformation of PDA and PMDA into PPPI.

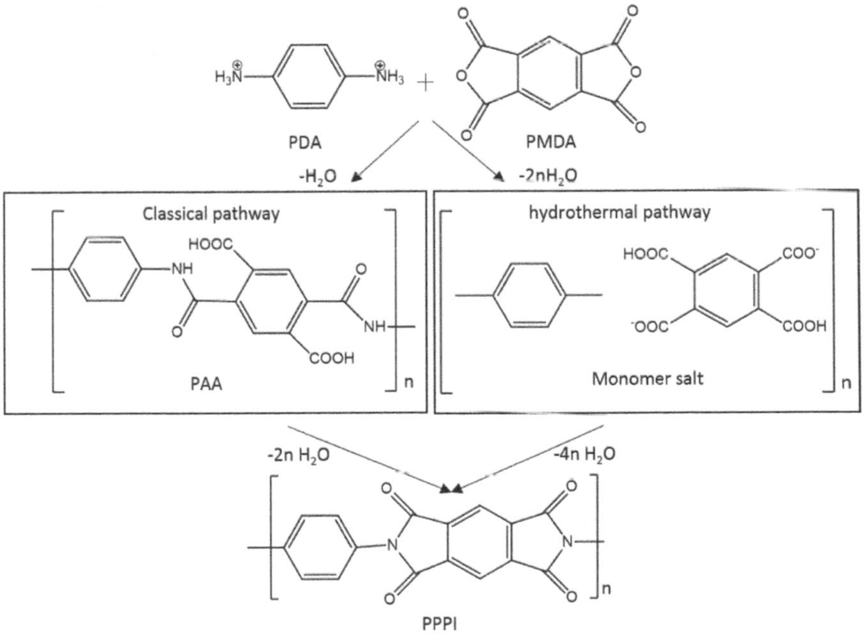

FIGURE 11.10 Different route for the synthesis of the PPPI.

11.3.3 FABRICATION OF POLYMERIC MEMBRANES

Nowadays, green technologies have been extensively preferred for their role in enhanced properties like less reaction time, high selectivity which they offer along with ecofriendly approach. With the recent development in techniques in the fabrication of membrane, now it is feasible to fabricate it in a different and more environmentally friendly method like solvent casting, ultrasonication assist, electrospinning, and so on.

Recently, to improve the physical properties of polymers, they are hybridized with various inorganic materials. The hybridized composite contains two phases:

the matrix phase constituting polymer and copolymers,

the inorganic phase, which is dispersed at the nanoscale level in the polymer matrix.

The hybridized material exhibits optical transparency and high electrical conductivity. It is biocompatible and shows considerable bioactivity and presents more significant opportunity to high tune pure polymers for their use in gas separation processes for the separation and purification of various gas mixtures.[14,15] Table 11.6 summarized the approaches adopted for the fabrication of the membrane via a greener path.

TABLE 11.6   Summary of the Various Methods Discussed above and the Benefits of these Methods Present Over Conventional Techniques.

Method used	Precursors used	Polymer used	Advantage
Ultrasonication assisted	ZnO-PVA	PI	Selective permeability and biodegradability
Ultrasonication assisted	TiO_2-CNF	PI	Selective permeability and enhance thermal stability
Electrospinning	CA and SNP	PAA-F with PB	Superhydrophobic and superoleophilic
Electrospinning	TiO_2	PI84	Enhanced mechanical stability
Electrospinning	TiO_2	PI	Enhanced gas barrier

The use of biocomposites (BC) is also can be an attractive alternative to reduce the carbon footprint of the membrane, and it will also reduce the expanse of polymer used. Incorporation of BC in membrane enhances the performance of the membrane by making it more selective toward filtration of gas and liquid. It also affects the permeability and thermomechanical and physicochemical properties of membrane. By considering this, it will be a good idea to amalgamate the green fabrication method with BC for the fabrication of membrane.

11.3.3.1 ULTRASONICATION ASSISTED FABRICATION OF POLYIMIDE MEMBRANE

There are very few works of literature available in the scientific world which are focusing on the ecofriendly method of membrane fabrication such as ultrasonication assisted, solution free, and green solvent casting with electrospinning and others. As an illustration, few authors mention the use of ultrasonic irradiation method for the BC fabrication such as interaction of zinc oxide and polyvinyl alcohol (ZnO–PVA)[16] nanocomposite with PI and in other example interaction of titanium dioxide-cellulose nanofiber (TiO$_2$-CNF) with PI.[17] On the main advantages of using this method, the energy of sonication may increase the possibility of interaction of modified particles with the polymer matrix. The high-frequency energy of ultrasonication activates the functional group present in the molecules, and it helps in smooth and robust interaction with polymer matrix. This method prevents the use of chemicals which should have been used in the functionalization process of the particle or for a polymer matrix; in addition to this, ultrasonic is used for achieving the dispersion of nano or microparticles by converting an electrical signal into a physical vibration to break substances apart, and converting the big agglomerated particle in the small dispersed and distributed particles by the virtue of which the size to surface area ratio increases; and as ratio increases, it helps in reduction of the reaction time and helps in higher yield of the product. The author shows the use of ultrasonic irradiation is a good alternative focusing toward the greener approach. It has been demonstrated by the authors that the membrane fabricated by the help of the ultrasonic irradiation had formed an excellent intermolecular interaction between polymer and bionanocomposite (BNC). In addition to the interaction, results proved better dispersion of BNC into the polymer matrix, and with the better dispersion of BNC, the thermal stability of the composite enhanced considerably with all selective permeability and biodegradability of the membrane and also improved with increase in the concentration of BNCs.

11.3.3.2 FABRICATION OF POLYIMIDE MEMBRANES BY THE ELECTROSPINNING METHOD

Ma et al. have explained the use of BC in the electrospinning method for the fabrication of membrane for the separation of oil–water mixture.[18] The author used PAA with heavily fluorinated polybenzoxazine (F-PB) for the in-situ polymerization of PI with cellulose acetate (CA) in a combination with silica nanoparticle (SNP). The fiber was coaxial electrospun with an outer layer of CA and inner layer of PI, where the role of PI is to provide the strength to the fiber where the CA has been used to introduce the roughness on the fiber. The fibers were coaxial electrospun by setting up the two syringe-like apparatus and with a different diameter needle coaxially placed inside an outer diameter. Results suggested the use of SNP as a filler, and CA, as an outer surface of fiber membrane, which makes the membrane superhydrophobic and superoleophobic. TEM image and thermogravimetric analysis (TGA) curve of the fiber has been shown in Figure 11.11 which explains the formation of by layer fiber.[18] Another author Lijo et al. describes the use of NBC in electrospinning which enhances the mechanical stability of the fiber without compromising with the membrane performance, and that has been proved using an FTIR analysis. The PI membrane was fabricated using electrospinning of PI84 solution and simultaneously electrospraying the TiO_2 dispersion over the membrane.[19]

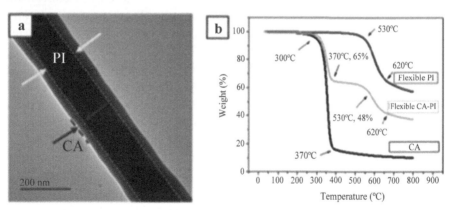

FIGURE 11.11   (a) TEM image of a PI/CA nanofiber; (b)TGA curve obtained on CA, flexible PI, and flexible PI/CA nanofiber membrane.

Source: Reprinted with permission from Ref. [18]. © 2017 Elsevier.

11.3.3.3   *FABRICATION OF POLYIMIDE MEMBRANES BY THE SOLVENT CASTING TECHNIQUE*

Few fragments of literature used TiO_2 as an NBC for the fabrication of PI membrane using solvent casting method in which they mainly focus on the interaction of PI with nanofiller (NF), permeability, and gas barrier properties of membrane. Schematic illustration of the solvent casting for the preparation of polymeric membrane is shown in Figure 11.12. With the incorporation of NF in the membrane, the water contact angle reduces and with respect to that the hydrophilicity of the membrane, increases. Results also reflected the permeability of the biocomposite membrane was enhanced by 20% when compared to the PI membrane, with a high barrier to gases like H_2 and O_2. Same has been proven by the SEM images shown in Figure 11.13.[20] The use of NF enhanced the compaction resistance which prevents the porous structure from collapsing and improves the mechanical and thermal stability.[20,21]

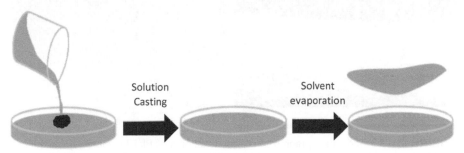

FIGURE 11.12 Schematic illustration of the preparation of the polymeric membrane using solvent-casting technique.

11.4 CROSS-LINKING IN POLYIMIDE MEMBRANE

The cross-linking involves the process of forming covalent bonds or relatively short sequences of chemical bonds to join two polymer chains together.

There are several benefits of cross-linking such as enhanced mechanical strength, control on porosity, and so on in PIs but the most important is the suppression of plasticization which occurs in gas phase separation reaction especially where natural gas is involved.[22]

The plasticization may refer to the swelling of polymer chains mostly faced in the separation of highly condensable gases which may hamper the selectivity as well as the permeability of the polymeric membrane and thus deteriorating the separation performance of the membrane. There are various ways in which cross-linking may be introduced in the PI membranes some of which are UV cross-linking, thermal cross-linking, and chemical cross-linking among others.

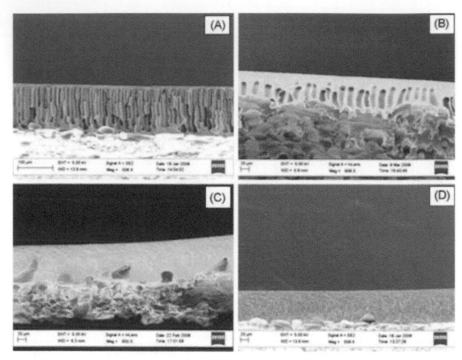

FIGURE 11.13 SEM micrograph of cross-sectional area of PI/TiO$_2$ composite membrane (a) M5 (0 wt% TiO2 in dope), (b) M6 (1 wt% TiO$_2$ in dope), (c) M7 (3 wt% TiO$_2$ in dope), and(d) M8 (5 wt% TiO$_2$ in dope).

Source: Reprinted with permission from Ref. [20]. © 2009 Elsevier.

A thermal method is considered as a green method to introduce cross-linking in the PI membranes.[23] The method includes heating of the sample in a controlled inert environment, keeping it at the required temperature for a given period and then allowing it to cool down naturally.

Also, instead of cross-linking the membrane afterward, the cross-linker can be added during the phase inversion section itself. This allows us to eliminate both the requirement of extra step and other solvent which in turn saves time and is seen as a comparatively greener approach than introducing the cross-linking after the PI is fabricated.[24]

Cross-linking may also be introduced by using a modified vapor technique the apparatus for which is shown in Figure 11.14. The cross-linked PI membrane developed by this route has proved to be an attractive class to separate biofuels and organic solvents.[25]

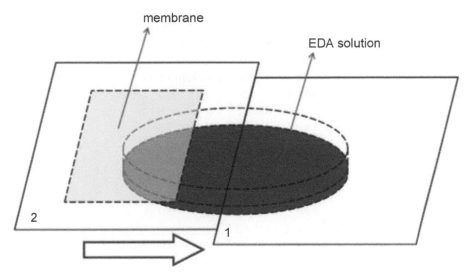

FIGURE 11.14   Apparatus for vapor cross-linking using EDA.

11.5   CONCLUSION

Solvent selection parameters, harmful effects of conventional solvents, properties of green solvents, and their use in the fabrication of PI membranes are discussed in detail. Hansen parameter which is a method based on which the choice of solvent could be made has been explored, and it can be of great help to the researcher.

Examples and mechanism for the synthesis process in the fabrication of PI polymers suggested green methodologies reducing the waste and minimizing the effect on the surrounding environment. The chapter comprises the facile and scalable methods based on environment-friendly techniques such as hydrothermal autoclave method which itself uses a clean strategy for the synthesis of PI in aqueous media, directional freezing method, and thermal condensation for the PI synthesis, whereas various other techniques such as electrospinning and solvent casting have been contemplated with examples for the fabrication of PI membranes.

Though the PIs are widely used in vast applications, still the methods and materials in most are not ecofriendly. The future perspectives should be the increased use of naturally derived monomers and ecofriendly routes for the synthesis processes. Also, the method discussed in the chapter called for the purpose of a solvent, which needs to be chosen carefully; it would be advisable if a solvent-free process could be adopted in future. It should also be taken care that the waste reduction is minimal and biodegradable.

Keeping all these things in mind, sustainability with the development should always move together so that developing new methods does not compromise our environment.

KEYWORDS

- **polyimide**
- **membrane**
- **greener approach**
- **eco-friendly membrane**
- **membrane fabrication**
- **solvents**
- **electrospinning**

REFERENCES

1. Microfiltration, M.; Ro, B.; Ultrafiltration, U.; Filtration, M. Membrane Filtration. 1–12.
2. Jiang, S.; Ladewig, B. P. Green Synthesis of Polymeric Membranes: Recent Advances and Future Prospects. *Curr. Opin. Green Sustainable Chem.* **2019,** 21, 1–8. https://doi.org/10.1016/j.cogsc.2019.07.002.
3. Jwo, S.; Whang, W.; Liaw, W. Effects of the Solubility Parameter of Polyimides and the Segment Length of Siloxane Block on the Morphology and Properties of Poly(Imide Siloxane). *J. Appl. Polym.* **1999,** 2832–2847.
4. Figoli, A.; Marino, T.; Simone, S.; Nicolò, E. Di; Li, X.; He, T.; Tornaghi, S. Towards Non-Toxic Solvents for Membrane Preparation : A Review. *Green Chem.* **2014,** *16* (9), 4034–4059. https://doi.org/10.1039/c4gc00613e.
5. Rama Koteswararao, P.; Tulasi, S. L.; Pavani, Y. Impact of Solvents on Environmental Pollution. *JCHPS Spcl.* **2014,** *3*, 132–135.
6. Yuejie, L.; Wang, W.; Zhao, Q. *J. Bioact Compat. Polym.: Biomed. Appl.* **2016.** https://doi.org/10.1177/0883911514554146.
7. Soroko, I.; Bhole, Y.; Livingston, A. G. Environmentally Friendly Route for the Preparation of Solvent Resistant Polyimide Nanofiltration Membranes. *Green Chem.* **2011,** *13*, 162–168. https://doi.org/10.1039/c0gc00155d.
8. Chakrapani, V. Y.; Gnanamani, A.; Giridev, V. R.; Madhusoothanan, M.; Sekaran, G. Electrospinning of Type I Collagen and PCL Nanofibers Using Acetic Acid. *J.Appl. Polym. Sci.* **2012,** *125* (4). https://doi.org/10.1002/app.

9. Gao, J.; Sun, S.; Zhu, W.; Chung, T. Green Modification of Outer Selective P84 Nanofiltration (NF) Hollow Fiber Membranes for Cadmium Removal. *J. Membr. Sci.* **2016**, *499*, 361–369. https://doi.org/10.1016/j.memsci.2015.10.051.

10. Jiang, S.; Hou, H.; Agarwal, S.; Greiner, A. Polyimide Nanofibers by "Green" Electrospinning via Aqueous Solution for Filtration Applications. *ACS Sustain. Chem. Eng.* **2016**, *4* (9), 4797–4804. https://doi.org/10.1021/acssuschemeng.6b01031.

11. Chem, J. M. Facile Green Synthesis of Crystalline Polyimide Photocatalyst for Hydrogen Generation from Water4. *J. Materials Chem.* **2012**, *22* (31), 15519–15521. https://doi.org/10.1039/c2jm32595k.

12. Wu, P.; Zhang, B.; Yu, Z.; Zou, H.; Liu, P. Anisotropic Polyimide Aerogels Fabricated by Directional Freezing. *J. Appl. Polym. Sci.* **2018**, *47179*, 1–9. https://doi.org/10.1002/app.47179.

13. Baumgartner, B.; Bojdys, M. J.; Skrinjar, P.; Unterlass, M. M. Design Strategies in Hydrothermal Polymerization of Polyimides. *Macromol. Chem. Phys.* **2016**, *217* (3), 485–500.

14. Novak, B. B. M. Hybrid Nanocomposite Materials: Between Inorganic Glasses and Organic Polymers. *Adv. Mater.* **1993**, *5* (6).

15. Smffl, M.; Noble, R. D. Organic–Inorganic Gas Separation Membranes: Preparation and Characterization. *J. Membr. Sci.* **1996**, *116* (2), 211–220.

16. Taylor, P.; Mallakpour, S. *Polymer-Plastics Technology and Engineering A Green Route for the Synthesis of Alanine-Based Poly(Amide-Imide) Nanocomposites Reinforced with the Modified ZnO by Poly(Vinyl Alcohol) as a Biocompatible Coupling Agent*; 2014. https://doi.org/10.1080/03602559.2014.996907.

17. Ahmadizadegan, H. Gas Permeation, Thermal, Mechanical and Biodegradability Properties of Novel Bionanocomposite Polyimide/Cellulose/TiO_2 Membrane. *J. Colloid Interface Sci.* **2016**. https://doi.org/10.1016/j.jcis.2016.11.043.

18. Ma, W.; Guo, Z.; Zhao, J.; Yu, Q.; Wang, F.; Han, J.; Pan, H.; Yao, J.; Zhang, Q.; Samal, S. K.; et al. Polyimide/Cellulose Acetate Core/Shell Electrospun Fibrous Membranes for Oil–Water Separation. *Separation Purif. Technol.* **2017**. https://doi.org/10.1016/j.seppur.2016.12.032.

19. Lijo, F.; Marsano, E.; Vijila, C.; Barhate, R. S.; Vijay, V. K.; Ramakrishna, S.; Thavasi, V. Electrospun Polyimide/Titanium Dioxide Composite Nanofibrous Membrane by Electrospinning and Electrospraying. *J. Nanosci. Nanotechnol.* **2011**, *11* (2), 1154–1159. https://doi.org/10.1166/jnn.2011.3109.

20. Soroko, I.; Livingston, A. Impact of TiO_2 Nanoparticles on Morphology and Performance of Crosslinked Polyimide Organic Solvent Nanofiltration (OSN) Membranes. *J. Membr. Sci.* **2009**, *343* (1–2), 189–198. https://doi.org/10.1016/j.memsci.2009.07.026.

21. Sun, H.; Ma, C.; Yuan, B.; Wang, T.; Xu, Y.; Xue, Q.; Li, P. Cardo Polyimides/TiO_2 Mixed Matrix Membranes : Synthesis, Characterization, and Gas Separation Property Improvement. *Sep. Purif. Technol.* **2014**, *122*, 367–375. https://doi.org/10.1016/j.seppur.2013.11.030.

22. Vanherck, K.; Koeckelberghs, G.; Vankelecom, I. F. J. Crosslinking Polyimides for Membrane Applications: A Review. *Prog. Polym. Sci.* **2013**, *38* (6), 874–896. https://doi.org/10.1016/j.progpolymsci.2012.11.001.

23. Chen, C. C.; Qiu, W.; Miller, S. J.; Koros, W. J. Plasticization-Resistant Hollow Fiber Membranes for CO_2/CH_4 Separation Based on a Thermally Crosslinkable Polyimide. *J. Membr. Sci.* **2011**, *382* (1–2), 212–221. https://doi.org/10.1016/j.memsci.2011.08.015.

24. Vanherck, K.; Cano-Odena, A.; Koeckelberghs, G.; Dedroog, T.; Vankelecom, I. A Simplified Diamine Crosslinking Method for PI Nanofiltration Membranes. *J. Membr. Sci.* **2010,** *353* (1–2), 135–143. https://doi.org/10.1016/j.memsci.2010.02.046.

25. Mangindaan, D. W.; Min Shi, G.; Chung, T. S. Pervaporation Dehydration of Acetone Using P84 Co-Polyimide Flat Sheet Membranes Modified by Vapor Phase Crosslinking. *J. Membr. Sci.* **2014,** *458,* 76–85. https://doi.org/10.1016/j.memsci.2014.01.030.

CHAPTER 12

Dynamic Molecular Phenomena in Polyimides Investigated by Dynamic Mechanical Analysis

MARIANA CRISTEA[1*], DANIELA IONITA[1], MANUEL GARBEA[2], and MARIANA-DANA DAMACEANU[1]

[1]*"Petru Poni" Institute of Macromolecular Chemistry, 41A Grigore Ghica Voda Alley, 700487, Iasi, Romania*

[2]*Microsin SRL, Bucharest, Romania*

*Corresponding author. E-mail: mcristea@icmpp.ro

ABSTRACT

The formation of polyimides by the combination of dianhydrides with diamines results in a large diversity of structures having various mobilities, inter/intra-chain interactions, and topologies. Generally, the processing of polyimides in the imidized form is complicated even because of molecular rigidity and strong inter-chain interactions that entails poor organosolubility. The molecular dynamics of polyimides is the result of the balance between molecular packing and chain motions. The phenomenon has effects also on the application of polyimide, for example, in the membrane field. Dynamic mechanical analysis (DMA) is a convenient method to investigate molecular dynamic in polyimides. The secondary relaxations of polyimides can be accurately defined by DMA, along with their influence on the whole behavior of the polymer. As a polyimide is prepared by thermal imidization of its precursor, poly(amic acid) (PAA), the result of a temperature sweep DMA gives also information on the evolution of the system during heating. In addition, the accurate assessment of the glass transition region cannot be established without considering the occurrence of imidization by increasing temperature. Even more challenging is the situation when, during imidization, the analyst deals with cross-linking, thermal rearrangement processes, or solvent evaporation. The chapter will frame all the aspects above by considering a couple of examples for illustration.

12.1   INTRODUCTION

A closer look to the history of polymers reveals that the appearance and development of a specific characterization method keeps the pace with the need to find answer for a specific polymeric behavior.[1,2] In the early 1920s, Staudinger introduced to the scientific world his theory about the high macromolecular compounds, covalently bonded, that were named *macromolecules*. Few years later, in 1929, rheology was established as a new branch of science when scientists became aware of the special nature of polymers. They are viscoelastic materials having simultaneously solid-like and liquid-like properties.[3–5] Since then, the face of the world has changed as polymers invaded literally our lives.[6] For example, we are surrounded by gadgets that help us in our work and keep us informed and connected to each other more than never before. One of the polymeric components of these electronic devices is a polyimide that usually functions as isolator.[7]

Polyimides belong to the class of high-performance polymers.[8,9] Herein high-performance refers to an outstanding thermal behavior combined with excellent mechanical and electrical properties that come from the attribute of the stable aromatic ring (imide) present in the polymer backbone. Unfortunately, these rings confer rigidity to the backbone and induce high intermolecular associations. This is why it is a troublesome task to process the polymer from the melt or to find solvents for casting. In this sense, the effort was concentrated to the manufacture of stable and processable polyimides, without sacrificing the desired properties. The major parameters that govern the processes involved in the preparative steps are *temperature* and *time profile*. They decisively influence the properties of the final polyimide material. The trade-off between flexibility and thermomechanical properties is established by investigation of the main relaxational phenomena that characterize the polymer. Dynamic mechanical analysis (DMA) is one of the most reliable methods for ascertaining transitions in polymers.[10,11] It is part of rheology and is based on the application of an oscillating stress (or strain) to a sample and on the recording of the material response to this stimulus. DMA tests the viscoelastic properties of a material (modulus and damping) and it can also be used to follow the temperature/frequency (and therefore time) dependence of these transitions.

This chapter will start with a general view on polyimides: how they are obtained, why the preparative step is challenging, what are the factors that determine the process, and how can they be controlled. Some representative structures will be presented. The second part will describe succinctly the DMA method and how it can describe the relation structure-property in polymers. Then, the next two parts will intend to put forward the utility of the method as a research instrument in the field of polyimides, either for the characterization of the polyimide materials, or for understanding and optimization of the synthesis steps. DMA exploits the concept of *time–temperature equivalence*, that is, the analogy between the transformations determined by temperature and frequency changes. The last parts will overview when this principle was applied to polyimides and with what purposes.

12.2 POLYIMIDES—OVERVIEW

The typical procedure to prepare polyimides consists in the polycondensation reaction of a dianhydride with a diamine (Fig. 12.1), in dipolar aprotic solvents like dimethyl sulfoxide (DMSO), dimethylacetamide (DMAc), dimethyl formamide (DMF), and N-methyl pyrrolidone (NMP). The scheme indicates that the synthesis is a two step method. The first step involves the preparation of the soluble polyimide precursor, the poly(amic acid) (PAA). In the second step, the cyclodehydratation of PAA is carried out, *thermally or chemically*, with the formation of polyimide.

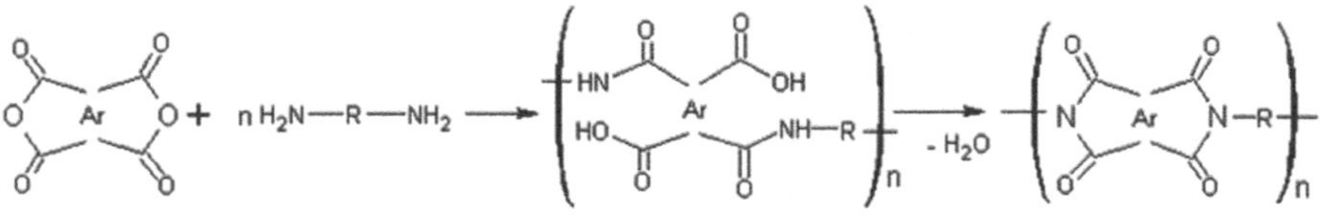

polyamic acid intermediate

FIGURE 12.1 Schematic representation of the two-step synthesis of polyimides.

Thermal imidization is performed on the PAA *in solid form* (a film cast or spin-coated on a glass plate)[12–16] or in solution.[17–21] The first approach demands a rigorously controlled curing profile in order to avoid voids and film shrinkage. It is reported that the *imidization in solution* avoids the side-reaction because cyclodehydratation takes place at relatively low temperature.

There are reports that describe the chemical imidization of PAA by using chemical dehydrating agents in combination with basic catalysts, like acetic anhydride and pyridine.[14,18,22,23]

Imidization in solution is possible to be realized in one step. It is used when the solubility of monomers is very low. The reaction may take place in m-cresol in the presence of benzoic acid and isoquinoline.[22,24,25] The polyimides with six-membered imide rings are mainly characterized by low solubility. In order to improve the solubility of monomers, Damaceanu et al. added LiCl in NMP and used benzoic acid as catalyst in the synthesis of co/poly(peryleneimides).[26,27]

Actually, it is the structure of the monomers that plays a decisive role in choosing the appropriate methodology for imidization. The outstanding thermal behavior combined with excellent mechanical and electrical properties come from the attributes of the stable imide and other aromatic rings. This aromaticity confers rigidity to the backbone and induces high intermolecular associations. To circumvent these drawbacks, the incorporation of kinks represented by asymmetric, noncoplanar, and flexible linkages between the aromatic rings or bulky pendent groups in the polymer chain was adopted. The main goal when working with polyimides was to find a good equilibrium between the flexibility, that dictates the processability, and the characteristics that confer high-performance properties. The papers published in the field report are a whole hierarchy of different commercial and synthesized monomers used to obtain various polyimide materials. There is such a large array of structures associated with polyimides that the reviews in the field focused only on a specific aspect of the polyimide domain. It is impossible to capture every single aspect of polyimides even in one book.

There are some publications that summarize briefly the main facets of polyimides: synthesis, structures, properties, and applications.[9,28,29] Other publications are focused on the main aspects of a specific class of polymides: isomeric polyimides,[30] polyimides with aliphatic/alycyclic segments,[31] polyimides containing 1,3,4-oxadiazole ring,[32] dendritic polyimides,[33] or cardo-polyimides.[34] Nevertheless, the list is not exhaustive. Given the complex phenomena that overlap during the synthesis, early studies were directed to find evidences of the polyimide formation mechanisms.[35] Ghosh et al. authored a review concerning the approaches that have been considered in order to improve the processability of polyimides.[36] It is meaningful that a substantial body of research has been devoted to the applications of polyimides in optoelectronic and memory devices,[37–40] membrane[41–43] and fiber technology,[44,45] or as carbon precursors.[46]

Aside from the general structural requirements concerning starting monomers, outlined at the beginning of the section, the polyimides for certain applications have additional requisites.

It is well-known that intra-/intermolecular charge transfer complex (CTC) are formed between the dianhydride segment (electron-acceptor) and the diamine segment (electron-donor). The formation of CTC is the cause of the deep-colored polyimide films, unwanted in optical applications. In order to obtain transparent polyimide films, a slight modification in the structure of the polyimide is needed, namely the joining of weak electron accepting dianhydride and a weak electron-donating diamine. In this way the formation of CTC is prevented. Fortunately, the strategies used for increasing the flexibility are also reliable in this situation: incorporation of fluorine, chlorine, sulfone groups; the presence of unsymmetrical and bulky pendent units; the use of alicyclic moieties in the structure of the starting monomers.

For obvious reasons, the molecular weight of the polyimide used in the formation of membranes as film or fibers should be enough high for reliable results.[47]

As for polyimide fibers, the dominating factor behind the option for the one stage or the two stage manufacturing process is the solubility of the polyimide. The spinning process (wet or dry) is included in the preparation method. When the polyimide is insoluble, the practitioners opt for the two-stage method.[48,49] Practically, the spinning is performed by using the PAA solution and the PAA fibers undergo thermal curing. Nevertheless, the challenge that has to be surpassed is to avoid the fusion of the filaments.

Recently, polyimides that are unprocessable were transformed into a hybrid poly(imide-imine)s by combining imide links with dynamic imine bonds. The new polymer is malleable and can be processed due to the reversible cleavage-reformation of the imine bond under external stimuli.[50]

These great varieties of structures with many intra/intermacromolecular interactions influence the flexibility of the polymer chains with conclusive effects on the properties of materials. DMA is one of the main methods used to investigate the dynamic molecular phenomena in polyimides.

12.3 DYNAMIC MECHANICAL ANALYSIS—MECHANICAL METHOD *VS* THERMAL METHOD

The science of the 19th century had described dichotomously the behavior of materials as solids and liquids. Simply said, solids respect the law of Hooke, that is, the deformation is proportional to stress and it is fully recoverable when the stress is removed. Liquids obey the Newton's model where the deformations are proportional to the rate of deformation and are permanent. Everything between these two categories represented exceptions. The emergence of polymers at the beginning of the 20th century reversed this statement. Solid- and liquid-type behaviors are particular cases because polymeric materials are viscoelastic. Mechanics, in the original sense repre-sents the study of the behavior of physical systems under the action of forces. The term *rheology* was introduced in 1929 in order to comprehend this dual behavior and it is the science of deformation and flow.[4,5]
In practice, rheology is seen as part of continuum mechanics, to characterize the flow of materials, having both solid and liquid characteristics.

Also, in polymer science the informal language uses the term mechanical properties to indicate the stress–strain relationship for polymeric materials. However, this relation neglects the viscous contribution and the elastic modulus is determined under the supposition that the material has only elastic behavior. Unlike many other materials, the properties of polymers (mechanical, electrical, optical) depend on time because the polymer intrinsic properties are set by their long-chained structure. The time-dependent characteristics of polymers are the output of experiments performed with discontinuous stress or strain levels (creep and stress relaxation). Neverthe-less, the time dependence may be investigated by the application of a small oscillatory perturbation to the samples, mainly sinusoidal.[10,11]
When we refer to DMA and the perturbation is the stress, it will be expressed as:

$$\sigma = \sigma_0 \sin \omega t, \tag{12.1}$$

where σ is the stress at time t, σ_0 is the amplitude of stress, and ω is the angular frequency.
The resulting strain will be also sinusoidally shaped:

$$\gamma = \gamma_0 \sin(\omega t + \delta), \tag{12.2}$$

where γ is the strain at the time t, γ_0 is the amplitude of strain, and δ is the phase difference between the stress signal and the strain signal. The phase lag δ is 0 for an ideal elastic solid and 90° for the perfect viscous fluid. The viscoelastic materials have $0° < \delta < 90°$.

The complex modulus is defined as a real part, E', and an imaginary part, E".

$$E^* = E' + iE''$$ (12.3)

The real part E' is the elastic (storage) modulus, while the imaginary part E" is the viscous (loss) modulus.

$$E' = E^* \cos \delta$$ (12.4)

$$E' = E^* \sin \delta$$ (12.5)

The elastic modulus represents the elastic contribution of the viscoelastic behavior and is associated with the rigidity of the sample. The viscous modulus denotes the viscous portion of the viscoelastic behavior and is related to the flowing properties of the sample. The loss factor tan δ is expressed as the ratio between the two moduli:

$$\tan \delta = \frac{E''}{E'}$$ (12.6)

Practically, it denotes the ratio between the energy dissipated and the energy stored elastically.

Rheometers used in rheology perform both rotational and oscillatory tests. Dynamic mechanical analysis (DMA) belongs to oscillatory tests conducted mainly at constant temperature with the particularity that it is mainly done on load-bearing samples.[51] Tests that are performed with varying temperature are also called dynamic mechanical thermal analysis (DMTA) (Fig. 12.2).

When the samples are liquids any test is usually regarded as being part of rheology. In this sense, it is important to mention that the rheological tests are mainly done by using the shear mode of deformation. For solid samples few loading modes are available (Fig. 12.3), each of them being adapted to the shape and rigidity of the samples: tension (films, fibers), single/dual cantilever and three point bending for bars, compression for soft samples, and shear.[52–55]

FIGURE 12.2   Dynamic mechanical analysis: mechanical and thermal method.

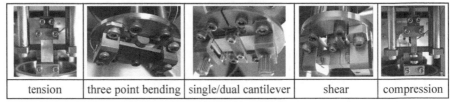

| tension | three point bending | single/dual cantilever | shear | compression |

FIGURE 12.3   Loading-types possible for DMA instruments.

A novel configuration was developed in order to facilitate the DMA investigations of powders: material pocket. The material is included in a rectangular stainless open container, which is fixed in the dual cantilever attachment. It was initially intended for pharmaceuticals but the procedure was also applied to other systems.[56–58]

The viscoelastic properties of polymers comprise also a mathematical apparatus intended for the analysis and design of proper polymer structures. As the polymers have become engineering structural materials, scientists with background in mechanics have oriented to the understanding of the polymer science. Detailed and proper explanations of the mathematical framework behind the polymer viscoelasticity can be found in some reference works.[10,59,60]

The most common DMA experiment is the isochronal temperature scanning. Figure 12.4 represents schematically the variation of the elastic modulus with temperature for a DMA experiment conducted at a single frequency, on an amorphous polymer. As the temperature increases, the magnitude of molecular motion is amplified and four characteristic regions are defined.[61–63]

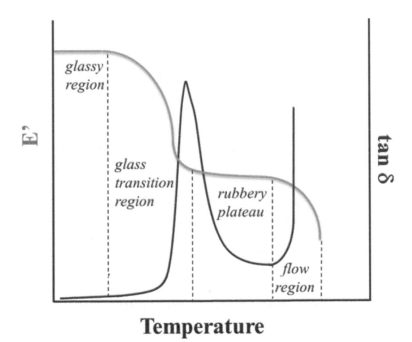

Temperature

FIGURE 12.4 Schematic representation of the variation of the elastic modulus E' and damping factor tan δ*vs.* T.

In the *glassy region,* the polymer chains do not have the possibility to move, they are practically frozen. Usually the elastic modulus E' takes values over 10^9Pa, this measurement order being characteristic for glassy polymers. Only secondary relaxations are possible in the glassy regions (β, γ) and they correspond to local motions (vibrations of side groups and bending/stretching of the main chain linkages). The secondary relaxation are detected when the frequency of the experiment is enough low (equivalent with long time) to track their relaxation time. Unlike DMA, differential scanning calorimetry (DSC) is not able to detect the faint secondary relaxation.

As the temperature increases, large-scale segmental movements become active. The polymer reaches the *α-relaxation* that is typically associated with *the glass transition region.* Herein, the modulus E' decreases substantially two-three orders of magnitude, until a plateau is reached. This method expresses better than many others the fact that the glass transition temperature of polymers is not a single value but an interval. Nevertheless, usually a single temperature value is reported as the onset of the E' drop or the peak of the E" or tan δ curves. No matter which manner is used to report the glass transition temperature, it is important to mention the experimental conditions (loading type, frequency, heating rate) and to maintain the procedure for the series of polymers under investigation.[10]

On the *rubbery plateau,* there is an extensive molecular movement; still the polymer chains are held together by cross-linking or other types of inter-macromolecular interactions. Processes like melting, cold crystallization, or curing can be noticed in this region, for appropriate polymer structures. As polymer interconnections mitigate, the chains acquire the ability to slip one past another in the flow region. Therefore, the polymer samples lose their load-bearing capacity.

As a matter of fact, all the phenomena that takes place in polymers overlap and a good knowledge of polymer synthesis and structure helps in the understanding of the viscoelastic behavior. In this sense, a correlation of the DMA results with other thermal methods is beneficial to the research.

12.4   SECONDARY RELAXATIONS IN POLYIMIDES—HOW IMPORTANT THEY ARE

The application of DMA to identify the sub-glass transition relaxations (β, γ) is justified by both theoretical and practical reasons.

The β-relaxation is generated by rotational vibrations of polyimide groups that depend on both the chemical composition and conformational features. There are multitudinous discussions in polyimide literature as regard the definite identification of specific groups responsible for β-relaxation.[64–70] The nature of β-relaxation—cooperative or noncooperative—was established by applying the Starkweather theory or by checking the deviations from the Arrhenius line.[71]

The temperature of β-relaxation is related to how easy the rotational vibration of the group is, while its magnitude can be correlated to the number of groups/unit length. The broad temperature range of β-relaxation may be attributed to the big differences between the activation temperature of the vibration groups.[64]

Li et al. investigated by DMA, the correlation between molecular packing and chain motions in polyimides obtained from 4,4'-(hexafluoroisopropyli-dene) diphthalique anhydride (6FDA) and 4,4'-diamino-2,2' disubstituted bisphenyls.[65] With regard to β-relaxation, the substituted groups introduce steric hindrance, hampering the local motion of small groups, and fragments on the repeating unit. Contrary, the effect is opposite for α-relaxation; the steric hindrance results in looser molecular packing, therefore it enhances the segmental motion. According to Starkweather procedure, the origin of β-relaxation was attributed to noncooperative motions in diamine for the disubstituted groups smaller than a ($-CF_3$) group, while the molecular motion evolved into a cooperative process for disubstituted groups larger than ($-CF_3$). The results of Arnold et al. demonstrated that, with increasing crystallinity and orientation, the noncooperative nature of the motion may be gradually lost.[66] Hougham et al. also reported that β-relaxation temperature shifted to higher temperatures with increasing fluorine content in the diamine unit, at the same time with decreasing the glass transition temperature.[72] Nevertheless, the bulky groups linked to the central moiety of the dianhydride residue had a smaller effect on the thermal transitions of polyimides than did meta/para isomerism.[73]

Ragosta et al. investigated a series of three polyimides differing for their molecular structures PMDA-ODA (PMDA—pyromellitic anhydride, ODA—4,4'-diaminodiphenyl ether), 6FDA-ODA, and 6FDA-6FpDA (6FpDA—4,4'-(hexafluoroisopropylidene).[74] They identified in the β process two relaxation processes (β' and β''), where β' and β'' were associated with the local motions of the diamine and dianhydride, respectively.

More recently, it was established that there is a relation between the coefficients of volumetric thermal expansion and the β-relaxation temperature.[75] In the same direction, Qian et al.[76] used the secondary relaxation behavior of a pendant group (the rotation of the meta-terphenyl unit that belongs to the pendant group) to obtain more free volume in the bulk. Using this strategy, the dielectric constant of the polymer was reduced.

The other secondary relaxation, γ-relaxation is situated at temperatures lower than $-50°C$ and is caused by the presence of water in polyimides.[77] Bas et al. made a correlation of this relaxation temperature with the microstructural parameters.[18] The γ-relaxation disappears by drying the samples and its temperature does not depend on the water content. Studies on semialiphatic polyimide structures[78] proved that the shift of this relaxation toward higher temperatures was attributed to the increase of chain packing (Fig. 12.5).

One more γ-relaxation is reported when the number of $(CH_2)_n$ groups included in the main chain of the polyimide is higher than four.[79] Both γ-relaxations were also found for other classes of polymers.[80–82]

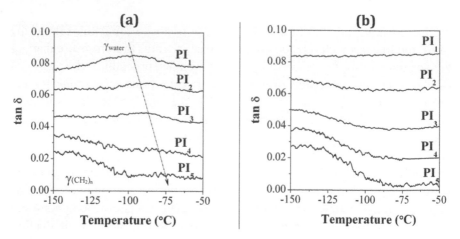

FIGURE 12.5 The γ-relaxations for semialiphatic copolyimides. (a) first heating, (b) second heating. Reprinted with permission from Ref. [78]. Copyright 2019 Elsevier.

12.5 SPECIFIC ASPECTS OF THE GLASS TRANSITION REGION IN POLYIMIDES

The glass transition temperatures (T_g) of polyimides (associated with α-relaxation) have values over 200°C. The results of a dynamic temperature sweep experiment offer for polyimides more than simply the determination of the relaxation temperatures. It is an opportunity to investigate the evolution of the system with heating.[83–86]

As it was presented in Section 12.2, a polyimide is prepared from its precursor, the PAA, that is transformed mostly thermally into the corresponding polyimide. Usually, the completion of thermal imidization is checked by IR spectroscopy. However, the overlapping of some characteristic absorption bands of PI and PAA can happen and this can impede the correct evaluation of the imidization reaction. Long-term practitioners in the synthesis of polyimides confirm that the imidization of PAA is never fully accomplished. Always there are some amic acid groups that are not imidized. Whenever a PAA or an incomplete cured PI is examined by DMA, the occurrence of the imidization in the DMA oven cannot be excluded.[85,87,88] A representative DMA for such a sample is represented in Figure 12.6.

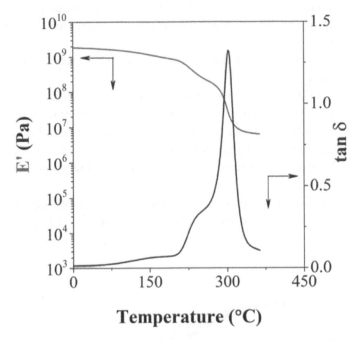

FIGURE 12.6 Representative DMA result for an incompletely imidized PAA. Reprinted with permission from Ref. [78]. Copyright 2019 Elsevier.

The two steps of the glass transition region are evident in the figure, one corresponds to the PAA and the other to the PI formed in the DMA oven. Additional imidization during the experiment increases the Tg of the initial sample. Gilham stated for the first time that what count in a scan temperature experiment of such a system is not the absolute temperature (T) of the sample at a certain moment (t) but the relation between this temperature and the T_g of the partially imidized PAA with the degree of imidization at that moment.[89] Lee and Goldfarb introduced a reduced parameter $(T-T_g)$ that expresses better the variable of the experiment for partially cured polymers.[90] According to them, when both thermal relaxation effects and imidization take place, there are four stages during the thermal experiment. When there is no additional imidization during the time scale of the experiment, $dT_g/dt=0$ (stage I). In stage II, the imidization rate is lower than the scanning rate, $dT_g/dt<dT/dt$; in stage III the imidization rate is higher than the scanning rate, $dT_g/dt>dT/dt$. A most frequent mistake in interpreting the DMA results of a polyimide is to associate the glass transition range of the incompletely transformed PAA with the β-relaxation. A strategy to avoid this is to perform the multifrequency experiment and to calculate the activation energies (E_a) of the relaxations.[86] The sub-glass transition relaxations have the E_a lower than 100 kJ/mol.

DMA can also highlight the processes that take place after the glass transition region.

It is known that fibers produced by conventional spinning showed very high tensile strength. Heat treatment helps in aligning the PI macromolecules parallel to each other without the need for mechanical stretching. Yang et al. demonstrated that during heating orientation/crystallization occur in the PAA.[91] The E' plateau registered a sudden increase at the temperature where these processes started.

In the same direction, Comer et al.[17] evidenced in a multifrequency DMA experiment high temperature thermal rearrangements (Fig.12.7).

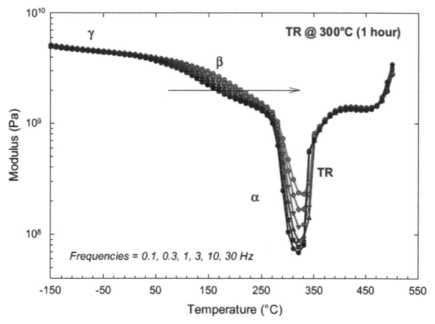

FIGURE 12.7   Thermal rearrangement evidenced by a multifrequency DMA experiment. Reprinted with permission from Ref. [17]. Copyright 2019 Elsevier.

For conclusive results, the variation of E', E" and tan δ should be correlated among them and, in the same time, with other methods of analysis.

12.6 TIME–TEMPERATURE SUPERPOSITION PRINCIPLE APPLIED FOR POLYIMIDES

The long-term exposure of polyimide to pressure, temperature, and moisture results in alteration of the mechanical properties, crucial issues not only for their design but also for their safe operation. Therefore, studying and understanding the evolution of the mechanical long-term properties is mandatory. While desirable, experimentally characterization of polyimide under the time and temperature conditions similar to those found in end-use applications is not always feasible. Reasons for these include the limited frequency range of the instruments or at best, time consuming and costly experiments. A comprehensive characterization of a magnetic tape substrates based on polyimide implies a 10–12 decade frequency range.[92] Several studies describing the effects of time and temperature upon the viscoelastic behavior of polyimide films (Kapton, Matrimid) have been published.[68,93] According to these studies, the effect of long-term behavior can be predicted in a very short time using time–temperature superposition (TTS) principle. It is based on the observation that for many polymer characteristics resulted from stress relaxation, creep, or DMA tests a shift in time is related with a shift in temperature. The relation between these two variables can be written as a mathematical equation:

$$E'(t,T_1) = E'(t/a_T, T_2) \tag{12.7}$$

where E' is the storage modulus, a_T is the shift factor, t is time and T_1, T_2 are the temperatures at the moment t and t/a_T.

TTS assumes a horizontally and/or vertically shifting, either at the left or at the right of the viscoelastic experimentally data points corresponding to a material property, at a series of temperatures, along the time (frequency) axis to create a continuous curve—*master curve*—at a reference temperature (Fig. 12.8).

The amount of shifting in a TTS plot is general described by two types of equation: Williams-Landel-Ferry equation (eq. 12.8) if the temperature is situated below the glass transition region or the Arrhenius equation (eq. 12.9) for temperatures well above the glass transition region.[94,95]

$$\log a_T = \frac{-C_1(T-T_r)}{C_2+(T-T_r)} \tag{12.8}$$

$$\log a_T = \frac{E_a}{2.303R}\left(\frac{1}{T} - \frac{1}{T_r}\right) \tag{12.9}$$

where a_T is the shift factor, C_1 and C_2 are material dependent parameters, T_r is reference temperature, T is the sample temperature, E_a is the activation energy and R is universal gas constant.

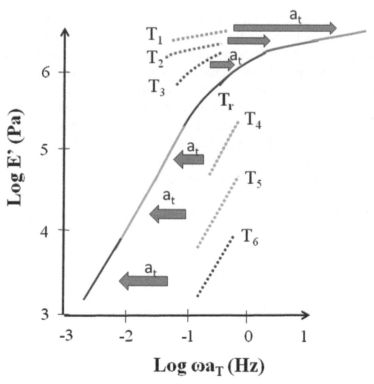

FIGURE 12.8 Schematical construction of a master curve.

Once the shift factors are known a master curve is created; thus the data generated at higher temperatures and shorter times can be used to predict behavior at lower temperatures and much longer time periods. Tsai et al.[96] studied the high temperature lifetime of a polyimide modified with p-amino-phenyltrimethoxysilane (5000-PIS) using TTS principle. The applications of TTS method also include evaluation of the dynamic fragility of the Matrimidbackbone[68] and monitoring of hydrothermal effects upon stress-relaxation of Kapton film.[97]

The most limiting requirement in the application of TTS principle relates to thermorheological behavior of the polyimide; this is divided into simple and complex. The thermorheological simple behavior is proved by generating a continuous master curve using horizontal shifting of the data obtained at different temperatures with respect to time. For complex thermorheological behavior the time–temperature superposition fails.

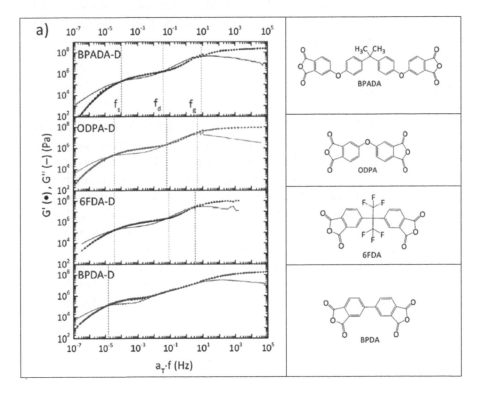

FIGURE 12.9 The elastic and viscous moduli master curve of four polyimides with different linkers connecting the two phtalic units of the dianhydride. Reprinted with permission from Ref. [99]. Copyright 2019 American Chemical Society.

The complexity can be caused by overlapping of primary and secondary relaxations, chemical reactions, or other type of physical changes that can modify the structure or morphology during measurement causing changes in the molecular mobility.[17,98] Although TTS has been used for thermorheological simple polyimide, the literature mentions its use in analyzing the behavior of more complex systems like self-healing polyimides.[99,100] Susa et al. have generated a master curve using the data gathered from frequency sweep experiments (from 0.1 to 10 Hz), performed at temperatures between 10° and 110°C in parallel plate geometry, in step of 5°C, at a reference temperature corresponding to healing temperature.[99]

The results shown in Figure 12.9 evidence the effect of dianhydride architecture on polymer dynamics over the extended frequency range obtained by superposition (from 10^{-7} Hz to 10^5 Hz). Moreover some correlations have been made between the crossover frequency in the flow relaxation and the degree of macroscopic healing.

12.7   CONCLUSIONS

DMA as part of rheological methods allows the determination of the visco-elastic properties of polymers as a function of temperature and in time. When referring to polyimides, it provides detailed information about the sub-glass transition relaxations and the glass transition temperature. All these information allow the formulation of detailed structure-properties correlation. Also, it is one of the best investigation methods that indicates the occurrence of various processes with increasing temperature: imidization process (thus, suggesting incomplete transformation during synthesis), crystallization/orientation, thermal rearrangement. By the application of the time–temperature equivalence, it is possible to estimate the behavior of one polyimide system after a period of time, much longer than the duration of an experiment.

ACKNOWLEDGMENT

The financial support of European Social Fund for Regional Development, Competitiveness Operational Programme Axis 1 – Project "Petru Poni Institute of Macromolecular Chemistry - Interdisciplinary Pol for Smart Specialization through Research and Innovation and Technology Transfer in Bio(nano)polymeric Materials and (Eco)Technology", InoMatPol, ID P_36_570, Contract 142/10.10.2016, cod MySMIS: 107464, is gratefully acknowledged.

KEYWORDS

- polyimide
- dynamic mechanical analysis
- viscoelastic properties
- relaxations
- structure-property correlation
- time–temperature equivalence

REFERENCES

1. Ebewele, R. O. *Polymer Science and Technology*; CRC Press: Boca Raton, FL, 2000.
2. Teegarden, D. *Polymer Chemistry: Introduction to an Indispensable Science*; National Science Teachers Associations: Arlington, Virginia, 2004.
3. Mülhaupt, R. Herman Staudinger and the Origin of Macromolecular Chemistry. *Angew. Chem. Int. Ed.* **2004**, *43*, 1054–1063.
4. Reiner, M. The Deborah Number. *Phys. Today* **1964**, *17*, 62–62.
5. Morrison F.A. What is Rheology Anyway? *The Industrial Physicist* **2004**, *10*, 29–31.
6. Freinkel, S. *Falling in and Out of Love: Our Troubled Affair with Plastics.* In *The Age of Plastic: Ingenuity and Responsibility;* Madden, O., Charola, A. E., Kobb, K. C., DePriest, P. T., Koestler, R. J., Eds.; Smithsonian Institution Scholarly Press: Washington D.C., 2017; pp 165–171.
7. Kim, D.; Shin, G.; Kang, Y. J.; Kim, W.; Ha, J. S. Fabrication of a Stretchable Solid-State Micro-Supercapacitor Array. *ACS Nano* **2013**, *7*, 7975–7982.
8. Baklagina, Y. G.; Milevskaya, I. S. *Supermolecular Structure of Polyamic Acids and Polyimides.* In *Polyamic Acids and Polyimides. Synthesis, Transformations, and Structure*; Bessonov, M. I., Zubkov, V. A., Eds.; CRC Press: Boca Raton, FL, 1993; pp 197–280.
9. Sroog, C. E. Polyimides. *Prog. Polym. Sci.* **1991**, *16*, 561–694.
10. Chartoff, P. R.; Menczel, D. J.; Steven, D. H. *Dynamic Mechanical Analysis.* In *Thermal Analysis of Polymers. Fundamentals and Applications*; Menczel, D. J., Bruce Prime, R., Eds.; John Wiley and Sons, Inc.: Hoboken, NJ, 2009; pp 387–495.
11. Sepe, M. P. *Dynamic Mechanical Analysis for Plastics Engineering*; Plastics Design Library: Norwich, NY, USA, 1998.
12. Hasegawa, M.; Kaneki, T.; Tsukui, M.; Okubo, N.; Ishii, J. High-Temperature Polymers Overcoming the Trade-Off between Excellent Thermoplasticity and Low Thermal Expansion Properties. *Polymer* **2016**, *99*, 292–306.
13. Kim, K.; Yoo, T; Kim, J.; Ha, H.; Han, H. Effects of Dianhydrides on the Thermal Behavior of Linear and Crosslinked Polyimides. *J. Appl. Polym. Sci.* **2015**, *132*, 41412 (1–9).

14. Ragosta, G.; Abbate, M.; Musto, P.; Scarinzi, G. Effect of the Chemical Structure of Aromatic Polyimides on their Thermal Aging, Relaxation Behavior and Mechanical Properties. *J. Mater. Sci.* **2012**, *47*, 2637–2647.

15. Kim, G.; Byun, S.; Yang, Y.; Kim, S.; Kwon, S. Film Shrinkage Inducing Strong Chain Entanglement in Fluorinated Polyimide. *Polymer* **2015**, *68*, 293–301.

16. Chen, W. J.; Chen, W; Zhang, B.; Yang, S.; Liu, C. Y. Thermal Imidization Process of Polyimide Film: Interplay Between Evaporation and Imidization. *Polymer* **2017**, *109*, 205–215.

17. Comer, A. C.; Ribeiro, C. P.; Freeman, B. D.; Kalakkunnath, S.; Kalika, D. S. Dynamic Relaxation Characteristics of Thermally Rearranged Aromatic Polyimides. *Polymer* **2013**, *54*, 891–900.

18. Bas, C.; Tamagna, C.; Pascal, T.; Alberola, N. D. On the Dynamic Mechanical Behavior of Polyimides Based on Aromatic and Alicyclic Dianhydrides. *Polym. Eng. Sci.* **2003**, *43*, 344–355.

19. Serbezeanu, D.; Butnaru, I.; Varganici, C. D.; Bruma, M.; Fortunato, G.; Gaan, S. Phosphorous-Containing Polyimide Fibers and their Thermal Properties. *RSC Adv.* **2016**, *6*, 38371–38379.

20. Butnaru, I.; Bruma, M.; Gaan, S. Phosphine Oxide Based Polyimides: Structure-Property Relationships. *RSC Adv.* **2017**, *7*, 50508–50518.

21. Bejan, A. E.; Constantin, C. P.; Damaceanu, M. D. n-Type Polyimides with 1,3,4-Oxadiazole-Substituted Triphenylamine Units—an Innovative Structural Approach. *J. Phys. Chem. C* **2019**, *123*, 15908–15923.

22. Meis, D.; Tena, A.; Neumann, S.; Georgopanos, P.; Emmler, T.; Shishatskiy, S.; Rangou, V.; Filiz, V.; Abetz, V. Thermal Rearranged of Ortho-Allyloxypolyimide Membranes and the Effect of the Degree of Functionalization. *Polym. Chem.* **2018**, *9*, 3987–3999.

23. Sava, I.; Resmerita, A. M.; Lisa, G.; Damian, V.; Hurduc, N. Synthesis and Photochromic Behavior of New Polyimides Containing Azobenzene Side Groups. *Polymer* **2008**, *49*, 1475–1482.

24. Hu, X.; Yan, J.; Wang, Y.; Mu, H.; Wang, Z.; Cheng, H.; Zhao, F.; Wang, Z. Colorless Polyimides Derived from 2R, 5R,7S, 10S-Naphthanetetracarboxylic Dianhydride. *Polym. Chem.* **2017**, *8*, 6165–6172.

25. Sidra, L. R.; Chen, G.; Mushtaq, N.; Xu, L.; Chen, X.; Li, Y.; Fang, X. High $_{Tg}$ Melt Processable Copolyimides Based on Isomeric 3,3' and 4,4'-Hydroquinone Diphthalic Anhydride (HQDPA). *Polymer* **2018**, *136*, 205–214.

26. Rusu, R. D.; Damaceanu M.-D; Marin, L.; Bruma, M. Copoly(peryleneimide) s containing 1,3,4-oxazole rings: synthesis and properties. *J. Polym. Sci. Pol. Chem.* **2010**, *48*, 4230–4242.

27. Damaceanu, M. D.; Constantin, C. P.; Bruma, M.; Pinteala, M. Tuning of the Color of Emitted Light from New Polyperyleneimides Containing Oxadiazole and Siloxane Moieties. *Dyes Pigments* **2013**, *99*, 228–239.

28. Bessonov, M. I., Zubkov, V. A., Eds.; *Polyamic Acids and Polyimides. Synthesis, Transformations, and Structure*; CRC Press: Boca Raton, FL, 1993.

29. Liaw, D. J.; Wang, K. L.; Huang, Y. C.; Lee, K. R.; Lai, J. Y.; Ha, C. S. Advanced Polyimide Materials: Synthesis, Physical Properties and Applications. *Prog. Polym. Sci.* **2012**, *37*, 907–974.

30. Ding, M. Isomeric Polyimides. *Prog. Polym. Sci.* **2007**, *32*, 623–668.

31. Zhuang, Y.; Seong, J. G.; Lee, Y. M. Polyimides Containing Aliphatic/Alicyclic Segments in the Main Chains. *Prog. Polym. Sci.* **2019**, *92*, 35–88.

32. Schulz, B.; Bruma, M. Brehmer, L. Aromatic Poly(1,3,4-Oxadiazole)s as Advanced Materials. *Adv. Mater.* **1997,** *9,* 601–613.
33. Jikei, M.; Kakimoto, M. A. Dendritic Aromatic Polyamides and Polyimides. *J. Polym. Sci. Pol. Chem.* **2004,** *42,* 1293–1309.
34. Korshak, V. V.; Vinogradova, S. V.; Vygodskii, Ya. S.; Nagiev, Z. M.; Urman, Ya. G.; Alekseeva, S.G.; Slonium, I. Ya. Synthesis and Investigation of the Properties of Cardo-Copolyimides with Different Microstructure. *Makromol. Chem.* **1983,** *184,* 235–252.
35. Sacher, E. A Reexamination of Polyimide Formation. *J. Macromol. Sci. B.* **1986,** *25,* 405–418.
36. Ghosh, A.; Sen, S. K.; Banerjee, S.; Voit, B. Solubility Improvements in Aromatic Polyimides by Macromolecular Engineering. *RSC Adv.* **2002,** *2,* 5900–5926.
37. Tapaswi, P. K.; Ha, C. S. Recent Trends on Transparent Colorless Polyimides with Balanced Thermal and Optical Properties: Design and Synthesis. *Macromol. Chem. Phys.* **2019,** *220,* 1800313 (1–33).
38. Tsai, C. L.; Yen, H. J.; Liou, G. S. Highly Transparent Polyimide Hybrid for Optoelectronic Applications. *React. Funct. Polym.* **2016,** *108,* 2–30.
39. Ni, H. J.; Liu, J. G.; Wang, Z. H.; Yang, S. Y. A Review on Colorless and Optically Transparent Polyimide Films: Chemistry, Process and Engineering Applications. *J. Ind. Eng. Chem.* **2015,** *28,* 16–27.
40. Kurosawa, T.; Higashihara, T.; Ueda, M. Polyimide Memory: a Pithy Guideline for Future Applications. *Polym. Chem.* **2013,** *4,* 16–30.
41. Vanherck, K.; Koeckelberghs, G.; Vankelecom, I. F. J. Crosslinking Polyimides for Membrane Applications: A Review. *Prog. Polym. Sci.* **2013,** *38,* 874–896.
42. Favvas, E. P.; Katsaros, F. K.; Papageorgiou, S. K.; Sapalidis, A. A.; Mitropoulos, A. Ch. A Review on the Latest Development of Polyimide Based Membranes for CO$_2$ Separations. *React. Funct. Polym.* **2017,** *120,* 104–130.
43. Xiao, Y.; Low, B. T.; Hosseini, S. S.; Chung, T. S.; Paul, D. R. The Strategies of Molecular Architecture and Modification of Polyimide-Based Membranes for CO$_2$ Removal from Natural Gas-A Review. *Prog. Polym. Sci.* **2009,** *34,* 561–580.
44. Zhang, M. Y.; Niu, H. Q.; Wu, D. Z. Polyimide Fibers with High Strength and High Modulus: Preparation, Structures, Properties and Applications. *Macromol. Rapid. Commun.* **2018,** *39,* 1800141 (1–14).
45. Ding, Y.; Hou, H.; Zhao, Y.; Zhu, Z.; Fong, H. Electrospun Polyimide Nanofibers and their Applications. *Prog. Polym. Sci.* **2016,** *61,* 67–103.
46. Inagaki, M.; Ohta, N.; Hishiyama, Y. Aromatic Polyimides as Carbon Precursors. *Carbon* **2013,** *61,* 1–21.
47. Chisca, S.; Barzic, A. I.; Sava, I.; Olaru, N.; Bruma, M. Morphological and Rheological Insights on Polyimide Chain Entanglements for Electrospinning Produced Fibers. *J. Phys. Chem. B* **2012,** 116, 9082–9088.
48. Goponenko, A. V.; Hou, H.; Dzenis, Y. A. Avoiding Fusion of Electrospun3,3',4'4'-Biphenyltetracarboxylic Dianhydride 4,4'-Oxydianiline Copolymer Nanofibers During Conversion to Polyimide. *Polymer* **2011,** *52,* 3776–3782.
49. Bubulac, T. V.; Serbezeanu, D. Innovative Materials Based on Polyimides. Book of Abstracts, Humboldt Kolegg. Science Without Borders: Alexander von Humboldt's Concept in Today's World, Varna, Bulgaria, September 18–21, 2019.
50. Lei, X. F.; Jin, Y. H.; Sun, H. L.; Zhang, W. Rehealable Imide-Imine Hybrid Polymers with Full Recyclability. *J. Mat. Chem. A.* **2017,** *40,* 21140–21145.

51. Mezger, T. G. *The Rheology Handbook*; Vincentz Network: Hannover, Germany, 2006.
52. Damaceanu, M. D.; Rusu, R. D.; Cristea, M.; Musteata, V. E.; Bruma, M.; Wolinska-Grabczyk, A. Insight into the Cahin and Local Mobility of Some Aromatic Polyamides and their Influence on the Physicochemical Properties. *Macromol. Chem. Phys.* **2014,** *215*, 1573–1587.
53. Ionita, D.; Cristea, M.; Banabic, D. Viscoelastic Behavior of PMMA in Relation to Deformation Mode. *J. Therm. Anal. Calorim.* **2015,** *120*, 1775–1783.
54. Xie, F.; Long, Y.; Chen, L.; Li, L. A New Study of Starch Gelatinization under Shear Using Dynamic Mechanical Analysis. *Carbohyd. Polym.* **2008,** *72*, 229–234.
55. Meyvis, T. K. L.; Stubbe, B. G.; Van Steenbergen, M. J.; Hennink, W. E.; De Smedt, S. C.; Demeester, J. A Comparison between the Use of Dynamic Mechanical Analysis and Oscillatory Shear Rheometry for the Characterization of Hydrogels. *Int. J. Pharm.* **2001,** *244*, 163–168.
56. Royal, P. G.; Huang, C. Y.; Tang, S. W. J.; Duncan, J.; Van-de-Velde, G.; Brown, M. B. The Development of DMA for the Detection of Amorphous Content in Pharmaceutical Powdered Materials. *Int. J. Pharm.* **2005,** *301*, 181–191.
57. Mahlin, D.; Wood, J.; Hawkins, N.; Mahey, J.; Royall, P. G. A Novel Powder Sample Holder for the Determination of Glass Transition Temperatures by DMA. *Int. J. Pharm.* **2009,** *371*, 120–125.
58. Cicala, G; Mannino, S.; Latteri, A.; Ognibene, G.; Saccullo, G. Effects of Mixing di- and tri-Functional Epoxy Monomers on Epoxy/Thermoplastic Blends. *Adv. Polym. Technol.* **2018,** *37*, 1868–1877.
59. McCrum, N. G.; Read, B. E.; Williams, G. *Anelastic and Dielectric Effects in Polymeric Solids*; John Wiley: New York, 1967.
60. Brinson, H. F.; Brinson, L. C. *Polymer Engineering Science and Viscoelasticity. An Introduction*; Springer: New York, 2015.
61. Menard, K. P. *Dynamic Mechanical Analysis. A Practical Introduction*; CRC Press: Boca Raton, FL, 1999.
62. Cristea, M. *Dynamic Mechanical Analysis in Polymeric Multiphase Systems*. In *Multiphase polymer systems. Micro- to Nanostructural Evolution in Advanced Technologies*; Barzic, A. I., Ioan, S., Eds.; CRC Press: Boca Raton, FL, 2017; p 173.
63. Sperling, H. L. *Introduction to Physical Polymer Science*; Wiley-Interscience: Hoboken, NJ, 2006.
64. Tang, H.; Dong, L.; Zhang, J.; Ding, M. Feng, Z. Study on the β Relaxation Mechanism of Thermosetting Polyimide/Thermoplastic Polyimide Blends. *Eur. Polym. J.* **1996,** *32*, 1221–1227.
65. Li, F.; Ge, J. J.; Honigfort, P. S.; Chen, J. C; Harris, F. W.; Cheng, S. Z. D. Diamine Architecture Effects on Glass Transitions, Relaxation Processes and Other Material Properties in Organo-Soluble Aromatic Polyimide Films. *Polymer* **1999,** *40*, 4571–4583.
66. Arnold, F. E. Jr.; Bruno, K. R.; Shen, D.; Eashoo, M; Lee, C. J.; Harris, F. W.; Cheng, S. Z. D. The Origin of β Relaxations in Segmented Rigid-Rod Polyimide and Copolyimide Films. *Polym. Eng. Sci.* **1992,** *33*, 1373–1380.
67. Sun, Z.; Dong, L.; Zhuang, Y.; Cao, L.; Ding, M.; Feng, Z. Beta Relaxation in Polyimides. *Polymer* **1992,** *33*, 4728–4731.
68. Comer, A. C.; Kalila, D. S.; Rowe, B. W.; Freeman, B. D.; Paul, D. R. Dynamic Relaxation Characteristics of Matrimid® Polyimide. *Polymer* **2009,** *50*, 891–897.

69. Okada, T.; Ishige, R.; Ando, S. Effects of Chain Packing and Structural Isomerism on the Anisotropic Linear and Volumetric Thermal Expansion Behaviors of Polyimide Films. *Polymer* **2018**, *146*, 386–395.

70. Nicholls, A. R.; Kull, K.; Cerrato, C.; Craft, G.; Diry, J. B.; Renoir, E.; Perez, Y.; Harmon, J. P. Thermomechanical Characterization of Thermoplastic Polyimides Containing 4,4'-Methylenebis(2,6-Dimethylaniline) and Polyetherdiamines. *Polym. Eng. Sci.* **2019**. DOI: 10.1002/pen.25226.

71. Starkweather, H. W .Jr. Aspects of Simple, Non-Cooperative Relaxations. Polymer **1991**, *32*, 2443–2448.

72. Hougham, G.; Tesoro, G.; Shaw, J. Synthesis and Properties of Highly Fluorinated Polyimides. *Macromolecules* **1994**, *27*, 3642–3649.

73. Coleman, M. R.; Koros, W. J. The Transport Properties of Polyimide Isomers Containing Hexafluoroisopropylidene in the Diamine Residue. *J. Polym. Sci. Polym. Phys.* **1994**, *32*, 1915–1926.

74. Ragosta, G.; Abbate, M.; Musto, P.; Scarinzi, G. Effect of the Chemical Structure of Aromatic Polyimides on their Thermal Aging, Relaxation Behavior and Mechanical Properties. *J. Mater. Sci.* **2012**, *47*, 2637–2647.

75. Ando, S.; Sekiguchi, K.; Mizoroki, M.; Okada, T.; Ishige, R. Anisotropic Linear and Volumetric Thermal-Expansion Behaviors of Self-Standing Polyimide Films Analysed by Thermomechanical Analysis (TMA) and Optical Interferometry. *Macromol. Chem. Phys.* **2017**, *219*, art. 1700354 (1–10).

76. Qian, C.; Bei, R.; Zhu, T.; Zheng, W.; Liu, S.; Chi, Z.; Aldred, M. P.; Chen, X.; Zhang, Y.; Xu, J. Facile Strategy for Intrinsic Low-k Dielectric Polymers: Molecular Design Based on Secondary Relaxation Behavior. *Macromolecules* **2019**, *52*, 4601–4609.

77. Bernier, G. A.; Kline, D. E. Dynamic Mechanical Behavior of a Polyimide. *J. Appl. Polym. Sci.* **1968**, *12*, 593–604.

78. Cristea, M.; Ionita, D.; Hulubei, C.; Timpu, D.; Popovici, D.; Simionescu, B. C. Chain Packing versus Chain Mobility in Semialiphatic BTDA-Based Copolyimides. *Polymer* **2011**, *52*, 1820–1828.

79. Willbourn, A. H. The Glass Transition in Polymers with the $(CH_2)_n$ Group. *Trans. Faraday Soc.* **1958**, *54*, 717–729.

80. Cristea, M.; Ionita, D.; Doroftei, F.; Simionescu, B. C. Effect of Long-Term and Short-Term Dynamic Mechanical Evaluation of Networks Based on Urethane and Soybean Oil. *J. Mech. Behav. Biomed. Mat.* **2013**, *17*, 317–326.

81. Cristea, M.; Ionita, D.; Simionescu, B. C. Dynamic Mechanical Analysis on Regenerated Cellulose. *Rev. Chem. (Bucuresti)* **2008**, *59*, 1088–1091.

82. Cristea, M.; Ibanescu, S.; Cascaval, C. N.; Rosu, D. Dynamic Mechanical Analysis of Polyurethane-Epoxy Interpenetrating Polymer Networks. *High Perform. Polym.* **2009**, *21*, 608–623.

83. Gillham, J. K.; Hallock, K. D.; Stadnicki, S. J. Thermomechanical and Thermogravimetric Analyses of Systematic Series of Polyimides. *J. Appl. Sci.* **1972**, *16*, 2595–2610.

84. Gillham, J. K., Gillham, H. C. Polyimides: Effect of Molecular Structure and Cure on Thermomechanical Behavior. *Polym. Eng. Sci.* **1973**, *13*, 447–454.

85. Cho, D.; Choi, Y.; Drzal, L. T. Simultaneous Monitoring of the Imidization and Cure Reactions of LaRC PETI-5 sized on a Brided Glass Fabric Substrate by Dynamic Mechanical Analysis. *Polymer* **2001**, *42*, 4611–4618.

86. Cristea, M; Ionita, D. G., Bruma, M; Simionescu, B. C. Thermal Behavior of Aromatic Polyamic Acids and Polyimides Containing Oxadiazole Rings. *J. Therm. Anal. Calorim.* **2008**, *93*, 63–68.

87. Seo, Y. Modeling of Imidization Kinetics. *Polym. Eng. Sci.* **1997**, *37*, 772–776.

88. Kim, K; Ryou, J. H.; Kim, Y.; Ree, M.; Chang, T. Thermal Imidization Behavior of Aromatic Poly(Amic dialkyl Ester) Precursors Derived from Biphenyltetracarboxylic Dianhydride). *Polym. Bull.* **1995**, *34*, 219–226.

89. Gillham, J. K. A Semimicro Thermomechanical Technique for Characterizing Polymeric Materials: Torsional Braid Analysis. *Am. Inst Chem. Engineers* **1974**, *20*, 1066–1079.

90. Lee, C. Y. C., Goldfarb, I. J. Glass Transition Temperature (Tg) Determination of Partially Cured Thermosetting Systems. *Polym. Eng. Sci.* **1981**, *21*, 787–791.

91. Yang, W.; Liu, F.; Zhang, J.; Zhang, E.; Qiu, X.; Ji, X. Influence of Thermal Treatment on the Structure and Mechanical Properties of one Aromatic BPDA-PDA Polyimide Fiber. *Eur. Polym. J.* **2017**, *96*, 429–442.

92. Weick, B. L.; Bhushan, B. The Tribological and Dynamic Behavior of Alternative Magnetic Tape Substrates. *Wear* **1995**, *190*, 28–43.

93. Huo, P. P.; Cebe, P. New TPI Thermoplastic Polyimide: Dielectric and Dynamic Mechanical Relaxation. *Polymer* **1993**, *34*, 696–704.

94. Williams, M. L.; Landel, R. F.; Ferry, J. D. The Temperature Dependence of Relaxation Mechanisms in Amorphous Polymers and Other Glass-Forming Liquids. *J. Am. Chem. Soc.* **1955**, *77*, 3701–3707.

95. Ferry, J. D. *Viscoelastic Properties of Polymers*, 3rd ed.; John Wiley & Sons: New York, 1980.

96. Tsai, M. H.; Whang, W. T. High Temperature Lifetime of Polyimide/Poly(Silsesquioxane)-Like Hybrid Films. *J. Polym. Res.* **2001**, *8*, 77–89.

97. Harper, B. D.; Rao, J. M.; Henner, V. H.; Popelar, C. H. Hygrothermal Effects Upon Stress Relaxation in a Polyimide Film. *J. Electron. Mater.* **1997**, *26*, 798–804.

98. Robertson, G. C.; Monat, J. E.; Wilkes, G. L. Physical Aging of an Amorphous Polyimide: Enthalpy Relaxation and Mechanical Property Changes. *J. Polym. Sci. Polym. Phys.* **1999**, *37*, 1931–1946.

99. Susa, A.; Mordvinkin, A.; Saalwachter, K.; Van der Zwaag, S., Garcia, S. J. Identifying the Role of Primary and Secondary Interactions on the Mechanical Properties and Healing of Densely Branched Polyimides. *Macromolecules* **2018**, *51*, 8333–8345.

100. Montano, V.; Picken, S. J., van der Zwaag, S.; Garcia, S. J. A Deconvolution Protocol of the Mechanical Relaxation Spectrum to Identify and Quantify Individual Polymer Feature Contributions to Self-Healing. *Phys. Chem. Chem. Phys.* **2019**, *21*, 10171–10184.

Index